SOCIETY OF GENERAL PHYSIOLOGISTS SERIES

Volume 32

Cell and Tissue Interactions

Society of General Physiologists Series

Published by Raven Press

SOCIETY OF GENERAL PHYSIOLOGISTS SERIES

Volume 32

Cell and Tissue Interactions

Editors:

James W. Lash, Ph.D.
Department of Anatomy
University of Pennsylvania
The School of Medicine
Philadelphia, Pennsylvania

Max M. Burger, M.D., Ph.D.
Department of Biochemistry
Biocenter, University of Basel
Basel, Switzerland

Raven Press ▪ New York

035105

The publication of this volume was supported by grants from the National Institutes of Health and the National Science Foundation

Made in the United States of America

Library of Congress Cataloging in Publication Data
Main entry under title:

Cell and tissue interactions.

   (Society of General Physiologists series ; v. 32)
   Includes bibliographies and index.
   1. Cell interaction. 2. Extracellular space.
3. Tissues. I. Lash, James W., 1929-   II. Burger,
Max M. III. Society of General Physiologists. Society
of General Physiologists series ; v. 32 [DNLM:
1. Cells. 2. Histology. W1 S0872G v. 32 / QS504
C3926]
QH604.2.C44   574.8'75   75-25111
ISBN 0-89004-180-6

QH
604.2
.C44
1977

Preface

Nature is nowhere wont to reveal her innermost secrets more openly than away from the beaten path, where she shows faint traces of herself.

The interphase between two liquid metals was where the alchemist searched in vain for his gold. It was the interphase between opposing "humours" where medieval physicians believed that life and death decisions were reached, for the detriment or benefit of their suffering patients. In modern science interphases are still important, it is the interphase between similar or disparate substances where we study the physical and chemical processes which give rise to new substances. In these interaction zones, chemical and physical communication determines the static and dynamic nature of all tissues. A multidisciplinary analysis of tissue interaction zones should thus lead to a better understanding of the ways and means by which tissues communicate with each other. The study of cell and tissue interactions has become very popular, and has lately been given a lot of credit, perhaps too much credit, for singlehandedly solving the principle questions in morphogenesis and embryonic development.

While a large number of biologists are seeking answers to how the genetic apparatus of cells function, an equally large number are investigating the circumstances that provoke cells to activate or regulate their genetic apparatus. One of the most fruitful areas of investigation is this rapidly growing field of cell and tissue interactions. The importance of these interactions, which invariably occur at interphases, has been recognized for many years, but only recently have technological advances made it possible to probe them at the molecular level. In just the past few years, results on cell and tissue interactions have produced exciting leads toward our understanding of many interactions, with particular emphasis on developmental events. It is now obvious that the intercellular matrix and the cell surface are implicated in most developmental and regulatory events.

When focusing with greater resolution on the zones of tissue interaction, it is realized that tissue interaction ultimately means cell interaction, and of equal importance, that these interactions are mediated through the interphase of the extracellular matrix. There is a structural and functional relation between the cell periphery, the surface membrane, and the extracellular matrix. Interactions take place directly via the plasma membrane, or via the extracellular matrix components (collagens and proteoglycans), which are intimately associated with the surface membrane.

The chapters in this volume are the result of the Thirty-second Annual Symposium of the Society of General Physiologists, which convened at the Marine Biological Laboratory in Woods Hole, September 12–16, 1976. We felt it would be fruitful to bring together a group of specialists on the extracellular matrix (Section II) with another group whose interests are directed towards the plasma membrane (Section III). As mediators, developmental biologists were helpful to form a tie between the specialists (Section I), reminding the participants that the purpose of both groups of specialists, those concerned with extracellular matrix and those concerned with plasma membranes, was to focus on developmental problems.

Tissue and cell interactions undoubtedly evoke changes in metabolic processes, thus a group of contributors was invited to deal with some basic cellular functions (cell migration and invasion, hormone response, etc.) that will have to be considered in any thorough analysis of tissue interactions (Section IV). The last two contributions illustrate how far we are from understanding the most complicated tissue organization in development, the neural network (Section V). There are very few concepts yet to explain such simple connections as those between nerve and target muscle, let alone the intricate wiring system in the brain.

It is impossible to cover all facets of the broad area of cell and tissue interactions in one volume, although the field is in its infancy and just beginning to unfold. A considerable, and some times arbitrary, selection in subject matters had to be made. Thus this volume should be useful as a primer for biomedical researchers and students who desire an introduction to the biochemistry of matrices and cell surfaces, and how they are implicated in developmental events. For those actively working in one of the many areas covered in these chapters, the bibliographies offer up to date reference material.

The meeting was a delightful educational experience for all who attended as the result of the wide variety of disciplines which were represented. In spite of the wide differences in the disciplines and training of the participants, the ultimate truth of scientific investigation was exemplified; all biological problems are related, and there are still new questions to be asked. Not only chemical, physical, and even biological interphases are necessary for the creation of new products, structures and forms, but the same principles of interphase activity also applies to the creation of new concepts and insights in science. We hope that this book will stimulate the reader to appreciate the usefulness of the interdisciplinary approach, as it did the participants.

Jay Lash
Max M. Burger

Acknowledgments

We gratefully acknowledge the assistance of the staff and community of the Woods Hole Marine Biological Laboratory for helping with the arrangements for the symposium.

We are also pleased to acknowledge the financial assistance of the National Institute of General Medical Sciences of the National Institutes of Health, the National Science Foundation, and the International Society of Developmental Biology. Most importantly, we are grateful to the contributors for their efforts in preparing yet another symposium paper, and for the prompt submission of their manuscripts. In many ways the counsel and generous assistance of Natalie N. Lash contributed to the success of the symposium.

Lastly, we thank the Society of General Physiologists for providing us with the opportunity to organize this topical symposium.

Contents

Contributors

M. Abercrombie
Strangeways Research Laboratory
Cambridge, England

Robert Auerbach
Department of Zoology
University of Wisconsin
Madison, Wisconsin 53706

G. Beattie
Department of Cancer Biology
The Salk Institute for
Biological Studies
San Diego, California 92112

C. R. Birdwell
Department of Cancer Biology
The Salk Institute for
Biological Studies
San Diego, California 92112

Anna Brownell
Department of Biochemistry and
Microbiology-Immunology
University of Southern California
Los Angeles, California 90007

K. W. Brunson
Department of Developmental and
Cell Biology
University of California
Irvine, California 92717

Max M. Burger
Department of Biochemistry
Biocenter, University of Basel
CH 4056 Basel, Switzerland

G. A. Dunn
Strangeways Research Laboratory
Cambridge, England

Robert Durr
Department of Biology
The Johns Hopkins University
Baltimore, Maryland 21218

I. J. Fidler
Basic Research Program
NCI-Frederick Cancer Center
Frederick, Maryland 21701

S. Filosa-Parisi
Institute of Histology and
Embryology
The University of Naples
Naples, Italy

Gerald D. Fischbach
Department of Pharmacology
Harvard Medical School
Boston, Massachusetts 02115

Eric Frank
Department of Pharmacology
Harvard Medical School
Boston, Massachusetts 02115

Luis Glaser
Department of Biology and
Biomedical Sciences
Washington University
St. Louis, Missouri 63110

David I. Gottlieb
Department of Biological Chemistry
Division of Biology and
Biomedical Sciences
Washington University
St. Louis, Missouri 63110

R. E. Hausman
Departments of Biology, Pathology, and
The Committee on Developmental
Biology
The University of Chicago
Chicago, Illinois 60637

Elizabeth D. Hay
Department of Anatomy
Harvard Medical School
Boston, Massachusetts 02115

J. P. Heath
Strangeways Research Laboratory
Cambridge, England

J. Jumblatt
Department of Biochemistry
Biocenter, University of Basel
CH 4056 Basel, Switzerland

A. Kaji
Department of Microbiology
University of Pennsylvania
The School of Medicine
Philadelphia, Pennsylvania 19104

James W. Lash
Department of Anatomy
University of Pennsylvania
The School of Medicine
Philadelphia, Pennsylvania 19104

N. M. Le Douarin
Institut d'Embryologie du Centre
National de la Recherche
Scientifique et du College de France
49 bis, Avenue de la Belle-Gabrielle
94130 Nogent-sur-Marne, France

C. Le Lièvre
Institut d'Embryologie du Centre
National de la Recherche
Scientifique et du College de France
49 bis, Avenue de la Belle-Gabrielle
94130 Nogent-sur-Marne, France

Jack Lilien
Department of Zoology
University of Wisconsin
Madison, Wisconsin 53706

E. R. Macagno
Department of Biological Sciences
Columbia University
New York, New York 10027

Ronald Merrell
Department of Biological Chemistry
Division of Biology and
Biomedical Sciences
Washington University
St. Louis, Missouri 63110

Edward J. Miller
Department of Biochemistry and
Institute of Dental Research
University of Alabama Medical Center
Birmingham, Alabama 35294

Alberto Monroy
Stazione Zoologica
Villa Comunale
80121 Naples, Italy

A. A. Moscona
Department of Biology, Pathology and
The Committee on Developmental
  Biology
The University of Chicago
Chicago, Illinois 60637

Helen Muir
Biochemistry Division
Kennedy Institute of Rheumatology
Bute Gardens
London W6 7DW England

M. Muto
Department of Microbiology
University of Pennsylvania
The School of Medicine
Philadelphia, Pennsylvania 19104

G. L. Nicolson
Department of Developmental and
Cell Biology
University of California
Irvine, California 92717

Minoru Okayama
Developmental Biology Laboratory
Departments of Medicine and Anatomy
Harvard Medical School at
Massachusetts General Hospital
Boston, Massachusetts 02114

Roslyn W. Orkin
Developmental Biology Laboratory
Departments of Medicine and Anatomy
Harvard Medical School at
Massachusetts General Hospital
Boston, Massachusetts 02114

E. Parisi
Stazione Zoologica
Villa Comunale
80121 Naples, Italy

B. De Petrocellis
C.N.R. Laboratory of
Molecular Embryology
Arco Felice
Naples, Italy

Howard Rasmussen
Departments of Internal Medicine and
Cell Biology
Yale University School of Medicine
New Haven, Connecticut 06510

J. C. Robbins
Department of Cancer Biology
The Salk Institute for Biological Studies
San Diego, California 92112

Stephen Roth
Department of Biology
The Johns Hopkins University
Baltimore, Maryland 21218

Richard Rutz
Department of Zoology
University of Wisconsin
Madison, Wisconsin 53706

Roger Santala
Department of Biological Chemistry
Division of Biology and Biomedical
  Sciences
Washington University
St. Louis, Missouri 63110

Lauri Saxén
Third Department of Pathology
University of Helsinki
SF-00290 Helsinki, Finland

Harold C. Slavkin
Laboratory for Developmental Biology
Ethel Percy Andrus Gerontology Center
University of Southern California
Los Angeles, California 90007

Barry D. Shur
Department of Developmental Genetics
Sloan-Kettering Institute for Cancer
  Research
New York, New York 10021

Nino Sorgente
School of Dentistry
University of Southern California
Los Angeles, California 90007

M. A. Teillet
Institut d'Embryologie du
Centre National de la Recherche
Scientifique et du Collège de France
49 bis, Avenue de la Belle-Gabrielle
94130 Nogent-sur-Marne, France

Bryan P. Toole
Developmental Biology Laboratory
Departments of Medicine and Anatomy
Harvard Medical School at
Massachusetts General Hospital
Boston, Massachusetts 02114

Gary N. Trump
Department of Biochemistry and
Microbiology-Immunology
University of Southern California
Los Angeles, California 90007

N. S. Vasan
Department of Anatomy
University of Pennsylvania
The School of Medicine
Philadelphia, Pennsylvania 19104

M. Yoshimura
Department of Microbiology
University of Pennsylvania
The School of Medicine
Philadelphia, Pennsylvania 19104

Cell and Tissue Interactions, edited by
J. W. Lash and M. M. Burger. Raven
Press, New York, 1977.

Directive Versus Permissive Induction: A Working Hypothesis

Lauri Saxén

Third Department of Pathology, University of Helsinki, SF-00290 Helsinki, Finland

Embryonic induction can be considered one of the most central and fascinating problems of developmental biology during the 1930s, when many leading schools of embryology followed in the footsteps of Hans Spemann. Soon, however, the enthusiasm died, and "induction" became almost a dirty word. The decline of this field was due to the slow progress made and the puzzling, seemingly contradictory results obtained by the different groups: Whatever the competent gastrula ectoderm, the classic target of these studies, was exposed to, neural differentiation resembling normal development resulted. Almost any heterotypic tissue, killed tissue fragments, cell-free fractions, various chemical compounds, and even traumatizing physical treatment of the target tissue triggered its neuralization, and it was not possible to find any common denominators in these artificial inducers (see ref. 21). The magic "organizine" transmitting inductive messages from the archenteron roof to the overlying ectoderm remained undiscovered, and many of the outstanding scientists involved in these studies could not resist the temptation to shift their interest to more clearly defined and more easily approachable problems in developmental biology.

Something of the ghost of the "organizine" still seems to affect our thinking, and we are confronted by the constant demand of isolating and characterizing the signal substances transmitting morphogenetic messages in various interactive events. Undoubtedly, this will be our ultimate goal as well as the clarification of the mode of action of such compounds, but in my opinion, our knowledge of the biology of most of the inductive interactions is still too fragmentary for a meaningful molecular approach. Should we search for actual informative, transmissible molecules ultimately interacting with the genome of the target cell? Or would it be more feasible to look for less specific factors which act as nutrients or growth stimulators? Or should we rather consider extracellular compounds with certain spheric configurations responsible for cell orientation and arrangement within a tissue? In most interactive situations, these questions remain unanswered, and there is no reason to believe that various morphogenetic interactions lumped together under the common epithet "induction" should operate through similar mechanisms.

1

Since the organizers of this section have entitled it "Evidence for Cell-Tissue Interactions" and because it will be followed by many biochemical and molecular approaches to the problem, I have felt it appropriate to open this session with a short review of some recent work on morphogenetic interactions as a biological phenomenon, and to present some evidence for their varying characteristics.

DIRECTIVE VERSUS PERMISSIVE INFLUENCES

To emphasize the biological diversity of various interactive events during embryogenesis, I have formulated an oversimplified and perhaps naive scheme suggesting two basic types of inductive influences. When an embryonic cell possesses more than one developmental option, the choice between them is affected by extracellular factors, which thus exert a true *directive* action on differentiation. A *permissive* action, on the other hand, refers to a step of development, in which the cell has become committed to a certain pathway, but still requires an exogeneous stimulus to express its new phenotype. According to the scheme (Fig. 1), these two types of influences may alternate during progressive differentiation of the embryonic tissues associated with a gradual restriction of the number of developmental options. Whenever a cell is committed to a new pathway, a directive influence is required and the developmental options become restricted. The following step, the expression of the new phenotype, might require permissive influences, which, however, do not further canalize differentiation. For this, a new directive message is necessary and will, again, be followed by a permissive condition necessary for the expression and stabilization of the cellular phenotype, the growth and proliferation of the tissues, and the maintenance of the organotypic organization. All these interactions fulfill most of the classic definitions for "embryonic induction" or "morphogenetic tissue interactions," but are, indeed, very different in their biological consequences and most probably in their molecular mechanisms.

In what follows, I will illustrate my postulate with some examples of recent work from my own and from several other laboratories.

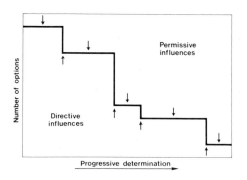

FIG. 1. A scheme of the relative importance of directive and permissive interactive events during progressive differentiation as a function of the number of developmental options.

DIRECTIVE INFLUENCES

Primary Embryonic Induction

The confusing results of the 1930s mentioned above may be taken as an example of permissive effects, since the cells of the gastrula ectoderm seem to have developed a neurogenic bias during their early development, which can be brought about by a great variety of triggers. Whether all these different treatments act by influencing the intracellular ion compartment as suggested by Barth and Barth (2) or through some other common mechanisms cannot yet be decided. Subsequent studies have, however, conclusively shown that these cells can also be experimentally converted into various cell types normally derived from mesoderm or entoderm (21), thus demonstrating their multiple developmental options. A subsequent restriction of these options during embryogenesis is shown in the following experiment: During early gastrulation, cells of the competent ectoderm can be converted into either neural or mesodermal derivatives by exposing them to various heterotypic inductors or other artificial inducers. At an early neurula stage, they have become committed toward neural elements and can no longer be geared off this pathway. However, they still have several options in the building of various regions of the central nervous system (CNS). If these cells from the anterior region of the neural plate are cultured *in vitro* without additional tissues, they will invariably follow their original destination and form forebrain structures. But if combined *in vitro* to cells of the axial mesoderm or to artificially mesodermalized cells, they develop into constituents of the caudal regions of the CNS (20,26). These options are, again, lost during neurulation and at stage 15 (Triturus), the cells of the neural plate are irreversibly regionalized (25).

Interactive processes during the later stages of neurogenesis are poorly understood, but some stages in the development of the neural crest are known to be guided by interactions of the directive type. One of the most striking examples is provided by the recent work of Le Douarin (*this volume*) showing how the mesenchymal environment directs the differentiation and functional maturation of the ganglioblasts of the autonomous nervous system.

Determination of the Derivatives of the Integument

Embryonic and, to a lesser degree, adult epidermis represent a tissue with remarkable flexibility and many developmental options, as illustrated by experiments in which the integumental epithelium is combined to various "inductor" tissues. Four examples will be given here. The development of the *cutaneous appendages* in mammalian, avian, and reptilian embryos have been thoroughly analyzed by Sengel and his group (5,22). Using heterospecific or interclass combination techniques, he concludes that the

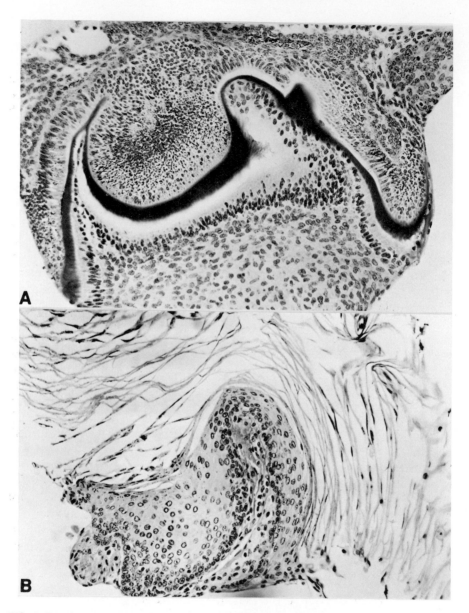

FIG. 2. Results of reciprocal combinations between the epithelium and mesenchyme of a tooth rudiment and of nondentogeneous gingival tissue. **A:** A combination of gingival epithelium with mesenchyme of the dental papilla results in the development of a well-shaped tooth rudiment and differentiation of both odontoblasts and ameloblasts. **B:** A combination of tooth epithelium and gingival mesenchyme shows no tooth development and the epithelium displays keratinization. (Courtesy of Dr. Irma Thesleff.)

appendages (hair, feathers, and scales) develop as a result of a two-step dermoepidermal interactive process. During the first step, a nonclass-specific inductive trigger of the dermis determines the size, shape, and distribution of the cutaneous appendages, but the ultimate class (hair, feather, scale) is determined by the genetic constitution of the epidermis. During the second step, which is class-specific, dermal influences guide the final morphogenesis of the appendages.

Vaginal morphogenesis is similarly guided by epitheliomesenchymal interactions in a directive way, as reported very recently by Cunha (3,4). Reciprocal combinations of uterine and vaginal tissue components demonstrated that, during the early neonatal period, the epithelial component still differentiates according to the origin of the mesenchymal component combined to it. Subsequently, the epithelia lose their competence to respond to this directive influence, but a permissive type of influence from the mesenchyme is still required for the maintenance of the structure and function of the vaginal epithelium (7).

Lentoid bodies developing at heterotypic sites have been detected long ago in Amphibian embryos (24). We have recently reopened this question in regard to Avian embryos and have shown that trunk epidermis of a 2-day-old chick embryo will respond *in vitro* to a directive influence of an optic cup by forming a lentoid body with advanced histodifferentiation. The epidermal origin of these lenses was confirmed by the use of biological nuclear markers and in transfilter experiments (Karkinen-Jääskeläinen, *unpublished*).

Tooth morphogenesis has also been shown to be guided by an epitheliomesenchymal interaction between the dental mesenchyme and the enamel epithelium. Differentiation of the latter seems to be triggered by a typical directive induction, as first shown by Kollar and Baird (11), and recently confirmed in organ culture experiments by Thesleff (23) in my laboratory. Reciprocal combinations were made of the presumptive enamel epithelium and dental mesenchyme of a bell stage tooth rudiment, and the epithelium and mesenchyme from nondentogeneous gingival tissue. Tooth epithelium combined to gingival mesenchyme failed to undergo enamel differentiation and the cultures displayed keratinizing squamous epithelium. When gingival, nondentogeneous epithelium was combined to the mesenchyme of the dental papilla, a well-differentiated enamel organ with secretion of enamel proteins was seen (Fig. 2).

PERMISSIVE INFLUENCES

Induction of Kidney Tubules

The interactive events behind the formation of the metanephric secretory tubules seem to represent a typical permissive influence. As shown by

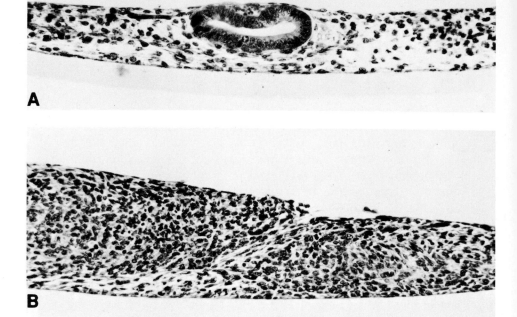

FIG. 3. Results of combinations of two potent tubule inductors with nonkidney mesenchyme. **A:** Isolated ureter bud combined with salivary mesenchyme for 5 days *in vitro*. **B:** Spinal cord associated with gastric mesenchyme and cultivated for 4 days. No signs of tubule formation are seen in these heterotypic mesenchymes. (From ref. 17.)

Grobstein and his school (1,10,27), a great variety of embryonal tissues can trigger this morphogenesis in the metanephric mesenchyme, which consequently can be considered "predetermined." Accordingly, other than kidney mesenchyme should not respond to these inductors by tubule formation and this has, in fact, been shown (17). Embryonic salivary, pulmonary, and gastric mesenchymes combined to either the normal tubule inductor, the ureter bud, or to another potent inducer, the spinal cord, failed to show any morphological changes suggestive of tubule formation (Fig. 3). This permissive induction is of relatively short duration and, hence, can be considered a trigger-type stimulus. Judging by our transfilter studies (14) and by recent observations of the development of cell contacts apparently required for this inductive interaction (19), an intercellular communication of 16 hours seems to be sufficient for the triggering of tubule formation, detectable only after some additional 24 hr (*unpublished*). This short interaction, however, seems to be followed by another permissive interaction between the pretubular condensates and the uninduced mesenchyme ensuring the elongation of the tubules (8).

Epitheliomesenchymal Interactions

Branching, morphogenesis, and cytodifferentiation of the epithelial component of various glandular organs have repeatedly been demonstrated to be dependent on the presence of their stromal mesenchyme. However, experiments in which the organs are at relatively advanced stages have also shown that the homotypic mesenchyme can be replaced by heterotypic ones: Salivary epithelium will undergo normal morphogenesis when combined to lung and other mesenchymes (4,12), and pancreatic mesenchyme at the 30-somite stage can be replaced by various heterotypic mesenchymes and by partially purified tissue extracts (15,16). The latter experiments seem to provide a good example of a permissive influence stimulating proliferation of the target cells necessary for the expression of their phenotype determined during the earlier stages. A word of caution, however, is appropriate here, since all of these experiments demonstrating a permissive influence were done *in vitro*. As pointed out by Lawson (12), some of these effects might primarily be required for the tissues to develop in the artificial *in vitro* conditions and consequently might not necessarily reflect the actual *in vivo* situation. This has been shown by Ellison and Lash (6) in another permissive event: triggering of somite chondrogenesis. Minor improvements in the tissue-culture conditions could enhance somite chondrogenesis *in vitro* and mimic true cartilage induction.

PROGRESSIVE DIFFERENTIATION AND STABILIZATION

My original scheme postulated a gradual restriction of the developmental options during embryogenesis followed by a shift from predominantly directive to predominantly permissive influences. The above examples and many others reviewed in the literature (see refs. 15,18) suggest that inductive interactions might be separated into these two major categories, and there are also convincing studies on the restriction of the developmental options of cells in early embryos (see ref. 9). Systemic "follow-up" studies on the fate of cells within a given tissue and the interactive requirements are few. Primary induction, beginning with a general neuralization and followed by the loss of competence to respond to such stimuli and subsequently showing time-limited responsiveness to regionalizing influences, can be taken as one example. The sequential interactive events in liver development may be taken as another example. Presumptive liver endoderm becomes programmed for hepatocyte differentiation at the 5-somite stage chick embryo, this being due to contact with the homotypic hepatocardiac mesenchyme. Expression of the hepatic cell phenotype, however, requires continuous contact with the hepatic mesenchyme, but this permissive influence is less specific, and various mesenchymes derived from the lateral plate can experimentally replace the homotypic liver mesenchyme (13). Similar examples of directive and permissive influences alternating during the organogenesis of a given tissue are known from the development of the lung, the pancreas, and the thyroid (see ref. 19).

CONCLUSIONS

There seems to be no doubt that a great variety of interactive events guide the ultimate fate of each embryonic cell and tissue. In the search for the mechanism of such interactive events, their diversity should always be borne in mind. Generalizations, although tempting, should not be made, for it is possible that common denominators will never be found.

REFERENCES

1. Auerbach, R. (1972): The use of tumors in the analysis of inductive tissue interactions. *Dev. Biol.,* 28:304–309.
2. Barth, L. G., and Barth, L. J. (1968): The role of sodium chloride in the process of induction by lithium chloride in cells of the *Rana pipiens* gastrula. *J. Embryol. Exp. Morphol.,* 19:387–396.
3. Cunha, G. R. (1972): Support of normal salivary gland morphogenesis by mesenchyme derived from accessory sexual glands of embryonic mice. *Anat. Rec.,* 173:205–212.
4. Cunha, G. R. (1976): Stromal induction and specification of morphogenesis and cytodifferentiation of the epithelia of the Mullerian ducts and urogenital sinus during development of the uterus and vagina in mice. *J. Exp. Zool.,* 196:361–370.
5. Dhouailly, D. (1975): Formation of cutaneous appendages in dermo-epidermal recombinations between reptiles, birds and mammals. *Wilhelm Roux's Arch.,* 177:323–340.
6. Ellison, M. L., and Lash, J. W. (1971): Environmental enhancement of *in vitro* chondrogenesis. *Dev. Biol.,* 26:486–496.
7. Flaxman, B. A., Chopra, D. P., and Newman, D. (1973): Growth of mouse vaginal epithelial cells *in vitro. In Vitro,* 9:194–201.
8. Gossens, C. L., and Unsworth, B. R. (1972): Evidence for a two-step mechanism operating during *in vitro* mouse kidney tubulogenesis. *J. Embryol. Exp. Morphol.,* 28:615–631.
9. Graham, C. F., and Kelly, S. J. (1977): Interactions between embryonic cells during the early development of the mouse. In: *Cell Interactions in Differentiation,* edited by M. Karkinen-Jääskeläinen, L. Saxén, and L. Weiss, pp. 45–57. Academic Press, London.
10. Grobstein, C. (1955): Inductive interaction in the development of the mouse metanephros. *J. Exp. Zool.,* 130:319–340.
11. Kollar, E. J., and Baird, G. R. (1970): Tissue interaction in embryonic mouse tooth germs. II. The inductive role of the dental papilla. *J. Embryol. Exp. Morphol.,* 24:173–186.
12. Lawson, K. A. (1974): Mesenchyme specificity in rodent salivary gland development: The response of salivary epithelium to lung mesenchyme *in vitro. J. Embryol. Exp. Morphol.,* 32:469–493.
13. Le Douarin, N. (1975): An experimental analysis of liver development. *Med. Biol.,* 53:427–455.
14. Nordling, S., Miettinen, H., Wartiovaara, J., and Saxén, L. (1971): Transmission and spread of embryonic induction. I. Temporal relationships in transfilter induction of kidney tubules *in vitro. J. Embryol. Exp. Morphol.,* 26:231–252.
15. Pictet, R. L., Filosa, S., Phelps, P., and Rutter, W. J. (1975): Control of DNA synthesis in the embryonic pancreas: Interaction of the mesenchymal factor and cyclic AMP. In: *Extracellular Matrix Influences on Gene Expression,* edited by H. C. Slavkin, and R. C. Greulich, pp. 531–540. Academic Press, New York.
16. Rutter, W. J., Wessells, N. K., and Grobstein, C. (1964): Control of specific synthesis in the developing pancreas. *Natl. Cancer Inst. Monogr.,* 13:51–65.

17. Saxén, L. (1970): Failure to demonstrate tubule induction in a heterologous mesenchyme. *Dev. Biol.,* 23:511–523.
18. Saxén, L., Karkinen-Jääskeläinen, M., Lehtonen, E., Nordling, S., and Wartiovaara, J. (1977): Inductive tissue interactions. In: *Cell Surface Interactions in Embryogenesis,* edited by G. Poste, and G. L. Nicolson, pp. 331–407. North-Holland Division of ASP Biological and Medical Press, Amsterdam.
19. Saxén, L., Lehtonen, E., Karkinen-Jääskeläinen, M., Nordling, S., and Wartiovaara, J. (1976): Are morphogenetic tissue interactions mediated by transmissible signal substances or through cell contacts? *Nature,* 259:662–663.
20. Saxén, L., Toivonen, S., and Vainio, T. (1964): Initial stimulus and subsequent interactions in embryonic induction. *J. Embryol. Exp. Morphol.,* 12:333–338.
21. Saxén, L., and Toivonen, S. (1962): *Primary Embryonic Induction.* Logos Press, Academic Press, London.
22. Sengel, P., and Dhouailly, D. (1977): Morphogenesis of amniote cutaneous appendages. In: *Cell Interactions in Differentiation,* edited by M. Karkinen-Jääskeläinen, L. Saxén, and L. Weiss, pp. 153–169. Academic Press, London.
23. Thesleff, I. (1977): Transfilter interactions in tooth development. In: *Cell Interactions in Differentiation,* edited by M. Karkinen-Jääskeläinen, L. Saxén, and L. Weiss, pp. 191–208. Academic Press, London.
24. Toivonen, S. (1945): Zur Frage der Induktion selbständiger Linsen durch abnorme Induktoren im Implantatversuch bei Triton. *Ann. Soc. Zool. Bot. Fenn. Vanamo,* 11:1–28.
25. Toivonen, S. (1967): Mechanism of primary embryonic induction. In: *Morphological and Biochemical Aspects of Cytodifferentiation,* edited by E. Hagen, W. Wechsler, and P. Zilliken, pp. 1–7. S. Karger, Basel.
26. Toivonen, S., and Saxén, L. (1968): Morphogenetic interaction of presumptive neural and mesodermal cells mixed in different ratios. *Science,* 159:539–540.
27. Unsworth, B., and Grobstein, C. (1970): Induction of kidney tubules in mouse metanephrogenic mesenchyme by various embryonic mesenchymal tissues. *Dev. Biol.,* 21:547–556.

Cell and Tissue Interactions, edited by
J. W. Lash and M. M. Burger. Raven
Press, New York, 1977.

Influence of the Tissue Environment on the Differentiation of Neural Crest Cells

N. M. Le Douarin, M. A. Teillet, and C. Le Lièvre

Institut d'Embryologie du Centre National de la Recherche Scientifique et du Collège de France, 49 bis, Avenue de la Belle-Gabrielle, 94130, Nogent-sur-Marne, France

The neural crest originates from the neural folds and appears early in embryogenesis, when the closure of the neural tube occurs. Called by Raven (18) one of the "primary organs" in the development of the vertebrate embryo, it has in fact a temporary existence, since its cells are soon dispersed throughout the body, differentiating into a number of tissues. From the moment they begin to migrate, the neural crest cells become indistinguishable from the tissues through which they move. For this reason, the details of the migratory process cannot be observed directly, and various indirect experimental methods have been applied to determine the developmental capabilities and normal fates of neural crest cells. They can be classified into three main procedures: ablation, explantation or heterotopic grafting in an embryonic area which does not belong to the neural primordium, and cell marking *in situ* or in combination with orthotopic grafting.

None of these techniques being fully satisfactory, controversies and uncertainties about the fates and migratory behavior of neural crest cells, may be attributed to ambiguities in the methods of analysis. Anyhow, convergent findings have shown that the neural crest gives rise to a number of differentiated cell types, such as mesenchymal derivatives, melanocytes, sensory ganglia, peripheral nerve cells of the autonomic system, calcitonin-producing cells of the ultimobranchial bodies, type I cells of the carotid body, Schwann cells, and also some supporting elements of the central nervous system (see refs. 4,10,21). This raises the problem of the time at which crest cells are committed to give rise to these various cell categories and also of the possible epigenetic influences which may control both the pattern of their migration and the expression of their differentiated phenotype.

In the study reported in this article, we have investigated the extent to which the developmental capabilities of neural crest cells are limited when they begin to migrate and we have tried to evaluate the importance for migratory behavior and differentiation of the environment the cells meet either during or at the end of their migration.

DEVELOPMENTAL FATE OF CREST CELLS AT THE VARIOUS LEVELS OF THE NEURAL AXIS

To trace the crest cell migration, we have used a stable cell-marking technique based on differences in the structure of the interphase nucleus in quail (*Coturnix coturnix japonica*) and chick (*Gallus gallus*) cells (7–9, 12). The general experimental procedure already described in previous articles (10) consists in grafting orthotopically fragments of quail neural primordium into chick embryos or conversely. The implanted neural crest cells migrate into the host according to a normal pattern and can be identi-

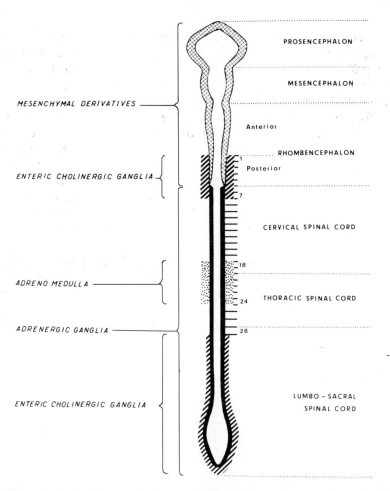

FIG. 1. Diagram shows the presumptive areas of the neural crest giving rise to the mesectoderm, the adrenomedulla, and the ganglion cells of the autonomic nervous system. This map has been inferred from the results of isotopic and isochronic grafts of fragments of neural primordium between quail and chick embryos.

fied due to their nuclear characteristics whatever the duration of the chimeric association may be.

From such transplantations, systematically carried out along the whole neural axis, a picture of the developmental fate of the crest cells at the various levels of the body has emerged (Fig. 1).

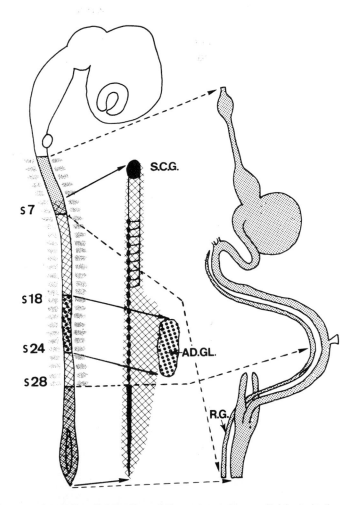

FIG. 2. Diagram shows the distribution of the autonomic ganglioblasts in the sympathetic system and the digestive tract. The vagal level of the neural crest (from somites 1 to 7) provides all the enteric ganglia of the preumbilical gut and contributes to the innervation of the postumbilical gut. The lumbosacral level of the neural crest gives rise to the ganglion of Remak **(RG)** and to most of the ganglia of the postumbilical gut. The orthosympathetic chain derives from the level of the neural crest posterior to the 5th somite and the adrenomedullary cells originate from the level of somites 18 to 24. **SCG:** Superior cervical ganglion. **AD. GL:** Adrenomedullary gland.

It appears that the neural crest can be divided into several areas with respect to the presumptive fate of its cells:

1. The potentiality to produce mesenchymal derivatives is restricted to the cephalic neural crest down to the level of the 4th to 5th somites (15).

2. On its whole length, the neural axis gives rise to melanoblasts; however, a number of observations suggest that presumptive melanocytes are more numerous in the cervicotruncal than in the cephalic area of the crest (13,19).

3. The autonomic ganglion cells of the sympathetic and parasympathetic systems originate from well-defined regions of the neural axis (Fig. 2).

The enteric ganglia are derived mostly from the "vagal" neural crest (level of somites 1 to 7). Cells originating from the *medulla oblongata* migrate ventrally in the branchial arches and become incorporated into the wall of the gut which is derived from splanchnopleural mesoderm. They are responsible for the formation of all the parasympathetic ganglia of the anteumbilical gut and participate also in the innervation of the postumbilical gut. However, at this level ganglion cells arise mainly from the lumbosacral neural crest (behind the level of the 28th somite), which also gives rise to the ganglion of Remak.

The adrenergic cells of the sympathetic system originate from the entire neural crest located behind the level of the 5th somite, and the adrenomedulla arises strictly from the area corresponding to somites 18 to 24.

Thus, regarding the origin of the autonomic ganglioblasts, the neural crest may be divided into several sections: one, at the "vagal" level (somites 1 to 5), gives rise only to cholinergic neurons; the next is cervicodorsal and only adrenergic cells develop (level of somites 8 to 28) from it; other parts have the ability to produce both adrenergic and cholinergic neurons (one at the level of somites 5 to 7 and the other corresponding to the lumbosacral spinal cord).

DIFFERENTIATING CAPABILITIES OF THE NEURAL CREST CELLS AT VARIOUS LEVELS OF THE NEURAL AXIS

The results of the isotopic–isochronic grafting experiments raise several questions:

Is the neural crest a mosaic of regions with limited differentiating abilities? Or, on the contrary, are the various potencies of the crest widely spread all over the neural axis? If the latter is the case, differentiation of crest cells could be under control of the environment they meet either during their migration or when they have already settled in their definitive site, or during both these processes. This problem was investigated in several ways, involving either the transfer of fragments of the neural primordium to heterotopic levels of the neural axis or the coculture of definite areas of the neural crest with various embryonic rudiments.

Capability for Connective Tissue Differentiation

The question first investigated concerns the capability of the neural crest to give rise to mesenchymal derivatives. As shown before, the cephalic part of the neural crest is the only one able to produce mesenchymal cells, the mesencephalon and rhombencephalon being the main sources of the facial and visceral structures (connective tissue, dermis, bone, cartilage, wall of the large arteries) (15).

Two types of experiments have been performed: (a) the transplantation of the mesencephalon and anterior rhombencephalon at the truncal level and (b) the transfer of the truncal neural crest at the level of mesencephalon or rhombencephalon.

In the first experimental series, the mesencephalon and anterior rhombencephalon from 4 to 9-somite quail embryos have been transplanted into chick embryos at the truncal level (somites 18 to 24) in order to see whether mesenchymal derivatives of neural crest origin can develop in the trunk. The host embryo sacrificed at 13 days of incubation showed a large mass of brain tissue that developed from the graft and prevented the cutaneous cicatrization, which occurs in isotopic transplants. At the level of the graft, induction of cartilage in the host sclerotome occurred, and vertebral structures abnormal in morphology were formed. In some cases, the vertebral cartilage was chimeric with nodules of quail cells included in the host vertebra (Fig. 3). Mesenchymal tissues arising from the neural crest were

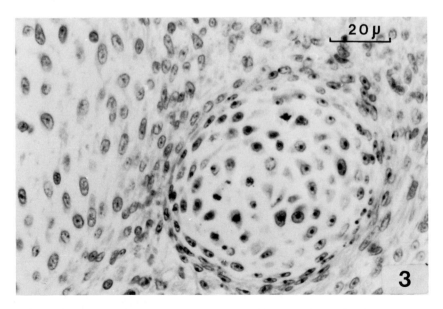

FIG. 3. Quail chondrocytes in the host vertebra following the graft of a quail mesencephalon at the "adrenomedullary" level of the chick neural axis. Feulgen-Rossenbeck's staining.

regularly found in the derivatives of the somite and the intermediate cell mass. Pieces of cartilage made up of quail cells have been encountered in all the cases observed inside or in the vicinity of the mesonephros and in the wall of the Müllerian duct. Connective cells of quail type participate in dermis, Müllerian duct wall, and also in the intertubular mesenchyme of the kidney (Figs. 4 and 5). No connective tissue of neural crest origin was found in the lateral plate derivatives (either somatopleura or splanchno-pleura).

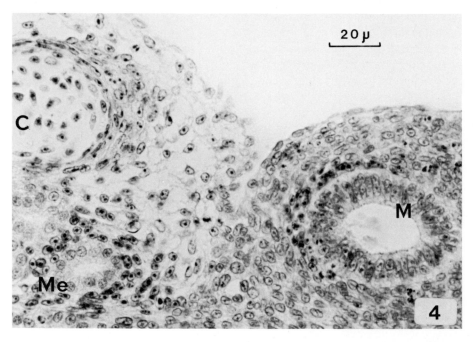

FIG. 4. Same experiment as in Fig. 3. Mesectodermal cells originating from the graft form connective tissue in the mesonephros **(Me)** and Müllerian duct **(M). C:** Piece of cartilage derived from the implant.

The graft of trunk neural primordium from 25-somite quail embryo at the cephalic level (mesencephalon, anterior or posterior rhombencephalon) of the chick neural axis has been performed. The external morphology of the head and neck of the host embryos was the same as that observed after the removal of the corresponding level of the neural tube. The operation resulted in severe deficiencies of facial and visceral arch structures (Fig. 6) (14), showing that the trunk neural crest is not able to replace efficiently the cephalic neural crest if transplanted into the head.

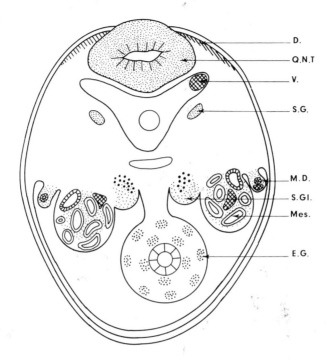

FIG. 5. Schematic representation of the sites where mesectodermal derivatives are found in the truncal structures of the host following the graft of a quail mesencephalon at the dorsal level of the chick neural axis. **D:** Dermis. **EG:** Enteric ganglia. **MD:** Müllerian duct. **Mes:** Mesonephros. **QNT:** Grafted quail neural tube. **SG:** Sympathetic ganglion. **S. Gl:** Suprarenal gland.

Capability for Autonomic Ganglia Formation

The isotopic–isochronic grafting experiments reported above have shown that in normal development the neural crest is precisely regionalized into areas from which either the cholinergic or the adrenergic neurons of the peripheral ganglia originate. It seemed interesting to investigate whether the developmental fate of the presumptive ganglion cells could be changed when their migration pathways and their definitive sites were experimentally modified.

Transplantation of the Cervicodorsal Region of the Neural Primordium at the Vagal Level

It has been previously established that, in normal development, the cells originating from the level of somites 8 to 28 do not migrate into the intestine and therefore do not participate in enteric ganglia formation. However, if the "adrenomedullary" level of the quail neural crest (somites 18 to 24) is

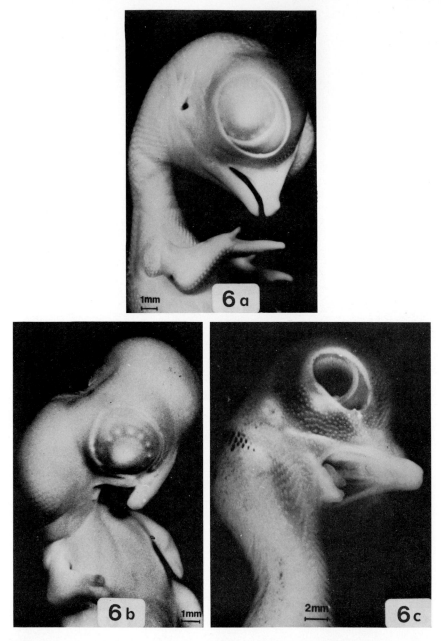

FIG. 6. a: Chick embryo at 9 days of incubation: control. **b:** Chick embryo at 9 days of incubation in which the mesencephalon and rhombencephalon have been excised. The lower jaw is not developed and the neck is reduced especially on its ventral side. These abnormalities result from the absence of the mesectodermal cells derived from mesencephalic and rhombencephalic neural crest. **c:** Chick embryo at 12 days of incubation in which the mesencephalon has been replaced by a fragment of the truncal neural primordium taken from a quail embryo. The lower jaw, the tongue, and the eyelids are reduced due to the lack of mesectodermal cells.

transplanted into the vagal region of a chick embryo, a massive colonization of the host gut by quail cells occurs. The enteric ganglia are entirely of the graft type down to the umbilicus, while in the postumbilical gut the parasympathetic innervation is derived from the lumbosacral crest of the host. Melanocytes originating from the graft migrate into the postumbilical gut and are always located either along the internal surface of the circular smooth muscle layer, like the cells of the Meissner's plexus do or are associated with the ganglia of Auerbach.

The neuroblasts originating from the vagal level of the neural crest do not differentiate into adrenergic cells according to their presumptive fate in normal development, but into cholinergic neurons as shown by the application of a physiological test for the cholinergic innervation of the gut smooth muscles (11).

These results show that the cervicodorsal neural crest can give rise to enteric ganglion cells if transplanted at the vagal level of the neural axis. This raises the question of whether changes of developmental significance take place either during the dorsoventral migration of the neural crest cells or after they have settled in the mesenchyme. To answer this question we have devised the following experiment.

Coculture of Truncal Neural Crest and Aneural Hind Gut

The transverse level of the embryo corresponding to the hind brain is characterized by a preferential migration route through which the crest cells are led to the developing foregut. From there, they migrate in a caudal direction and progressively invade the whole digestive tract. In normal development, cells of the vagal neural crest reach the posterior gut several days after the onset of migration, since they are seen in the large intestine and rectum only from the 7th day of incubation.

It is also at this stage that ganglion cells originating from the lumbosacral level of the crest colonize the posterior gut. Therefore, if the hind gut is taken from the embryo at 4 or 5 days of incubation, and grafted on the chorioallantoic membrane (CAM), it remains aneural, although otherwise its development proceeds normally. The hind gut of 5-day-old chick embryos is cultured for 10 days on the CAM in association with the neural primordium taken from quail embryos at the level of somites 18 to 24. The crest cells migrate into the gut mesenchyme and form ganglia of Auerbach's and Meissner's plexuses. Ganglion cells do not show any catecholamine content as revealed by the application of Falck's technique (2) but develop into neurons with a strong acetylcholinesterase activity and a positive reaction with the silver impregnation technique of Ungevitter (20). Thus, it appears that the factors controlling cholinergic differentiation of the autonomic ganglioblasts do not act significantly during the migration process of the cells but after they are localized in the gut mesenchymal environment.

Capability of the Cephalic Neural Crest to Differentiate into Autonomic Ganglia and Paraganglia

The interspecific transplantation of the cephalic neural crest at the truncal level (somites 18 to 24) results in colonization of the suprarenal gland of the

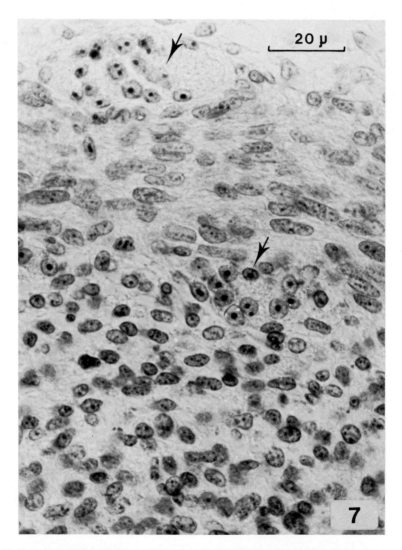

FIG. 7. Enteric ganglia made up of quail cells **(arrows)** in the hind gut of a chick embryo which has received the graft of a quail mesencephalon at the "adrenomedullary" level of the neural axis. Feulgen-Rossenbeck's staining.

host by cells originating from the graft. The latter differentiate into adrenomedullary cells exhibiting a high content in catecholamines and the typical electron dense secretory granules of epinephrine-producing cells (13). Neurons of graft origin are also seen along the ventrolateral sides of the aorta and in the location of the sympathetic ganglion chain. In addition, some neural crest cells migrate into the dorsal mesentery and penetrate the developing gut at the level of the graft. Later on they proceed caudally and are finally found forming the enteric ganglia of both Auerbach's and Meissner's plexuses of the hind gut (Fig. 7). The ganglion of Remak, however, is always of host type.

This observation brings up two new points: (a) that the capacity to give rise to autonomic ganglion cells is widely spread in the whole neural crest and (b) that cephalic crest cells can migrate into the gut if transplanted into the trunk, although the cells originating normally from this level of the neural primordium never penetrate the dorsal mesentery.

The most likely explanation of this fact is that the cephalic neural crest is made up of a much larger number of cells than the dorsal one. Thus, after a certain amount of cells have settled in the trunk, the others proceed ventrally, invade the gut, and differentiate into enteric ganglia.

INFLUENCE OF THE AXIAL STRUCTURES OF THE TRUNK ON THE DIFFERENTIATION OF CATECHOLAMINERGIC CELLS

In the embryo, the localization of the sympathetic adrenergic neurons and of the adrenalinogen paraganglion cells is strictly restricted to the dorsal region of cervix and trunk. No cells of that kind are actually found either in splanchnopleural or in somatopleural derivatives. The question arises then as to whether the dorsal structures (the neural tube, the notochord, or the somitic mesenchyme) are of any significance in promoting catecholamine synthesis in the autonomic ganglioblasts.

Previous works of Cohen (1) and Norr (17) have suggested that sympathoblast differentiation could be stimulated exclusively by the somitic mesenchyme. The latter, however, was supposed to acquire such a capacity from the influence of notochord and neural tube. We have reinvestigated this question in order to analyze the respective roles of the notochord and the somitic mesenchyme in this developmental process.

The neural primordium, taken from the adrenomedullary level of 25-somite quail embryos, was inserted together with the notochord into the hind gut of a 5-day chick and cultivated for 9 to 10 days on the CAM. In three out of seven explants, catecholamine-producing cells developed in the vicinity of the notochord. In addition, enteric ganglia differentiated as in the experiment reported above, i.e., association of the hind gut with the neural primordium and without the notochord.

This result shows that the gut wall mesenchyme is able to sustain adrenergic cell differentiation provided that the notochord is present in the cultures. This leads us to consider that the capacity for promoting adrenergic cell differentiation is not restricted to the somitic derivatives but rather depends on the specific influence of the notochord and maybe also the neural tube.

SHIFT IN NEUROTRANSMITTER METABOLISM IN THE GANGLION OF REMAK BY A CHANGE IN ITS TISSUE ENVIRONMENT

The ganglion of Remak, which develops in the dorsal mesentery, belongs partially to the cholinergic parasympathetic system, although it contains nerve fibers of postganglionic sympathetic neurons. If taken at 4 or 5 days of incubation and cultured on the CAM, it develops into an elongated ganglion in which no catecholamine content is detectable by Falck's procedure. In contrast, if the ganglion of Remak of a 4-day quail embryo is transplanted into the dorsal structure of a 2.5-day chick (Fig. 8), its further evolution is modified. The cells of the grafted ganglion do not remain associated together; they migrate and become distributed among the following sites: (a) the orthosympathetic ganglia, where they can be either mixed up with host sympatho-

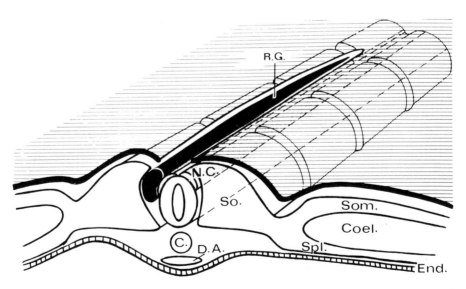

FIG. 8. Diagram showing the implantation of the ganglion of Remak taken from a 4-day quail embryo into a 2.5-day chick. **C:** Notochord. **Coel:** Coelomic cavity. **DA:** Dorsal aorta. **End:** Endoderm. **NC:** Neural crest. **SO:** Somite. **Som:** Somatopleure. **Spl:** Splanchnopleure.

blasts or form the whole ganglion, (b) the aortic plexus, and (c) the suprarenal gland. In this location, some cells participate in the formation of the
adrenomedullary strands, while some others are grouped in the vicinity of
the gland (Fig. 9).

The cytochemical procedure of Falck for catecholamine detection, followed by Feulgen's staining, shows that the cells located in the sympathetic

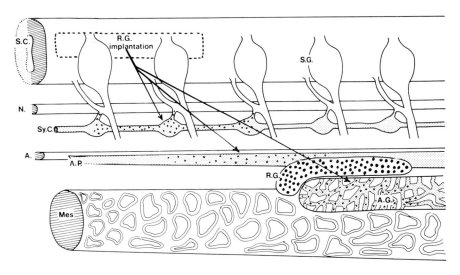

FIG. 9. Diagram shows the distribution of cells of the ganglion of Remak **(RG)** following
its transplantation at the dorsal level between the spinal cord **(SC)** and the somites (see
Fig. 8). Six days after the graft, the quail cells are distributed in the orthosympathetic
chain **(SyC)**, the aortic plexus **(AP)**, and the adrenomedullary gland **(AG)**. Large nerve
cells deriving from the ganglion of Remak are also found in the vicinity of the suprarenal
gland and of the mesonephros **(Mes)**. **SG:** Spinal ganglion.

ganglia, the aortic plexus, and the suprarenal gland contain both fluorogenic
amines and the quail nuclear marker. In the ganglion located close to the
suprarenal gland, only some cells are fluorescent, while the others with
large nucleus and perikaryon do not seem to have fluorogenic amine content
(Fig. 10).

CONCLUSIONS AND DISCUSSION

The experiments reported above show that, although the neural crest is
regionalized into well-defined areas giving rise to different cell types, most of
the developmental capabilities of crest cells are widely spread along the
neural axis. Such is the case, for example, for the autonomic ganglioblasts
that can be produced by the cephalic as well as by the trunk neural crest. In

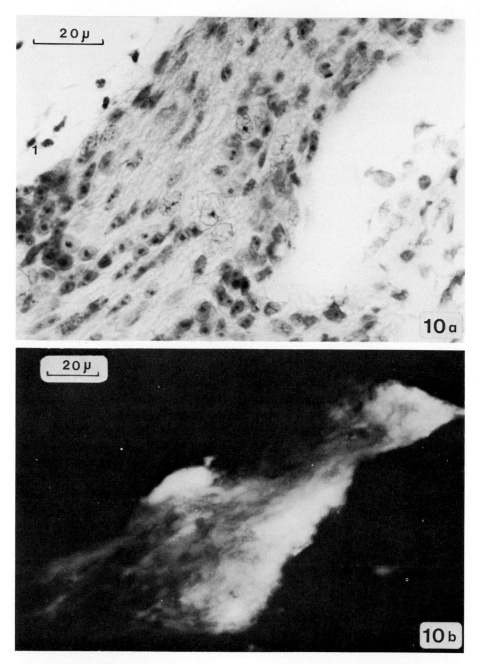

FIG. 10. Quail cells located in a ganglion of the sympathetic chain of an 8-day chick embryo host into which the ganglion of Remak of a 4-day quail embryo has been grafted in the dorsal region **(a)** (see Fig. 8). The application of Falck's technique for fluorogenic monoamine detection shows that the quail cells contain catecholamines **(b).** The two photographs represent the same section of the ganglion.

addition, whatever the level of origin of the autonomic neuroblasts, their differentiation into adrenergic and cholinergic neurons depends exclusively upon the site they reach at the end of their migration. The notochord seems, in this respect, to have a decisive role in inducing the catecholamine metabolic pathway in the differentiating neuroblasts.

The normal developmental fate of the ganglioblasts of the ganglion of Remak is to become cholinergic neurons. However, if their environment is changed early, that is, at 4 days of incubation, their metabolism can be oriented toward catecholamine synthesis.

By contrast, the cells that give rise to the mesectoderm are strictly localized in the head neural crest, since the truncal part of the neural axis does not seem to have capacity to produce mesenchymal derivatives. The fact that head neural crest transplanted at the truncal level gives rise to connective tissue and cartilage suggests the existence of a population of "mesenchymal crest stem cells" determined early and distinct from those responsible for the differentiation of the nonmesenchymal cell types (nerve cells, melanocytes, calcitonine cells, carotid body cells, etc.).

Another point that is evidenced by the observations described in this chapter concerns the migratory behavior of the neuroblasts. It is obvious that the site reached by crest cells depends on the transverse level of the embryo from which they start migrating. Heterotopic transplants of the crest in various places of the neural axis show that the cells follow the same routes and stop at the same sites whatever their primary origin is. For instance, at the hind brain level, the cells migrate ventrally, become incorporated into the intestine, and then proceed in a caudal direction. The first cells that migrate settle in the anterior gut, while the next successively stop in the medium and posterior gut.

It is obvious that the trunk neural crest contains a smaller number of cells than the head one. When a piece of truncal neural crest is transplanted at the hind brain level, quail ganglion cells are present only in the preumbilical part of the gut. Very likely this is due to the fact that crest cells are not numerous enough to invade the whole intestine. When the cephalic neural crest is grafted at the level of somites 18 to 24 it gives rise to enteric ganglia. The latter are located exclusively in the hind gut of the host, showing again that the migration of cells in the gut proceeds caudad. In addition, the cells of cephalic origin settle in the hind gut before the onset of migration of the lumbosacral neuroblasts. The latter are then prevented from participating in the posterior gut innervation. Therefore, it seems that preferential sites of arrest for neural crest cells are distributed according to a precise pattern in the embryonic primordia into which those cells migrate. Each site is able to house a limited amount of cells. If migrating cells arrive to a fully occupied site, they proceed further and settle in the first free location they encounter.

Regarding the evolution of the mesenchymal derivatives of the neural crest, some points should be noted. First, mesectodermal cells transplanted into the trunk differentiate into dermis and take part in germ feather formation when they are in contact with the ectoderm. In addition, they are sensitive to the vertebral morphogenetic field, just as mesodermal cells are. They are indeed induced to differentiate into cartilage and become incorporated in the vertebra of the host. The fact that cartilage of mesectodermal origin is also found in the derivatives of the intermediate cell mass may be related to the ability of this rudiment to differentiate into cartilage as reported previously by Lash (6). The tissue environment provided by various truncal structures is able to ensure chondrification of mesectodermal cells in the same way as the pharyngeal endoderm does as shown in Amphibian embryo by a number of works (3,5,16).

REFERENCES

1. Cohen, A. M. (1972): Factors directing the expression of sympathetic nerve traits in cells of neural crest origin. *J. Exp. Zool.*, 179:167–182.
2. Falck, B. (1962): Observations on the possibility of the cellular localization of monoamines by a fluorescence method. *Acta Physiol. Scand.*, 56 (Suppl. 197): 1–25.
3. Holtfreter, J. (1968): Mesenchyme and epithelia in inductive and morphogenetic processes. In *Epithelial-Mesenchymal Interactions*, edited by R. Fleischmajer, and R. E. Billingham, pp. 1–30. Williams & Wilkins, Baltimore.
4. Hörstadius, S. (1950): The neural crest. Its properties and derivatives in the light of experimental research. Geoffrey Cumberlege, Oxford University Press, London.
5. Hörstadius, S., and Sellman, S. (1946): Experimentelle Untersuchungen über die Determination des Knorpeligen Kopfskelettes bei Urodelen. *Nova Acta R. Soc. Sci. Uppsal.*, Ser. IV, 13:1–170.
6. Lash, J. W. (1963): Studies on the ability of embryonic mesonephros explants to form cartilage. *Dev. Biol.*, 6:219–232.
7. Le Douarin, N. (1969): Particularités du noyau interphasique chez la Caille japonaise (Coturnix coturnix japonica). Utilisation de ces particularités comme "marquage biologique" dans des recherches sur les interactions tissulaires et les migrations cellulaires au cours de l'ontogenèse. *Bull. Biol. Fr. Belg.*, 103:435–452.
8. Le Douarin, N. (1973): A biological cell labeling technique and its use in experimental embryology. *Dev. Biol.*, 30:217–222.
9. Le Douarin, N. (1973): A feulgen-positive nucleolus. *Exp. Cell Res.*, 77:459–468.
10. Le Douarin, N. (1974): Cell recognition based on natural morphological nuclear markers. *Med. Biol.*, 52:281–319.
11. Le Douarin, N., Renaud, D., Teillet, M. A., and Le Douarin, G. H. (1975): Cholinergic differentiation of presumptive adrenergic neuroblasts in interspecific chimeras after heterotopic transplantations. *Proc. Natl. Acad. Sci., U.S.A.* 72:728–732.
12. Le Douarin, N., and Rival, J. M. (1975): A biological nuclear marker in cell culture: Recognition of nuclei in single cells and in heterokaryons. *Dev. Biol.*, 47:215–221.
13. Le Douarin, N., and Teillet, M. A. (1974): Experimental analysis of the migration and differentiation of neuroblasts of the autonomic nervous system and of neurectodermal mesenchymal derivatives, using a biological cell marking technique. *Dev. Biol.*, 41:162–184.
14. Le Lievre, C. (1976): Contribution des crêtes neurales à la genèse des structures céphaliques et cervicales chez les Oiseaux. Thèse d'Etat, Nantes.
15. Le Lievre, C., and Le Douarin, N. (1975): Mesenchymal derivatives of the neural crest: Analysis of chimaeric quail and chick embryos. *J. Embryol. Exp. Morphol.*, 34:125–154.
16. Newth, D. R. (1954): Determination of the cranial neural crest of the Axolotl. *J. Embryol. Exp. Morphol.*, 2:101–105.
17. Norr, S. C. (1973): *In vitro* analysis of sympathetic neuron differentiation from chick neural crest cells. *Dev. Biol.*, 34:16–38.

18. Raven, C. P. (1942): *C. R. Soc. Neerl. Zool.,* quoted from ref. 4.
19. Teillet, M. A. (1971): Recherches sur le mode de migration et la différenciation des mélanoblastes cutanés chez l'embryon d'Oiseau: étude expérimentale par la méthode des greffes hétérospécifiques entre embryons de Caille et de Poulet. *Ann. Embryol. Morphol.,* 4:95–109.
20. Ungevitter, L. H. (1951): A urea silver nitrate method for nerve fibers and nerve endings. *Stain Technol.,* 26:73–76.
21. Weston, J. A. (1970): The migration and differentiation of neural crest cells. In *Advances in Morphogenesis,* Vol. 8, edited by M. Abercrombie, J. Brachet, and T. King, pp. 41–114. Academic Press, London.

Cell and Tissue Interactions, edited by
J. W. Lash and M. M. Burger. Raven
Press, New York, 1977.

Epithelial-Mesenchymal Interactions: Mesenchymal Specificity

Harold C. Slavkin, Gary N. Trump, Anna Brownell, and Nino Sorgente

Laboratory for Developmental Biology, Ethel Percy Andrus Gerontology Center, and Departments of Biochemistry and Microbiology-Immunology, School of Dentistry, University of Southern California, Los Angeles, California 90007

The regulatory processes operating during epithelial-mesenchymal interactions associated with epidermal organ development are as yet obscure. Examples of several extensively described epithelial-mesenchymal organ systems include salivary gland morphogenesis (11,12), kidney formation (22,24,32,33), pancreas development (23,26,28,30), limb development (31,54,55), feather pattern development (37), and tooth morphogenesis (19–21,29,45). A critical developmental question common to the study of each of these systems is: What is the nature of and how are inductive signals transmitted over short intercellular distances between heterotypic tissues?

The objective of this discussion is to describe a series of experiments designed to identify mechanisms by which close-range intercellular communication effects a specific epithelial phenotype. The value of the tooth organ system lies in its high degree of developmental programming for sequential mesenchymal and epithelial cell differentiation (6). It is evident that the lexicon of developmental "instructions" necessary to initiate ameloblast cell differentiation reside in part within mesenchymal cell surface molecules. The transmission of a signal, the receiving of the signal, and the translation of the signal into a specific phenotype appear to be mediated by interactions among molecules located within the outer surfaces of mesenchymal cell processes, matrix vesicles, and the inner enamel epithelia.

EPITHELIAL-MESENCHYMAL INTERACTIONS: MESENCHYME SPECIFICITY

The differentiation of epidermal organ systems into functional systems is dependent upon interactions termed epithelial-mesenchymal interactions. An understanding of such heterotypic tissue interactions is a prerequisite in studies designed to investigate normal and abnormal organogenesis. A series of insights into the principles associated with tissue interactions have been acquired primarily through experimental intervention during embryonic organ development: The developmental capabilities of tissues have been tested under conditions of *in vitro* isolation and in tissue recombination experiments of various heterotypic tissues. By these methods it has been repeatedly shown that, in general, an embryonic epithelium depends upon the influence of an underlying mesenchyme in order to undergo histological differentiation and morphogenesis (11,12,32,45). Numerous embryonic organ rudiments, such as the pancreas, salivary gland, tooth organs, kidney, thyroid, lung, and mammary gland, can be isolated and continue to differentiate and express phenotypic characteristics unique to a specific organ when cultured *in vitro*. The epithelia and mesenchyme can be dissociated, recombined, and will also continue to express the developmental characteristics of the organ system *in vitro*. Each isolated tissue interactant will not differentiate when cultured alone. In general, the organ specificity of the mesenchyme will influence a "responsive" epithelia to differentiate into an epithelium complementary to the mesenchymal specificity. For example, embryonic rabbit mammary gland mesenchyme has been shown to induce avian flank epithelium to differentiate into mammary gland-like epithelium complete with branching tubule formation (27). In numerous examples, mesenchymal specificity has been demonstrated through heterologous tissue recombinations to mediate developmental information across vertebrate species (21).

The results obtained from a number of studies indicate that mesenchyme produces permissive and/or growth rate-limiting factors that accommodate the phenotypic expression of protodifferentiated epithelia in tissue recombinants. Mesenchymal specificity in this sense is analogous to an "amplification phenomenon" in which already determined developmental events (protodifferentiated) were provided a microenvironment complementary to the expression of a particular phenotype. In such experiments, specific epithelial gene products were detectable in low concentrations before the experiment had begun. Evidence from yet other studies indicates that mesenchyme produces a specific set of molecular instructions that determines the cytodifferentiation of the epithelium. For example, it has been shown that mouse embryonic lung mesenchyme can elicit branching morphogenesis in tracheal epithelium, a tissue that does not normally branch (1). Embryonic mouse molar tooth organ mesenchyme has been shown to induce mouse foot pad epithelium to differentiate into ameloblasts and form a tooth complete with dentine and enamel matrices [see extensive review (21)].

In almost all published studies of epithelial-mesenchymal interactions, histogenesis and morphogenesis (e.g., branching tubule formation, glandular appearance, metachromatic extracellular matrix deposition, cessation of mitosis, cell elongation, etc.) have been used as the criteria to assess mesenchyme specificity. In a defined system, no evidence as yet has been reported that unequivocally demonstrated mesenchymal induction of *new* gene transcription, translation, posttranslational modifications, or secretion of unique macromolecules by a responding epithelium. Therefore, one of the central issues is to design experiments that might obtain evidence that could discriminate between (a) instructive developmental signals derived from the mesenchyme, and (b) permissive and/or growth rate-limiting factors that might amplify or retard the synthesis and secretion of epithelial gene products.

EMBRYONIC MAMMALIAN TOOTH ORGAN DEVELOPMENT: GENERAL CONSIDERATIONS

The embryonic tooth organ provides a unique opportunity to study epithelial-mesenchymal interactions *in vivo* and *in vitro*. In the course of embryonic development, cranial neural crest cells migrate into the forming craniofacial complex and differentiate into dental papilla mesenchyme among other cell type derivatives (16). The progenitor tooth organ mesenchyme differentiate in juxtaposition to an ectodermally-derived epithelium that will differentiate into the enamel organ epithelium [see extensive descriptions of tooth morphogenesis (10,45)]. These two tissue types interact throughout the formative phases of tooth organ morphogenesis. The interface between these heterotypic tissues provides a useful spatial and temporal orientation with which to describe relative features along the gradients of first dentinogenesis and then amelogenesis (i.e., dentine formation precedes enamel matrix secretion by approximately 48 hr in most mammalian species).

During mammalian embryogenesis (for example, 9.5 to 10th day of gestation in mice), the dental lamina is a band of thickening of the ectodermally derived oral epithelium, which indicates the future position of each dental arch (10). The dental lamina forms an ingrowth of oral epithelial cells into the adjacent ectomesenchyme (i.e., cranial neural crest-derived cells) of the jaw. Both tissues are separated by a continuous basal lamina associated with the undersurface of the epithelia (a metachromatic basement membrane is evident in the light microscope). Presumably through cell proliferation, the lamina first forms a bud that subsequently becomes invaginated and forms a bell-shaped epithelial structure called the "enamel organ epithelia." Through continued epithelial-mesenchymal interactions the enamel organ increases in size and morphological complexity until the basic configuration of the tooth organ is determined with the onset of odontoblast cell differentiation and dentine matrix deposition (10).

The inner surface of the enamel organ, inner enamel epithelium, is in juxtaposition to the differentiating odontoblast cells. The inner enamel epithelia is a proliferating sheet of cuboidal cells. Among the epithelial undersurface, facing the predentine forming between the tissue interactants, is a continuous basal lamina. The mesenchymal cells have differentiated into preodontoblasts, then odontoblasts, and synthesize and secrete dentine collagen, dentine phosphoprotein, and proteoglycans prior to the initiation of enamel protein synthesis and secretion (45). The odontoblast differentiation begins at the apex of the forming tooth organ and appears as a gradient moving in a lateral and distal direction. After 4 to 6 μm of predentine is secreted, the predentine begins to mineralize and forms dentine (45). At this time and place, amelogenesis is initiated in juxtaposition to that location in which dentine is first formed (10). Prior to the initiation of enamel synthesis and secretion, the epithelial basal lamina is discontinuous (17,25,38) and is then completely degraded. The preameloblasts stop cell division and acquire an elongated appearance into tall columnar ameloblast cells with an elaborate rough endoplasmic reticulum and a prominent supranuclear Golgi apparatus. Essentially all of the mitochondria are translocated to the infranuclear zone of the cell. Secretion of enamel matrix is through the lateral and apical surfaces of Tomes' process; each ameloblast has one Tomes' process.

Mesenchymal specificity in the embryonic tooth organ has been described in some detail (for example, see refs. 20,21,29,45). Various mammalian embryonic molar and incisor tooth organ rudiments (i.e., protodifferentiated) have been successfully cultured as explants, prior to either dentine or enamel matrix formation, on the chick chorioallantoic membrane (CAM), as intraocular grafts, subcapsular grafts in kidney, testes, or spleen, and as explants in semidefined organ culture medium *in vitro*. Such explants expressed both histogenesis and morphogenesis and the production of dentine and enamel matrices. Such embryonic tooth rudiments were found to be easily dissociated into enamel organ epithelium and dental papilla mesenchyme using proteases or cation chelating agents. When the two separated tissue components were recombined, they continued to differentiate and developed into tooth organs comparable to the intact explant cultures.

Whereas the artificial environments used in these studies were readily permissive to tooth organogenesis, separated tissue explants did not differentiate under these experimental conditions; isolated mesenchyme produced fibroblast-like outgrowths and the enamel organ epithelium lost its bud shape and transformed into a flat sheet of cells. The information within these isolated tissues was not sufficient to express a specific tooth phenotype (e.g., odontoblast or ameloblast) when cultured for extended periods of time in an artificial environment previously demonstrated to be suitable to support and maintain tooth morphogenesis. Koch (1967) extended these observations and demonstrated "embryonic transfilter induction" between enamel organ epithelium and dental papilla mesenchyme cultured in juxtaposition to a Millipore filter (0.45 μm pore size and 25 μm thick) (19). Dentine matrix formation was evident on one side of the filter and an "enamel-like" refractile substance was evident on the epithelial surface of the filter assembly. If the tissue recombinants were separated by cellophane or Millipore filters 70 to 80 μm thick, no differentiation was detectable (19).

To test the specificity of dental papilla mesenchyme, tooth mesenchyme was cultured in the presence of various foreign or heterologous epithelium. The foreign epithelium differentiated into ameloblasts and secreted enamel matrix (e.g., snout, foot pad, nontooth oral, and dorsal skin epithelium) (21,29,45).

CLOSE-RANGE INTERCELLULAR COMMUNICATION

We assume that mesenchymal cells are prerequisite to the initiation of enamel protein synthesis and secretion from ameloblast cells. Intimate, close-range interactions between mesenchymal cell processes, matrix vesicles, and the outer surfaces of the responding epithelium appear to be crucial determinants of tooth morphogenesis. During embryonic and neonatal mammalian tooth organ formation, mesenchymal cells produce matrix vesicles that accumulate along the undersurface of the inner enamel epithelium, adjacent to the continuous basal lamina, *prior to the discontinuity of the basal lamina* and degradation of the lamina (39–42). Direct cell-cell contacts between the mesenchymal cell processes and the inner enamel epithelium have not been observed in this region (43). We postulate that the basal lamina serves to insulate, in part, the outer cell surfaces of the epithelia from contacting the outer surfaces of the mesenchymal cell processes.

Subsequently, along the gradient of increasing cell differentiation (inner enamel epithelium→preameloblast→ameloblast), the basal lamina becomes discontinuous along the undersurface of the inner enamel epithelial and preameloblast cells (5,6,38,43). Epithelial microvilli project through these discontinuities. Numerous matrix vesicles were observed in proximity to regions in which the basal lamina was degraded (5,43). Following basal lamina degradation, numerous direct cell-cell interactions were observed between the preodontoblast and odontoblast cell processes and the adjacent preameloblast cells (5,17,25,38–43). These close-range, direct cell contacts were noted to be coincident with the initial dentine matrix mineralization, cessation of cell division within the epithelia, dramatic intracytoplasmic alterations which occurred (5,6,43) as the preameloblast elongated into the ameloblast, and the basal lamina degradation (5,6,43).

Matrix vesicles are discrete extracellular "organelles" approximately 0.1 μm in diameter and are specifically formed by "budding" from the plasma membrane of the preodontoblast cell processes (42). At least three different ultrastructural types of vesicles have been identified (39). It is also evident that both the matrix vesicle-limiting membrane and the adjacent mesenchymal cell process plasma membrane have H-2 antigens located on their respective outer surfaces (42,44). The most distal dimension of the extended preodontoblast and odontoblast cell processes were approximately 0.1 to 0.2 μm in diameter (43).

Experiments with many epithelial-mesenchymal organ systems indicate, to a first approximation, successful heterotypic cell-cell interactions when epithelium and mesenchyme were cultured in juxtaposition to Millipore and Nucleopore filters with porosities of 0.4 μm or more and filter thicknesses of 25 μm (12,22,24,32,33). Transfilter embryonic inductions were not successful using 0.1 μm pore size Nucleopore filters (33). Recently, Saxén (34) discussed the results of a large series of transfilter embryonic tooth tissue recombinant experiments in which she confirmed that 0.1-μm pore size Nucleopore filters prevented epithelial-mesenchymal interactions associated with tooth formation. Using embryonic mouse molar tooth organ tissues in juxtaposition to 0.4-μm pore size Nucleopore filters, in which the culture assembly was grown using the Trowell method, inner enamel epithelium differentiated into ameloblasts. These studies and others have concluded that epithelial-mesenchymal interactions require direct heterotypic cell-cell contact as a prerequisite to epithelial cell differentiation (22,33).

MODE OF ACTION OF THE MESENCHYMAL SIGNAL(S)

Our laboratory is investigating close-range, epithelial-mesenchymal interactions in the developing embryonic rabbit and mouse molar and incisor tooth organs *in vivo* and *in vitro*. One of the most useful experimental features of these selected stages of tooth morphogenesis is that development is continuous and the consequence of heterotypic tissue interactions can be identified as: (a) reciprocal cell differentiation into nondividing, terminally differentiated secretory cells, and (b) the deposition of two different extracellular matrices (dentine and enamel), in apposition to one another, which mineralize and calcify.

We have focused upon the initial formation of the extracellular matrix interposed between essentially single layers of interacting epithelial (inner enamel epithelium) and mesenchymal cells (preodontoblasts). In the "proliferative zone" of the developing tooth organ (after the convention of Kallenbach) (17), we have described a fibrous material (collagenase-labile) associated with the undersurface of the basal lamina along the outer and inner enamel epithelium in embryonic rabbit incisor tooth organs (e.g., cervical loop regions) (46), and in embryonic and neonatal mouse molar and incisor tooth organs *in situ* (43). In the proliferative zone, both epithelial and mesenchymal cells were dividing and had not as yet demonstrated a well-defined Golgi apparatus and rough endoplasmic reticulum (46). At this stage, the mesenchymal cells have not condensed to form a sheet of preodontoblast cells. Extracellular collagen fibrils and fibers were not readily observed in association with these cells. Subsequently, these cells differentiated into preodontoblast and then odontoblast cells which secreted Type I collagen and formed predentine. In this "early differentiation zone" the mesenchymal cells extended long cell processes toward the undersurface of the basal lamina and also formed matrix vesicles (5,6,39,45). The matrix vesicles appeared to accumulate along the undersurface of the basal lamina prior to the degradation of the lamina (43). During the basal lamina degradation, mesenchymal cell processes formed direct cell-cell contacts with the "exposed" undersurface of the preameloblast cells (43). In this "differentiation zone," odontoblast cells formed predentine and mineralizing dentine approximately 24 hr prior to the initiation of enamel protein synthesis and secretion as evaluated by electron microscopy and high-resolution autoradiography (47).

Assumptions

Based on the morphological descriptions previously discussed, and the results obtained from transfilter embryonic induction experiments, we formulated the following assumptions: (a) the fibrous material associated with the basal lamina in the proliferative zone was synthesized and secreted by the inner enamel epithelial cells prior to preodontoblast differentiation; (b) the fibrous material was collagen and served as a template to signal adjacent mesenchymal cells to secrete Type I dentine collagen (analogous to the synthesis and secretion of Type I collagen by the embryonic chick corneal epithelial cells prior to the appearance of mesenchymal cell synthesis and secretion of the primary corneal stroma) (2,15); (c) the inner surface cells of the enamel organ epithelia (i.e., inner enamel epithelium) synthesized and secreted Type IV basal lamina collagen and Type I collagen prior to the differentiation of preodontoblast cells (50); (d) matrix vesicles were formed by "budding" from the plasma membrane of the preodontoblast cells and were sequestered along the undersurface of the basal lamina; (e) the basal lamina degradation is the result of matrix vesicle and/or mesenchymal cell process protease activity; and (f) the degradation of the lamina "unmasks" the epithelial cell surface receptors, which form direct contacts with the other surfaces of the mesenchymal cell processes and thereby function to mediate the initiation of enamel protein synthesis and secretion.

Identification of Matrix Vesicle Collagenase Activity

On the basis of morphological evidence and an enlarging cytochemical and biochemical literature concerning matrix vesicles (42,48), several possible functions have been postulated: (a) the transport of unique molecules synthesized within the cells responsible for producing the matrix vesicles (e.g., preodontoblast, chondroblast, osteoblast), (b) a mechanism by which proteolytic enzymes and phosphatases can be transported within vesicles into the forming extracellular matrix milieu for subsequent functions; and (c) a mechanism for the concentration and transport of specific ions for subsequent calcification. Of these postulated functions for matrix vesicles, we have recently completed a series of experiments to evaluate the possible protease activity of these extracellular "organelles."

Since we assumed that matrix vesicles mediated the degradation of the basal lamina, and since the basal lamina was allegedly Type IV collagen, we designed experiments to assay for mammalian collagenase activity associated with physically isolated matrix vesicles. Fractions (matrix vesicle enriched, membrane particulate preparations) were isolated from embryonic New Zealand white rabbit incisor tooth organs in the void volume effluent of a Bio-Gel A-50 M column by procedures previously reported by our laboratory (40,48). For the collagenase activity assay, ^{14}C-labeled Type I rabbit skin collagen was reconstituted in fibrillar form and then incubated, in the presence of various protease inhibitors, with aliquots of matrix vesicles for 18 hr at 25°C (4). At the termination of the incubation, experimental and control groups were processed for scanning electron microscopy (Fig. 1), 200-μl aliquots were assayed for radioactive counts released into the reaction mixture supernatant, and the remaining supernatant was dialyzed, lyophilized, and then analyzed using 5% sodium dodecylsulphate-polyacrylamide gel electrophoresis (SDS-PAGE) to determine the degree of heterogeneity and relative molecular weights of the reaction products. The 5% SDS-PAGE patterns of the reaction products resulting from the collagenase activity of the matrix vesicles demonstrated two major products (P_1 and P_2) having molecular weights of 67,000 and 32,000 daltons, respectively (Fig. 2). The matrix vesicles unequivocally demonstrated a mammalian collagenase-like activity characterized by the enzymatic cleavage of native collagen into 3/4 and 1/4 fragments (49).

Morphological Evidence for Direct Heterotypic Cell-Cell Contacts Prior to Enamel Protein Synthesis and Secretion

Intercellular contacts at the epithelial-mesenchymal interface during prenatal development of the rat submandibular salivary glands have been described (7). Recent transfilter embryonic induction studies using Nucleopore filters have demonstrated direct cell-cell contacts during epithelial-mesenchymal interactions and have suggested that direct mesenchymal cell contact with an adjacent responding epithelium was crucial for the induction of cytodifferentiation (33,51,52). During tooth development, Pannesse (25), Kallenbach (17,18), Croissant et al. (6), and Slavkin and Bringas (43) have described direct cell-cell contacts between preodontoblast cell processes and the surface of the preameloblast cell membrane during basal lamina degradation (Fig. 3). These direct cell-cell contacts preceded

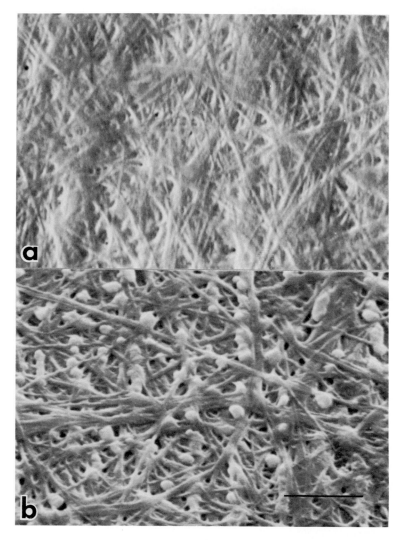

FIG. 1. Scanning electron microscopy (SEM) was used to compare the topographical characteristics of **(a)** the control using buffer only and the collagen substrate and **(b)** matrix vesicles incubated on native Type I collagen fibrils. The assay for collagenase activity in the matrix vesicles consisted of ^{14}C-labeled young lathyritic rabbit skin Type I collagen, 0.05 M Tris–0.005 M CaCl$_2$ (pH 7.6), the matrix vesicle fractions collected from the void volume of a Bio-Gel A-50M column, and 0.1 ml of protease inhibitors. The reaction mixture was incubated for 18 hr at 25°C and then processed for SEM. Line = 0.5 μm.

ameloblast cytodifferentiation. Our tentative interpretation of these observations is that epithelial cell enamel protein synthesis is stimulated by preameloblast plasma membrane interactions with either mesenchymal cell-derived matrix vesicles and/or cell processes. The data have also been inter-

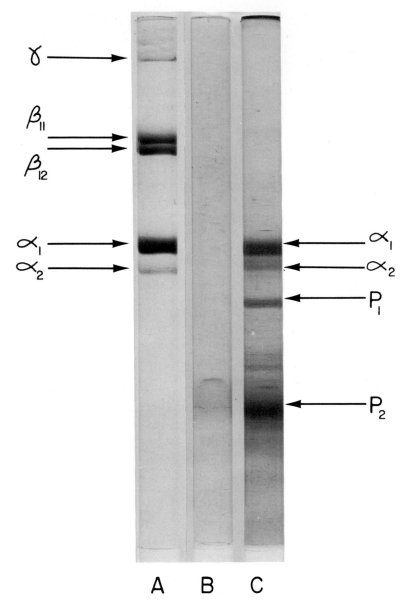

FIG. 2. The 5% SDS-PAGE patterns of the reaction products resulting from the collagenase activity of the ^{14}C-labeled Type I collagen standard **(A)**, bacterial collagenase-treated control **(B)**, and the matrix vesicles **(C)**. Two major reaction products (P_1 and P_2) resulted from the incubation of matrix vesicles with collagen; the degradation products were 67,000 and 32,000 daltons, respectively, and clearly demonstrated a mammalian collagenase activity associated with the vesicles.

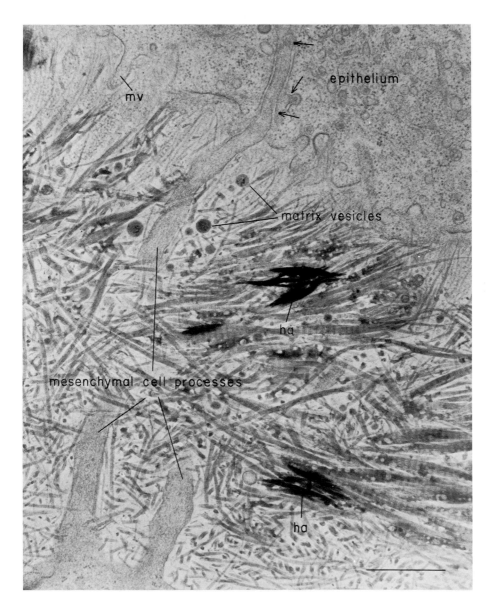

FIG. 3. Following the basal lamina degradation along the undersurface of the inner enamel epithelial cells, preodontoblast and odontoblast cell processes formed direct cell-cell contacts with the outer surfaces of the epithelial cells. Arrows indicate several regions of apposition between the outer surfaces of these heterotypic cell types. **MV,** epithelial cell microvilli; **ha,** forming calcium hydroxyapatite crystal formation in dentine matrix. Line = 0.5 μm.

preted to suggest that the initiation of enamel protein synthesis and secretion is timed by the transient stability of the condensed fibrous materials associated with the basal lamina.

Assays for Mesenchymal Induction of Enamel Matrix Secretion

Assays for inductive and/or permissive activities is essential for identifying, purifying, and characterizing the mesenchymal influence upon a responding epithelium. Morphological criteria have been defined that unequivocally indicate the secretion of enamel matrix proteins using scanning electron microscopy (3) (Fig. 4), transmission electron microscopy (Fig. 5), histochemistry capable of localizing sulphydryl groups (8), and light and electron microscopic autoradiography using isotopic precursors characteristic of embryonic enamel proteins (e.g., proline, histidine, tyrosine, cystine, tryptophan, fucose, glucosamine) (53).

In addition to morphological criteria, recent biochemical advances in the identification and partial characterization of newly secreted enamel proteins have been reported (9,13,14). These studies indicated a precursor/product relationship in which a precursor enamel protein of 65,000 daltons was degraded into increasingly smaller polypeptides of 50,000, 22,000, and 20,000 daltons. All of these polypeptides were glycosylated and the 65,000 and 22,000 polypeptides were found to be phosphorylated (14). Enamel protein(s) are antigenic. Recently, Schonfeld reported the production of alloantibodies directed against newly secreted enamel matrix proteins; embryonic 26-day New Zealand white molar extracellular matrices were used as the immunogen to produce alloantibodies in young adult female New Zealand white rabbits (35,36). Using indirect fluorescence microscopy and immunoelectrophoresis methods, the alloantisera were found to be specific for enamel matrix proteins in embryonic molar and incisor tooth organs. The rabbit antirabbit antisera was cross reactive with mouse enamel matrix.

It is evident that assays for mesenchymal induction of enamel matrix synthesis and secretion will require tissue culture methods. Our experiences in using the chick chorioallantoic membrane to culture tooth tissues have suggested that this *en ovo* method, albeit useful for obtaining valuable information for some questions, would not be useful to pursue the thesis of this discussion. Therefore, we modified several methods and have now been able to: (a) explant embryonic mouse molar tooth organs, prior to the characteristic cytodifferentiation of dentine-forming odontoblast and enamel-forming ameloblast expression, and in a defined medium permit the expression of dentinogenesis and amelogenesis (Fig. 6); (b) after dissociating embryonic molar tooth organs, we recombined enamel organ epithelium and adjacent dental papilla tissues and obtained dentinogenesis and amelogenesis; and (c) finally, we demonstrated that isolated epithelium or mesenchyme in this system did not differentiate in the absence of one another. In

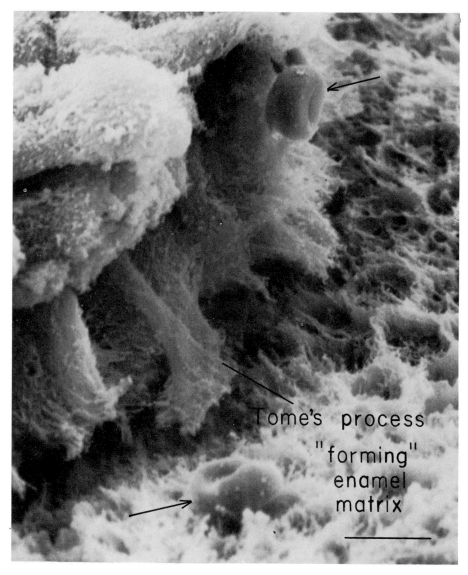

FIG. 4. Scanning electron micrograph showing a series of embryonic rabbit incisor secretory ameloblasts in the process of secreting enamel matrix proteins. Precursor enamel matrix proteins are secreted from the lateral and apical surfaces of Tome's processes. The degree of ameloblast cell elongation can be appreciated in comparison to red blood cells **(arrows)** in this preparation. Line = 7 μm.

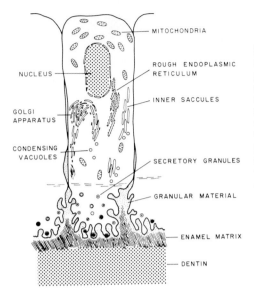

FIG. 5. Diagrammatic presentation of the embryonic secretory ameloblast. Enamel protein is synthesized on membrane-associated polysomes (rough endoplasmic reticulum), glycosylated in the Golgi apparatus, transported through the cisternae of the smooth endoplasmic reticulum, condensed into vacuoles, transported through the cytoplasm in secretory granules, which subsequently secrete their contents from lateral and apical surfaces of Tome's processes as a granular material. Phosphorylation presumably is mediated by a phosphokinase prior to secretion. The formation of intra- and intermolecular disulfide linkages presumably occurs during secretion.

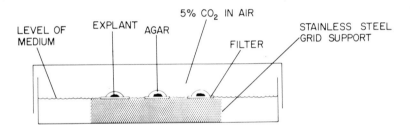

ORGAN CULTURE ASSEMBLY

FIG. 6. Diagram of the *in vitro* method found to be permissive for the expression of both dentine and enamel secretion. A modified Trowell procedure was used to cultivate explanted 16- and 17-day-old embryonic mouse manidublar first molar tooth organs. The explants were cultured on 0.45-μm pore size Millipore filters, overlayed with a 1% bactoagar to inhibit artifactual keratinization of the epithelia in a medium consisting of MEM in Eagle's salts (F-11), sodium bicarbonate (16 mM), glutamine (292 μg/ml), ascorbic acid (100 μg/ml), glycine (150 μg/ml), gentamycin (100 μg/ml), and 10% fetal calf serum. The initial pH of cultures in 5% CO_2 in 95% air at 37.5°C was 7.4. Experiments were generally terminated after 5 days and the pH was 6.9. Embryonic chick extract was found to inhibit enamel synthesis and secretion.

this culture system, we observed the same sequence of cytological events that were characteristic of tooth morphogenesis (Fig. 7).

Experiments are now in progress that have been designed to combine many of these methods and procedures in order to test and possibly validate our predictions.

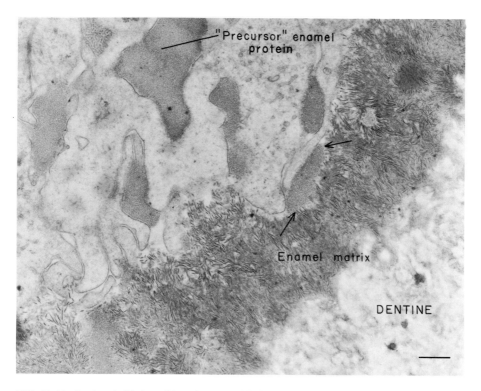

FIG. 7. Explants of 17-day-old embryonic CD-Swiss mouse molar tooth organs differentiated and secreted enamel matrix within 5 days of culture. The characteristics of the enamel matrix formed *in vitro* was analogous to that formed *in vivo*. Initially, a granular material was secreted prior to the enamel matrix *per se* being formed. Line = 1 μm.

ACKNOWLEDGMENTS

The authors wish to acknowledge the excellent technical assistance of Pablo Bringas, Jr., P. Matosian, P. Wilson, and W. Mino. The research investigations reviewed in this manuscript were supported by USPHS Research Grants DE-02848, DE-03569, and DE-03513 and Training Grant DE-00094 from the National Institute of Dental Research. The authors wish to express their appreciation to Mrs. Joanne Leynnwood for typing the manuscript.

REFERENCES

1. Alescio, T., and Cassini, A. (1962): Induction *in vitro* of tracheal buds by pulmonary mesenchyme grafted on tracheal epithelium. *J. Exp. Zool.,* 150:83.
2. Bard, J. B. L., Hay, E. D., and Meller, S. M. (1975): Formation of the endothelium of the avian cornea: A study of cell movement *in vivo. Dev. Biol.,* 42:334.

3. Boyde, A., and Jones, S. J. (1972): Scanning electron microscopic studies of the formation of mineralized tissues. In: *Developmental Aspects of Oral Biology,* edited by H. C. Slavkin and L. A. Bavetta, p. 243. Academic Press, New York.
4. Brownell, A., Sorgente, N., and Slavkin, H. C. (1976): Collagenase activity associated with extracellular matrix vesicle-enriched fractions. *J. Cell Biol.,* 70:271.
5. Croissant, R. D. (1975): Induction, extracellular matrix vesicles, and RNA: A study of epithelial-mesenchymal interaction during late embryonic rabbit incisor development. Ph.D. Thesis, University of Southern California.
6. Croissant, R. D., Guenther, H., and Slavkin, H. C. (1975): How are embryonic pre-ameloblasts instructed by odontoblasts to synthesize enamel? In: *Extracellular Matrix Influences on Gene Expression,* edited by H. C. Slavkin and R. C. Greulich, p. 515. Academic Press, New York.
7. Cutler, L. S., and Chaudhry, A. P. (1973): Intercellular contacts at the epithelial-mesenchymal interface during the prenatal development of the rat submandibular gland. *Dev. Biol.,* 33:229.
8. Everett, M. M., and Miller, W. A. (1974): Histochemical studies on calcified tissues. I. Amino acid histochemistry of foetal calf and human enamel matrix. *Calc. Tissue Res.,* 14:229.
9. Fukae, M., and Shimizu, M. (1974): Studies on the proteins of developing bovine enamel. *Arch. Oral Biol.,* 19:381.
10. Gaunt, W. A., and Miles, A. E. W. (1967): Fundamental aspects of tooth morphogenesis. In: *Structural and Chemical Organization of Teeth,* Vol. 1, edited by A. E. W. Miles, p. 151. Academic Press, New York.
11. Grobstein, C. (1967): Mechanisms of organogenetic tissue interaction. *Natl. Cancer Inst. Monogr.,* 26:279.
12. Grobstein, C. (1975): Developmental role of intercellular matrix. In: *Extracellular Matrix Influences on Gene Expression,* edited by H. C. Slavkin and R. C. Greulich, p. 9. Academic Press, New York.
13. Guenther, H. Croissant, R. D., Schonfeld, S., and Slavkin, H. C. (1975): Enamel proteins: Identification of epithelial-specific differentiation products. In: *Extracellular Matrix Influences on Gene Expression,* edited by H. C. Slavkin and R. C. Greulich, p. 387. Academic Press, New York.
14. Guenther, H., Croissant, R. D., Schonfeld, S., and Slavkin, H. C. (1977): Identification of four extracellular matrix enamel proteins during embryonic rabbit tooth organ development. *Biochem. J. (in press).*
15. Hay, E. D., and Revel, J. P. (1969): Fine structure of the developing avian cornea. In: *Monographs in Developmental Biology,* edited by A. Wolsky and P. S. Chen, pp. 1–144. Basel.
16. Johnston, M. C., and Listgarten, M. (1972): Observations on the migration interactions and early differentiation of orofacial tissues. In: *Developmental Aspects of Oral Biology,* edited by H. C. Slavkin and L. A. Bavetta, p. 53. Academic Press, New York.
17. Kallenbach, E. (1971): Electron microscopy of the differentiating rat incisor ameloblast. *J. Ultrastruc. Res.,* 35:508.
18. Kallenbach, E. (1976): Fine structure of differentiating ameloblasts in the kitten. *Am. J. Anat.,* 145:283.
19. Koch, W. E. (1967): *In vitro* differentiation of tooth rudiments of embryonic mice. I. Transfilter interaction of embryonic tissues. *J. Exp. Zool.,* 165:155.
20. Koch, W. E. (1972): Tissue interaction during *in vitro* odontogenesis. In: *Developmental Aspects of Oral Biology,* edited by H. C. Slavkin and L. A. Bavetta, p. 151. Academic Press, New York.
21. Kollar, E. (1972): Histogenetic aspects of dermal-epidermal interactions. In: *Developmental Aspects of Oral Biology,* edited by H. C. Slavkin and L. A. Bavetta, p. 125. Academic Press, New York.
22. Lehtonen, E., Wartiovaara, J., Nordling, S., and Saxen, L. (1975): Demonstration of cytoplasmic processes in Millipore filters permitting kidney tubule induction. *J. Embryol. Exp. Morphol.,* 33:187.
23. Levine, S., Pictet, R. L., and Rutter, W. J. (1973): Control of proliferation and cytodifferentiation by a factor reacting with the cell surface. *Nature [New Biol.]* 246:143.
24. Nordling, S., Miettinen, H., Wartiovaara, J., and Saxen, L. (1971): Transmission and

spread of embryonic induction. I. Temporal relationships in transfilter induction of kidney tubules *in vitro. J. Embryol. Exp. Morphol.*, 26:231.

25. Pannesse, E. (1962): Observations on the ultrastructure of the enamel organ. III. Internal and external enamel epithelia. *J. Ultrastruc. Res.*, 6:186.

26. Pictet, R. L., Filosa, S., Phelps, P., and Rutter, W. J. (1975): Control of DNA synthesis in the embryonic pancreas: Interaction of the mesenchymal factor and cyclic AMP. In: *Extracellular Matrix Influences on Gene Expression*, edited by H. C. Slavkin and R. C. Greulich, p. 531. Academic Press, New York.

27. Propper, A. (1975): Epithelial-mesenchymal interactions between developing chick epidermis and rabbit embryo mammary gland mesenchyme. In: *Extracellular Matrix Influences on Gene Expression*, edited by H. C. Slavkin and R. C. Greulich, p. 541. Academic Press, New York.

28. Rall, L. B., Pictet, R. L., and Rutter, W. J. (1975): Glucocorticoids potentiate the *in vitro* development of the embryonic rat pancreas. *J. Cell Biol.*, 67:350a.

29. Ruch, V. J., Karcher-Djuricic, V., and Gerber, R. (1973): Les determinismes de la morphogenese et des cytodifferenciations des ebauches dentaires de souris. *J. Biol. Buccale*, 1:45.

30. Rutter, W. J., Wessells, N. K., and Grobstein, C. (1964): Control of specific synthesis in the developing pancreas. *Natl. Cancer Inst. Monogr.*, 13:51.

31. Saunders, J. W., and Gasseling, M. T. (1963): Transfilter propogation of apical ectoderm maintenance factor in the chick embryo wing bud. *Dev. Biol.*, 7:64.

32. Saxen, L., Koskimies, O., Lahti, A., Miettinen, H., Rapola, J., and Wartiovaara, J. (1968): Differentiation of kidney mesenchyme in an experimental model system. In: *Advances in Morphogenesis*, edited by M. Abercrombie, J. Brachet, and T. J. King, p. 251. Academic Press, New York.

33. Saxen, L. (1975): Transmission and spread of kidney tubule induction. In: *Extracellular Matrix Influences on Gene Expression*, edited by H. C. Slavkin and R. C. Greulich, p. 523. Academic Press, London.

34. Saxen, I. (1977): Transfilter embryonic induction during tooth development. In: *Cell Interactions in Differentiation*, edited by L. Saxen and L. Weiss, Academic Press, London (*in press*).

35. Schonfeld, S., Trump, G. N., and Slavkin, H. C. (1977): Immunogenicity of two naturally occurring solid-phase enamel proteins. *Proc. Soc. Exp. Med. Biol.*, 115:111.

36. Schonfeld, S. (1975): Demonstration of an alloimmune response to embryonic enamel matrix proteins. *J. Dent. Res.*, 54:72.

37. Sengel, P. (1975): Feather pattern development. In: *Cell Patterning* (CIBA Foundation Series No. 29), p. 51. Elsevier-Excerpta Medica-North-Holland, New York.

38. Silva, D., and Kailis, D. (1972): Ultrastructural studies on the cervical loop and the development of the amelo-dentinal junction in the cat. *Arch. Oral. Biol.*, 17:279.

39. Slavkin, H. C., Bringas, P., and Croissant, R. D. (1972): Epithelial-mesenchymal interactions during odontogenesis. II. Intercellular matrix vesicles. *Mech. Ageing Dev.*, 1:139.

40. Slavkin, H. C., Croissant, R. D., and Bringas, P. (1972): Epithelial-mesenchymal interactions during odontogenesis. III. A simple method for the isolation of matrix vesicles. *J. Cell Biol.*, 53:841.

41. Slavkin, H. C., and Croissant, R. D. (1973): Intercellular communication during odontogenic epithelial-mesenchymal interactions: Isolation of extracellular matrix vesicles containing RNA. In: *The Role of RNA in Reproduction and Development*, edited by M. Niu and S. Segal, p. 247. North-Holland Publishing Co., Amsterdam.

42. Slavkin, H. C., Croissant, R. D., Bringas, P., Matosian, P., Wilson, P., Mino, W., and Guenther, H. (1976): Matrix vesicle heterogeneity: Possible morphogenetic functions for matrix vesicles. *Fed. Proc.*, 35:127.

43. Slavkin, H. C., and Bringas, P. (1976): Epithelial-mesenchymal interactions during odontogenesis. IV. Morphological evidence for direct heterotypic cell-cell contacts. *Dev. Biol.*, 50:428.

44. Slavkin, H. C., Trump, G. N., Mansour, V., Matosian, P., and Mino, W. (1974): Localization of H-2 histocompatibility alloantigens on epithelial and mesenchymal mouse embryonic tooth cell surfaces. *J. Cell Biol.*, 60:795.

45. Slavkin, H. C. (1974): *Embryonic Tooth Formation: A Tool for Developmental Biology*.

Oral Sciences Reviews, Vol. 4, edited by A. H. Melcher and G. A. Zarb. Munksgaard, Copenhagen.

46. Slavkin, H. C., Bringas, P., LeBaron, R. D., Cameron, J. C., and Bavetta, L. A. (1969): The fine structure of the extracellular matrix during epithelio-mesenchymal interactions in the rabbit embryonic incisor. *Anat. Rec.,* 165:237.

47. Slavkin, H. C., Mino, W., and Bringas, P. (1976): The biosynthesis and secretion of precursor enamel protein by ameloblasts as visualized by autoradiography after tryptophan administration. *Anat. Rec.* 185:289.

48. Slavkin, H. C. (1975): The isolation and characterization of calcifying and non-calcifying matrix vesicles from dentine. In: *International Colloquium on Physical Chemistry and Crystallography of Apatites of Biological Interest,* edited by G. Montel, p. 161. Centre National de la Recherche Scientifique, Paris.

49. Sorgente, N., Brownell, A., and Slavkin, H. C. (1977): Basal lamina degradation: The identification of mammalian-like collagenase activity in mesenchymal-derived matrix vesicles. *Biochem. Biophys. Res. Comm.,* 74:448.

50. Trelstad, R. L., and Slavkin, H. C. (1974): Collagen synthesis by the enamel organ epithelia of the embryonic rabbit tooth. *Biochem. Biophys. Res. Comm.,* 59:443.

51. Wartiovaara, J., Lehtonen, E., Nordling, S., and Saxen, L. (1972): Do membrane filters prevent cell contacts? *Nature,* 238:407.

52. Wartiovaara, J., Nordling, S., Lehtonen, E., and Saxen, L. (1974): Transfilter induction of kidney tubules: Correlation with cytoplasmic penetration into Nucleopore filters. *J. Embryol. Exp. Morphol.,* 31:667.

53. Weinstock, A. (1972): Matrix development in mineralizing tissues as shown by radio-autography: Formation of enamel and dentine. In: *Developmental Aspects of Oral Biology,* edited by H. C. Slavkin and L. A. Bavetta, p. 202. Academic Press, New York.

54. Wolpert, L., Lewis, J., and Summerbell, D. (1975): Morphogenesis of the vertebrate limb. In: *Cell Patterning* (CIBA Foundation Series No. 29), p. 95. Elsevier-Excerpta Medica-North-Holland, New York.

55. Zwilling, E. (1956): Interaction between limb bud ectoderm and mesoderm in the chick embryo. II. Experimental limb duplication. *J. Exp. Zool.* 132:173.

Cell and Tissue Interactions, edited by
J. W. Lash and M. M. Burger. Raven
Press, New York, 1977.

Toward a Developmental Theory of Immunity: Selective Differentiation of Teratoma Cells

Robert Auerbach

Department of Zoology, University of Wisconsin, Madison, Wisconsin 53706

Seven years ago, in the first paper of this series (3) I suggested "that the origin, development and function of immunocompetent cells may be considered analogous to the origin, development and function of germ cells." The basic tenet of the proposal for a "germinal theory" was that "there is a meaningful correlation between germ line and blood line of the embryo . . . and that there are fundamental similarities between germ plasm and immunoplasm development and function" (3). I drew special attention to the unique nature of linkage group IX of the mouse, and to the concept that cell-ligand-type materials involved in cell interactions in development and materials released in mixed lymphocyte cultures may have meaningful homologies. I should add, at this later date, that several years earlier, Dr. Haim Ginsburg, then at the Weizmann Institute of Science in Rehovoth, had made available to me a manuscript in which the correlation between sperm and lymphocyte biology was clearly outlined (15), and I am indebted to him for helping to focus my attention on this important association between the reproductive and lymphoid systems.

In the second paper of this series (4), developmental correlations between the immune response and other tissue interactive systems were outlined, while the third paper dealt with the problem of ontogeny of immune responsiveness and the concept of allosteric (stem cell) tolerance (8; see also 9).

In this, the fourth paper in the series, I wish to focus attention on the use of teratomas as models for the study of development of the immune response, by reviewing first some of the earlier studies I reported with respect to teratoma cell differentiation toward immunocompetence, next by reporting on some of the more recent studies carried out in my laboratory, and finally by using these studies as a framework for incorporating current information from other laboratories on surface antigens expressed on sperm, embryonic cells, teratomas, and neural tissue into a more generalized view of the development of specificity during ontogeny of immune responsiveness.

EARLIER STUDIES

Starting with teratoma cells that were multipotential (Tumor OTT6050 obtained from L. C. Stevens in 1970), we attempted to select or induce cell subpopulations that would be capable of participating in immunological reactions. To do this, freshly dissected tumors were prepared as cell suspensions and injected intravenously into adult, syngeneic, and lethally irradiated mice. Some of these mice, when examined a few days later, were found to have enlarged spleens filled with hematopoietic cells, suggesting that teratoma cells had settled there and had either provoked regeneration of residual radiation-resistant stem cells or, more likely, had differentiated into cells of the hematopoietic system (7). The histological appearance of the spleens of teratoma-injected animals resembled that of spleens obtained from animals injected with syngeneic bone marrow cells, although it should be pointed out that teratoma cells did not provide radiation protection similar to that afforded by bone marrow cells.

Since we were interested particularly in ontogeny of the immune system, we next combined as cell inoculum dissociated teratoma cells and freshly

TABLE 1. *Cooperation between thoracic duct lymphocytes and cells obtained from bone marrow or teratomas: response to sheep red blood cells*[a]

Cells[b]	PFC/0.05 spleen[c]	Mean PFC/spleen	Index[d]
None	0, 0, 0	0	
TDL	23, 10, 3	240	
Tumor	0, 0	0	
BM	9, 54	630	
TDL + BM	206, 203, 351	5,067	5.82
TDL + Tumor	52, 95, 367	3,447	14.36
TDL + BM + Tumor	91, 149, 56	1,973	2.27

[a] From ref. 5.
[b] For concentration see text.
[c] 1/20th of spleen suspension was assayed for PFC.
[d] Calculated as PFC of animals injected with combination of cells/sum of PFC of singly injected animals.

obtained thoracic duct lymphocytes. Sheep red blood cells were administered simultaneously in order to assess the restitution of immunocompetence that could be effected by tumor cells alone, by thoracic duct lymphocytes, or by a combination of the two cell types. After 6 days, the number of antibody-forming cells present in the spleen was determined by use of a plaque assay; the results of one experiment are given in Table 1, while the results obtained in nine consecutive experiments are reported in Table 2. Synergistic interaction between teratoma cells and thoracic duct lympho-

TABLE 2. *Cooperation between thoracic duct cells and teratoma cells in response to sheep red blood cells: summary of nine experiments[a]*

Experiment	TDL	Tumor	Both	Index
1	0	80	1380	17.3
2	25	26	80	1.6
3	20	27	192	4.1
4	0	7	80	11.4
5	130	10	760	5.4
6	10	55	530	8.2
7	867	10	1895	2.2
8	970	0	880	0.9
9	240	0	3447	14.4
n = 30				Mean index = 5.78

[a] From ref. 5.

cytes was apparent and indeed was comparable to the synergism seen in the usual bone marrow–thoracic duct cell combination.

The next experiments were designed to verify that the resulting antibody-forming cells were indeed turmor-derived. To do so, two different strains of mice were used for source of thoracic duct cells and tumor cells (BALB/c and 129/J, respectively), with F_1 (BALB/cx129/J) animals serving as hosts. The amount of synergism obtained in this combination was limited, however (a fact which can now be ascribed to the requirement for syngeneic cell cooperation), and it was not possible to obtain a definitive answer to this question at that time. Shortly thereafter, our OTT6050 subline lost its capacity to cause spleen enlargement *in vivo,* as well as its previous capacity to generate embryoid bodies intraperitoneally, and the studies were not then continued.

During the course of these studies, a variety of other experiments were also carried out, aimed at characterizing the broad spectrum of developmental capacities that could be manifested by teratoma cells. Of particular pertinence to more recent work which has been carried out was the demonstration that neurally biased teratoma cells show the capacity to act as embryonic inducer of kidney tubules. Specifically, teratoma cells obtained either from a line A neural teratoma, characterized by its capacity to form neural tubes *in vivo,* or from neural tissue obtained from cultured OTT6050 or 402C teratoma by selective trypsinization were combined with metanephrogenic mesenchyme from 11-day mouse embryos. Typical metanephric tubules were induced in the mesenchyme, indicating that these neuralized teratoma explants demonstrated an inductive capacity similar to that shown by dorsal spinal cord or medullary plate as shown previously by Grobstein (18) (Table 3). The more advanced C1300 neuroblastoma failed to show such inductive capability.

TABLE 3. *Effect of various tumor preparations on metanephrogenic mesenchyme* in vitro[a]

Tumor source	Characteristics	No. of cultures	No. with tubules	Time of appearance (hr)
Teratoma 402C	Embryonal-carcinoma (undifferentiated)	6	(1?)	
Teratoma 6050	Embryonal-carcinoma (undifferentiated)	3	0	
	Multiple differentiations (including brain)	6	2	72
	Trypsin-resistant (largely neural)	6	5	72
	Reaggregate (largely neutral)	3	3	48–72
Line A neural teratoma	Fresh explant	4	4	48
	Trypsin-resistant	6	6	36–48
	Cultures (2 weeks)	6	6	36–48
Neuroblastoma C1300	Fresh explant	6	0	
Dorsal spinal cord		14	14	36–48
None		10	0	

[a] From ref. 6.

CURRENT STUDIES

Having established that selection of teratoma subpopulations suitable for carrying out partially restricted functions (e.g., neural-type inductions, differentiation along lymphoid pathways), we embarked on an extensive program to select teratomas that maintained broad capabilities of growth and differentiation but that were at least partially restricted in probable phenotypic expression. My own interest in development of the immune response led to selection of spleen-seeking teratomas. When injected intraperitoneally, OTT6050 occasionally leads to spleen enlargement, suggesting that either there is a host response to the teratoma marked by splenomegaly or that tumor cells proceeded to settle in and proliferate within the host spleen. That the latter was at least in part correct was shown by intraperitoneal injection of a cell suspension prepared from such enlarged spleens; i.e., cells derived from enlarged spleens of teratoma-bearing mice gave rise to teratomas when transferred intraperitonally into a second syngeneic host animal. Most significant was the finding that the frequency of enlarged spleens in these second hosts was high (50% versus 5%) in unselected teratoma transfers. By selection it was further possible to obtain a subline of tumor that regularly preferred spleen as a growth site.

In cooperation with Dr. Brenda Kahan cultures were prepared from a spleen harboring these spleen-seeking teratoma cells and subcultures and clones were prepared for further study. At the present time, we have de-

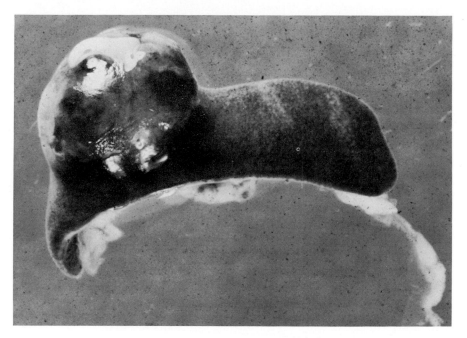

FIG 1. Mouse spleen showing massive "splenoteratoma."

rived from these cultures, which *in vitro* appear to be largely composed of embryonal carcinoma cells which while following intraperitoneal injection into syngeneic host animals form massive spleen-associated tumors (Fig. 1). Experiments aimed at characterizing the functional capacity of these "splenoteratomas" are currently in progress.

A different teratoma subpopulation has been isolated by Youniss and Kahan (*unpublished observations*) from a tumor originating in the LT line. This subpopulation as well as a series of clones isolated from this subpopulation tend to form secondary ovary-associated tumors when inoculated into LT female mice. Youniss (26) carried out preliminary studies designed to determine the relative adhesiveness of the ovary-seeking and spleen-seeking teratoma cells, using a modified Roth assay to test organ-specific adhesion to tissue slices (kidney versus ovary versus spleen). His results are still quite tentative, but he interprets his data to suggest that the ovary-seeking teratoma shows selective preference for ovary tissue slices, while the spleen-seeking teratoma shows no specificity preference. At the present time, moreover, Dr. Kahan has selected a teratoma subpopulation that appears to develop exclusively in the ovarian bursa when injected intraperitoneally; adhesion studies with this teratoma variant are now in progress.

Yet another teratoma (neural teratoma from line 129) has been tested recently to determine its developmental and inductive potential *in vitro*.

As with the neural teratoma studied previously, this tumor too showed excellent capacity to induce metanephric tubules from metanephrogenic mesenchyme (Table 4). Moreover, Dr. Vr. Muthukkaruppan recently examined the ability of such primitive neural teratomas to induce lens formation *in vitro* in a manner analogous to the *in vitro* lens induction previously described by him for the 10-day mouse embryo (21). His initial results, which should be considered preliminary in nature, indicate that indeed such neural teratomas can promote lens development *in vitro*.

Finally, we have initiated studies examining the ability of supernatants from teratoma cultures to induce metanephric tubules from metanephrogenic mesenchyme in the absence of tissue inducers such as dorsal spinal cord. Our earlier work (2) had indicated that such inductive activity could not readily be obtained from extracts of dorsal spinal cord, but we had suggested that it was attractive to consider the possibility that the materials involved in inductive tissue interactions and those involved in specific cell adhesion might be the same (2). For this reason we chose to use as collecting method a serum-free medium obtained from teratoma monolayer cultures following overnight incubation, as described for the collection of specific aggregation-promoting materials by Lilien and his co-workers (see Rutz and Lilien, *this volume*). Our work to date (see Table 4) indicates that supernatants from embryonal carcinoma and neural teratoma cultures can substitute at least in part for inductively active tissues in the metanephric system, while supernatants obtained from several other tumor sources are not active. Studies are now in progress to determine the nature and specificity of the inductively active materials.

TABLE 4. *Further studies on tumor-induced development of metanephric tubules*

	No. +/total	Strength of response
Positive control (dorsal spinal cord)	7/7	++++
Negative control (metanephrogenic mesenchyme only)	0/7	−
Stevens line 129 neural teratoma	4/4	+++
Cultured neural teratoma	4/4	++++
Standard culture medium from neural teratoma	0/3	−
Overnight collection, serum-free medium, neural teratoma	4/6	++
Overnight collection, serum-free medium, clone TG3	4/4	++++
C755 mammary tumor	0/6	−

GENERAL COMMENTS

At the present time we are rapidly acquiring information concerning the expression of surface antigens on teratoma cells and the relation of these antigens to those expressed on cells of the early embryo and on sperm cells

(10,13,14,17,19). It has been suggested, indeed, that the F9-teratoma-associated antigen may be equivalent to the normal antigen whose mutant form is represented by the t^{12} allele (1,12,25). Other studies suggest that cells from the developing nervous system share antigens with teratoma cells, and once again such antisera may also be able to detect surface antigens associated with germ cells (23,24). Thus, the relation between antigens of the early embryo, of teratomas, of germ cells, and of neural tissue seems to be an extremely close one and one which may have considerable developmental significance.

Against this background one may consider the recent studies of brain-associated theta antigen (22). These studies suggest that the prethymic cell may have on its surface an antigenic specificity not readily detected by anti-θ bodies but which does absorb antibrain-specific antibodies (BAT). Moreover, since anti-θ and BAT antibodies both detect the same population of θ-bearing cells further along the differentiative pathway to thymus differentiation, it is attractive to accept the speculation that the BAT-sensitive cells are cells that express precursors to θ antigen the latter representing a modification of the earlier antigen (see refs. 11, 20, 22) possibly by simple addition of terminal sugar residues.

Seen in this context the early prethymic stem cell, the germ line cells, and the cells of the early embryo represent a group of cells sharing surface properties detected by specific antisera. While the significance of this finding remains to be elusive, the observation helps to clarify why we can obtain by relatively low-pressured selection teratomas that tend to seek spleen or ovary or testis; why we can expect that our intravenous selection system (Table 1) favors "T"-cell functions; why a single segment of chromosome XVII appears to regulate simultaneously events of early differentiation and induction (16) of sperm cell antigens, of lymphocyte specificities, and of cell recognition factors associated with T-cell activation and T-B cell interactions.

ACKNOWLEDGMENTS

Research work reported in this paper was supported in part by grants from the American Cancer Society, the National Science Foundation, and the National Institutes of Health. The author is indebted to his colleagues in the laboratory, especially to Dr. Brenda Kahan, Mr. Louis Kubai, and Dr. Younan Sidky, for many helpful discussions.

REFERENCES

1. Artzt, K., Bennett, D., and Jacob, F. (1974): Primitive teratocarcinoma cells express a differentiation antigen specified by a gene at the T-locus in the mouse. *Proc. Natl. Acad. Sci. U.S.A.*, 71:811–814.
2. Auerbach, R., and Grobstein, C. (1958): Inductive interaction of embryonic tissues after dissociation and reaggregation. *Exp. Cell Res.*, 15:384–397.

3. Auerbach, R. (1970): Toward a developmental theory of antibody formation: The germinal theory of immunity. In: *Developmental Aspects of Antibody Formation and Structure,* edited by J. Sterzl, and I. Riha, pp. 23–33. Academic Press, New York.

4. Auerbach, R. (1971): Towards a developmental theory of immunity: Cell interactions. In: *Cell Interactions and Receptor Antibodies in Immune Responses,* edited by O. Makela, A. Cross, and T. U. Kosunen, pp. 393–398. Academic Press, New York.

5. Auerbach, R. (1971): Contacts and communications between cells and their relationship to morphogenesis and differentiation. In: *The Dynamic Structure of Cell Membranes,* 21st Symposium of the Gesellschaft fur Biologische Chemie, Mosbach, April 15–17, 1971, pp. 37–49. Springer-Verlag, Berlin.

6. Auerbach, R. (1972): The use of tumors in the analysis of inductive tissue interactions. *Dev. Biol., 28:304–309.*

7. Auerbach, R. (1973): Controlled differentiation of tumor cells. In: *Unifying Concepts of Leukemia,* edited by Kutcher, and Chieco-Bianchi, pp. 906–909. Karger, Basel.

8. Auerbach, R. (1974): Towards a developmental theory of immunity: Ontogeny of immunocompetence and the concept of allosteric tolerance. In: *Cellular Selection and Regulation in the Immune Response,* edited by G. Edelman, pp. 59–70. Raven Press, New York.

9. Auerbach, R. (1975): Development of immunity and the concept of stem cell tolerance. *Am. Zool., 15:209–213.*

10. Babinet, C., Condamie, H., Fellous, M., Gachelin, G., Kemler, R., and Jacob, F. (1975): Expression of a cell surface antigen common to primitive mouse teratocarcinoma cells and cleavage embryos during embryogenesis and spermatogenesis. In: *Teratomas and Differentiation,* edited by D. Solter, and M. Sherman, pp. 101–108. Academic Press, New York.

11. Barclay, A. N., Letarte-Muirhead, M., and Williams, A. F. (1976): Chemical characterization of the Thy-1 glycoproteins from the membranes of rat thymocytes and brain. *Nature,* 263:563–567.

12. Bennett, D., Artzt, K., Spiegelman, M., and Magnuson, T. (1977): Experimental teratomas as a way of analyzing the development of the T/t locus mutations. In: *Cell Interactions in Differentiation,* Proceedings of the Sixth Sigrid Juselius Symposium, edited by Karkinen and Jaaskelainen. Academic Press, New York (*in press*).

13. Boyse, E. A., and Bennett, D. (1974): Differentiation and the cell surface: Illustrations from work with T cells and sperm. In: *Cellular Selection and Regulation in the Immune Response,* edited by G. M. Edelman, pp. 155–176. Raven Press, New York.

14. Edidin, M., and Gooding, L. (1975): Teratoma-defined and transplantation antigens in early mouse embryos. In: *Teratomas and Differentiation,* edited by D. Solter, and M. Sherman, pp. 109–122. Academic Press, New York.

15. Ginsburg, H. (1965): The role of the thymus in immunology: A hypothesis by analogy with the testis. (*Unpublished manuscript.*)

16. Gluecksohn-Waelsch, S., and Erickson, R. P. (1970): The T locus of the mouse: Implications for mechanisms of development. *Curr. Top. Dev. Biol., 5:281–316.*

17. Gooding, L. R., Hsu, Y-C., and Edidin, M. (1976): Expression of teratoma-associated antigens of murine ova and early embryos. *Dev. Biol., 49:479–486.*

18. Grobstein, C. (1955): Inductive interaction in the development of the mouse metanephros. *J. Exp. Zool., 130:319–340.*

19. Jacob, F. (1977): Mouse teratocarcinoma and embryonic antigens. *Immunol. Rev.,* 33:3–32.

20. Morris, R. J., Letarte-Muirhead, M., and Williams, A. F. (1975): Analysis in deoxycholate of three antigenic specificities associated with the rat Thy-1 molecule. *Eur. J. Immunol.,* 5:282–285.

21. Muthukkaruppan, V. (1965): Inductive tissue interaction in the development of the mouse lens *in vitro. J. Exp. Zool., 159:269–287.*

22. Sato, V. L., Waksal, S. D., and Herzenberg, L. A. (1976): Identification and separation of pre-T cells from nu/nu mice: Differentiation by preculture with thymic reticuloepithelial cells. *Cell. Immunol., 24:173–185.*

23. Solter, D., and Schachner, M. (1976): Brain and sperm cell surface antigen (NS-4) on preimplantation mouse embryos. *Dev. Biol., 52:98–104.*

24. Stern, P. L., Martin, G. R., and Evans, M. J. (1975): Cell surface antigens of clonal teratocarcinoma cells at various stages of differentiation. *Cell, 6:455–465.*

25. Vitetta, E. S., Artzt, K., Bennett, D., Boyse, E. A., and Jacob, F. (1975): Structural similarities between a product of the T-locus isolated from sperm and teratoma cells, and H-2 components isolated from splenocytes. *Proc. Natl. Acad. Sci. U.S.A.*, 72:3215–3219.
26. Youniss, S. T. (1976): The characterization of an embryoid body-derived malignant endoderm cell: The role of intercellular adhesion in organ specific tumor cell metastasis. B. A. Thesis, University of Wisconsin, Madison.

Cell and Tissue Interactions, edited by
J. W. Lash and M. M. Burger. Raven
Press, New York, 1977.

The Shape and Movement of Fibroblasts in Culture

M. Abercrombie, G. A. Dunn, and J. P. Heath

Strangeways Research Laboratory, Cambridge CBI 4RN, England

A fibroblast in a fairly sparse culture has a characteristic shape, or rather, range of shapes, and most of the time it is performing active locomotion. Its shape and its locomotion are closely related. This chapter reviews some working hypotheses about them, restricting itself mostly to the conventional culture conditions of a plane substratum covered by a liquid medium, and mostly to the familiar chick embryo heart fibroblast. We shall be concerned with gross form. The generation and the relation to movement of the structures that make up the fine texture of the cell surface, such as the microvilli, microspikes, and so forth, are too little understood to be included. The main aspects that we shall have to consider are the cell's adhesion to the solid surface around it, the tensions that go with these adhesions, and the protrusive activity by which the cell makes an extension over its substratum.

THE ADHESIONS OF A CULTURED FIBROBLAST

The most extensive adhesions of a fibroblast in a sparse culture are to the substratum provided by the culture vessel. We can detect a good deal about these adhesions when the substratum is a plane sheet of glass, because the interference reflection microscope (13,32) will then reveal them in the living state. The well-known chick heart fibroblast, free from contact with other cells, assumes on such a substratum the shape that is characteristic of many kinds of fibroblasts isolated on any plane surface to which they can adhere. The cell, although some 50 μm long, is flattened on the substratum to a thickness of no more than 2 to 3 μm (over the nucleus). As it moves, its front end is a broad, thin sheet of cytoplasm (the leading lamella) tapering down to half a micrometer thick or less at the front edge. Its rear end is narrower in plan, its sides often a smooth gently concave curve, although sometimes rather less regular. In the interference reflection microscope, regions of very close approach of the underside of the cell to the substratum are almost black. They take the form of a varying number of small discrete areas, termed focal contacts (32), most of them elongated roughly in the direction in which the cell (or more strictly the nearby cell periphery) is moving (or in some cases has recently been moving). The size of each is

2 to 10 μm long by about 0.25 to 0.5 μm wide; and the gap between cell and substratum is about 10 to 15 nm (32). Most of them are usually found underneath the leading lamella, but some are clustered at the rear end, and some near the tip of any major projection from the cell (Fig. 1). The evidence that

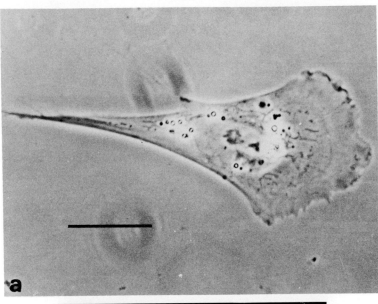

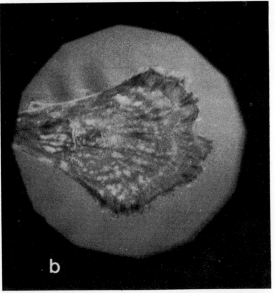

FIG. 1. a: Phase contrast micrograph of a living chick embryo heart fibroblast moving on a glass substratum. **b:** The same cell 5s later viewed with interference reflection microscopy showing focal contacts and gray areas under the leading lamella. Bar 20 μm.

these focal contacts are adhesions is good (1,32). They are, for instance, absolutely stationary in relation to the substratum, however fast or slow the cell is moving; when the cell retracts, they remain as the attachment of retraction fibers to the glass; when part of one cell intrudes between the substratum and another, it avoids the focal contacts.

One expects that these focal contacts will be apparent in electron microscope sections cut vertically to the substratum. They were in fact observed before the interference microscope had resolved them. Cornell (12) reported that close application of fibroblasts to a substratum was confined to small regions. Abercrombie et al. (5) identified small areas in which the bilayers of the cell membrane were separated from the substratum by less than 30 nm. They were mostly under the leading lamella. These areas corresponded in size, shape, and number with the focal contacts subsequently shown by the interference microscope, their identity with which is confirmed by evidence to be mentioned that both are attached to bundles of microfilaments within the cell.

Focal contacts in a moving cell are formed anew close to the anterior edge of the cell (39), and they fade away further posteriorly as the cell moves forward. Their length of life is inversely correlated with the speed of cell movement, and it is an interesting problem as to what detaches them at a roughly constant average position along the length of the cell.

These conspicuous attachments are probably not the only means by which a cell is held to the substratum. The interference microscope commonly shows a larger, less clear-cut area of uniform gray under the anterior part of the leading lamella where most of the focal contacts lie (32). It seems to change in shape and size as the cell moves. It represents a less close approximation of cell to glass than do the focal contacts (about 30 nm) but perhaps one close enough for adhesive forces to act. Fibroblasts may be found, especially amongst those that first emerge in culture from an explant on to the glass, with no clearly resolvable dark focal contacts, but only gray areas. Furthermore, we have seen only gray areas beneath polymorphs, macrophages, and a few varieties of sarcoma cells (most sarcoma cells, however, move very poorly on glass). The ingenious method by which Revel and Wolken (44) have applied the scanning electron microscope to the substratum side of fibroblasts confirms the occurrence of large flat areas of possible attachment (in L cells).

BUNDLES OF MICROFILAMENTS

A striking characteristic of focal contacts is that the plasma membrane is not only adherent on the outside to the substratum, but on the other, internal side it is attached to a cytoplasmic strand, which can often be seen in living cells (10,32). In the electron micrographs of vertical longitudinal sections of fibroblasts in which the focal contacts were seen as regions of close approach of cell to substratum (5), they also commonly showed as regions where there was an increased electron density, or plaque, on the inner surface of the plasma membrane, reminiscent of the similar plaques in smooth muscle cells to which actin filaments are attached (6,34,43). From a number of these densities in the fibroblasts a bundle of microfilaments was seen to proceed obliquely backward in the cell, rising from the ventral toward the dorsal surface. Doubtless these bundles of filaments are to be identified with the strands visible in living cells, referred to above. Similar plaques associated with attachment to the substratum and with microfilaments were found in glia cells by Brunk et al. (9).

A better view of these bundles can be gotten from electron micrographs of whole cells, especially when a high-voltage electron microscope is used on thick sections parallel to the substratum (Fig. 2). Correlation of pictures obtained in this way with pictures of the same cells before sectioning taken with the interference reflection microscope enables the relation of these bundles to the attachment plaques to be determined rather clearly. The great majority of the plaques under the leading lamella form the termination of a bundle. Sometimes a bundle links two plaques, but more usually the strand terminates anteriorly in a plaque while posteriorly it ends diffusely

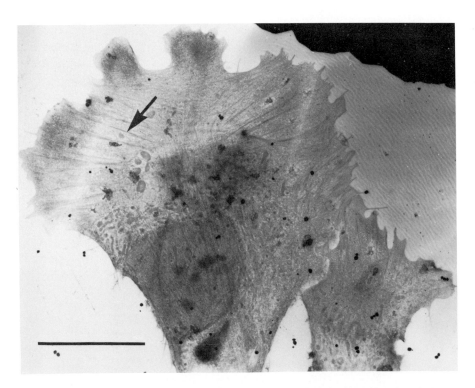

FIG. 2. High-voltage electron micrograph (700 kV) of a whole, fixed, and embedded chick heart fibroblast with many bundles of microfilaments **(arrow)** in the lamella. Bar 10 μm.

near the nucleus. The oblique bundles seen in longitudinal sections presumably belong to the latter category.

There is little doubt that these bundles of microfilaments are identical with those shown with heavy meromyosin to contain actin (31,40) and by much recent work with fluorescent antibodies to contain the major muscle proteins: actin (24), myosin (48), tropomyosin (35), and α-actinin (37). Lazarides and Burridge (37) find that the plaques themselves contain a concentration of α-actinin, and in that respect are similar to the Z-bands of striated muscle to which actin filaments are attached. A striking demonstration that the bundles in fibroblasts are contractile has recently been provided by Isenberg et al. (30): They isolated a bundle from a glycerinated cell by means of a laser microbeam and made it contract with ATP.

It seems a reasonable hypothesis that the bundles that have a single attachment to a plaque are linked at the other end to the fibrillar system of the cell. They are well placed, therefore, to draw more posterior parts of the moving cell forward toward the adhesions to the substratum. Fleischer and Wohlfarth-Bottermann (21) have found in experiments with the contractile cytoplasm of *Physarum* that definite bundles of parallel-disposed microfilaments only form when contraction is isometric. This suggests that the bundles in fibroblasts are contracting against a considerable resistance, which presumably comes partly from the focal contacts usually found at the rear of a migrating fibroblast. The rear end of the cell is indeed usually under tension. It often moves forward in a series of jerks, apparently as posterior adhesions part, and when it breaks or is broken (26) free from its adhesions it rapidly retracts. The resistance to the contraction of the bundles also comes from adhesions to other cells, which we now consider.

Besides its adhesions to the substratum, served by the rather elaborate apparatus described, a fibroblast may have one or more areas of adhesion to neighboring fibroblasts. These areas of adhesion usually influence cell shape, the outline of the cell being drawn toward the contact. The resulting cell process is, like the free rear end of a migrating cell, under tension. The tension develops very quickly in any part of the front end of a moving fibroblast that makes an adhesion with another fibroblast. It is one of the most obvious characteristics of the phenomena of contact inhibition (see, for instance, ref. 1). The tension persists, so that a sheet of fibroblasts *in vitro,* when freed from extraneous adhesions, contracts with a force that has been measured (33) and found to be about one-tenth of that exerted by a smooth muscle of similar cross-sectional area.

The initial adhesion formed between two fibroblasts that move into contact is in many ways similar to that formed between fibroblast and substratum. There are areas of fairly close apposition (probably closer than in the gray areas) and within them plaques to which bundles of microfilaments become attached (29). The plaques are developed by both cells, in pairs; a plaque appears on either side of a small region of very close apposition

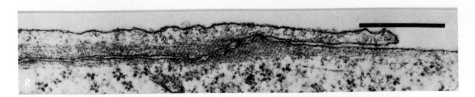

FIG. 3. Electron micrograph of a vertical section through an adhesion between two chick heart fibroblasts showing plaques and microfilament bundles associated with the two apposed membranes. Bar 0.5 μm. (From ref. 29.)

(Fig. 3). The plaques appear within 20 sec of the first contact between the cells. There is some suggestion in the work of Lazarides and Burridge (37) that they contain concentrations of α-actinin. Tight junctions are also formed quickly after contact (20).

The areas of adhesion between fibroblasts persist for varying times, which have not been systematically studied. They are very firm while they last: often firmer in aggregate than the adhesions to the substrate (26). They are, however, broken by the cell movements in culture, although whether this involves some intrinsic process of decay is unknown.

SKELETAL ELEMENTS

We have considered the fibroblast as if its shape came from the way it is, so to speak, pinned out by its adhesions. But are there no skeletal structures that provide the cell with an intrinsic resistance to deformation? The system of microtubules which permeates the cytoplasm (49), although there is not a high concentration of microtubules in any part of a cultured fibroblast, might be a candidate for the role. So, rather more convincingly because of their size, might be the system of microfilament bundles. But whatever the possibilities, there seems in fact to be no evidence that an effective skeletal system exists in fibroblasts. When the cell's adhesions to its surroundings are released, by whatever method, it starts to retract toward a spherical shape, and that is the shape it maintains in the absence of renewed adhesions. Microtubules seem to be reduced in number in rounded cells (23). The bundles of actin are lost (7,36) at any rate if the cells are freed by trypsin. Their disappearance is to be expected if one accepts the argument of Fleischer and Wohlfarth-Bottermann (21) that oriented microfilaments appear only when contraction is resisted. It seems that if shape is maintained by rigid structure, rigid structure is maintained by adhesions.

The rounded form taken up by fibroblasts in suspension is still sometimes regarded as a consequence of surface tension. This is, however, highly unlikely, because the surface area is far from being reduced to a minimum. A sphere is only its gross form: It has numerous submicroscopic irregularities such as microvilli, as has been known since the observations of Taylor

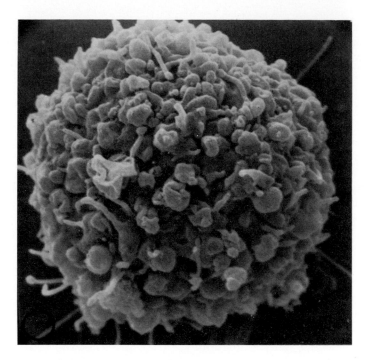

FIG. 4. A BHK21 cell rounded up by trypsinization. The surface is highly convoluted and covered with blebs and some microvilli. Bar 1 μm. (From ref. 18.)

(45) and Lesseps (38) (Fig. 4). Indeed, Erickson and Trinkaus (18) have estimated that there is no change of surface area in the transition from round (mitotic or suspended) to flattened form in BHK21 cells. Rounding up of cells is much more likely to be due to a subsurface tension, produced by the contraction of cortical actomyosin.

Cooke (11) has proposed a persuasive case that there is a different kind of skeletal system in smooth muscle fibers, the 100-Å filaments, that protects the cell against tensile stress. There seems no evidence against the view that the 100-Å filaments of fibroblasts have a similar function, but their attachments within the fibroblast are unknown as yet.

PROTRUSIVE ACTIVITY

Parts of the edge of a cultured fibroblast, as seen in plan, are relatively thick, smooth in outline, gently concave, and rather unchanging. They are probably regions of the edge that are under longitudinal tension, between areas of adhesion to the substratum or to other cells. Other parts of the edge, especially the broad leading edge, are different; they are flattened, somewhat irregular, and usually convex in outline, and they are undergoing continuous change, in part sufficiently rapid to be directly visible. We term this change protrusive activity, since its final effect is to extend that part of the cell across the substrate. The extension is not, however, a steady advance: the margin of the cell fluctuates backward and forward over a distance of several micrometers in a highly irregular way. It slowly gains ground on average, because it spends a longer time extending than withdrawing. The observed fluctuations are caused by the projection from the edge of the cell of a very thin sheet of cytoplasm, a lamellipodium, which usually does not touch the substratum (39); its withdrawal follows, and this is often accompanied by a folding of the lamellipodium away from the substratum, to form a ruffle projecting into the liquid medium.

The precise events during this strange protrusive activity have been described in some detail by Abercrombie et al. (2–5) [and for epithelia by Di-Pasquale (14)]. Our hypothesis was that new material is continually added to the cell surface at the anterior edge, in fluctuating amounts, an excess leading to the extension of the delicate folds of the surface membrane (the lamellipodia) in this region, a deficit to their withdrawal; the output of surface is compensated by its internalization behind the edge, with a backward flow on the cell surface and presumably a forward flow within the cytoplasm to complete a circulation (4). Harris (25) has linked the circulation with the contractile system of the cell, by supposing that actin attached to the cell surface is responsible for pulling it back, on both dorsal and ventral surfaces, thereby creating a tension at the extreme anterior edge, which is met by the insertion of new membrane there. An ingenious recirculation hypothesis, in which only the cell membrane lipids are involved, leaving the protein floating in the lipid bilayer, has recently been proposed by Bretscher (8); it has the great advantage of accounting satisfactorily for the phenomena of "capping." These hypotheses remain to be tested rigorously. They imply that a great deal of material must be moved forward (relative to the rest of the cell). It should, however, be noted that even if they are incorrect, such forward transport is still required. This is because the material making up and associated with the plaques moves backward in relation to the rest of the cell, and so must eventually be moved forward in relation to the rest of the cell. The driving force for this transport has not been identified. It may be noted that the direction of protrusion, and hence presumably of the flow of material, corresponds with the direction of the bundles of microfilaments attached to the nearby plaques. Whether there is dependence of one on the other is an interesting question.

The protrusive activity takes place most frequently at or near the edge of the cell, although lamellipodia are sometimes projected from the dorsal surface. A region of adhesions to the substratum is always found near any edge that is active (26). It is conceivable that this particular situation at the edge of a cell sets going protrusive activity and keeps it going. This would explain why a rounded cell, settling from suspension, starts to flatten by protrusive activity when it touches the substratum. [An isolated epithelial cell does not do so; it requires contact with another epithelial cell to initiate protrusive activity (42).] The suggestion that spreading of a cell is initiated by contact with an area of solid surface, and that phagocytosis by leukocytes may be basically the same reaction, but with a smaller area of solid surface that can be engulfed by the spreading, was current in the days when cells were regarded as simple liquid drops (19) and may still be valid.

Whether or not protrusive activity is promoted by adhesion to a solid substratum, it is inhibited by adhesion to another fibroblast. This is the phenomenon of contact inhibition (28). When two fibroblasts in culture collide, they usually adhere and the fine structure of the adhesions, as we have seen, is reminiscent of those between cell and substratum, with focal contacts connected to bundles of microfilaments as contraction develops. The effect, though, of the adhesion is quite different. It is to produce a localized cessation of protrusive activity (46), and a localized contraction which in its manifestation is more severe than any that occurs during uninhibited movement. The result is the formation of one of the mutual adhesions under tension, which has already been described.

Protrusive activity is therefore suppressed by contacts with other cells. It is clearly also controlled from within the cell. A fibroblast free of contacts with others, unless it has just settled from suspension, usually has its main protrusive activity limited to one part of its periphery, which is naturally the leading edge. Other minor regions may occur transitorily at other places, particularly at the tail end; occasionally there is extensive activity at both ends of a bipolar cell, which is thereby immobilized. If the leading protrusive region of a migrating cell is inhibited by contact inhibition, activity quickly develops elsewhere, often in an existing region of minor activity, and a new front end is formed. The impression is strong that there is some sort of competitive relation between protrusive regions, a major one tending to suppress any other. From time to time, independently of contact inhibition, a dominant region or a large part of one loses its activity, and the cell then undergoes an apparently spontaneous change of direction as a new region comes into activity. Such a competitive relation was first proposed by Weiss and Garber (50).

A rounded cell that settles from suspension onto a solid substratum spreads and flattens on the solid surface by protrusive activity (45,51). This activity is not exactly the same as that of the leading edge of a moving cell in a number of ways, most obviously in that it is not polarized, but occurs all round the periphery of the cell. The cell has evidently not immediately regained the postulated competitive relation, but it soon does, and becomes polarized in the usual fashion.

Little can be said about how this competitive relation is maintained. Drugs that abolish microtubules also abolish the usual polarity, but do not abolish the protrusive activity, which appears at many places around the periphery, so that the cell often comes to a standstill (22,47). This suggests that the competitive relation depends on microtubules, which perhaps canalize the direction of transport of the material that takes part in the protrusive activity. There is no effect of colchicine on the symmetrical spreading of cells deposited from suspension (22) nor on the protrusive activity of epithelial cells (15); in neither case can one say that there is a competitive relation. Microtubules are therefore clearly important for fibroblast shape even if they have no skeletal role.

VARIOUS SUBSTRATA

We have concentrated on the plane glass surface as a substratum because it has been so widely used. It is not necessarily the totally foreign environment that it has seemed to some critics, since the glass is usually coated with a layer adsorbed from the medium (9,44). In any case, the form that

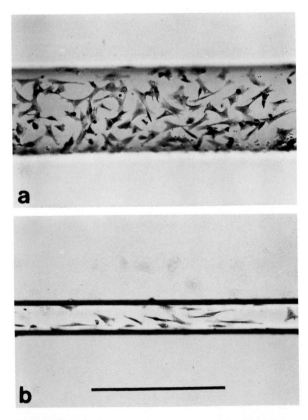

FIG. 5. Chick heart fibroblasts cultured for 24 hr on glass fiber substrata with radii of curvature of 127 μm **(a)** and 54 μm **(b)** demonstrating that only in **b** has the glass fiber a small enough radius to elicit an oriented response from the cells. Bar 200 μm.

fibroblasts take on any roughly plane surface, including a dense mat of collagen or the consolidated surface layer of a fibrin gel, is very similar to that on glass. With a large range of substrata, the gross shape of the surface they present, rather than their composition, is the most significant factor. There are, however, relatively nonadhesive substrata (hydrophobic or mechanically yielding hydrophilic; 41) on which cells are markedly less flattened (27), presumably because they cannot maintain many of their adhesions against the stress produced by contraction in the cell.

When the form of the substratum is not even roughly a plane, cell shape becomes different. Cells are very sensitive to curvature of substrata, and will move in the direction in which it is least (Fig. 5). This is probably the basis of the well-known contact guidance of Paul Weiss, and with suitable linearly oriented substrata it will produce strongly polarized cells. A hypothesis to account for contact guidance has recently been presented (16); it is that curvature of the cell in its long axis restricts the possibility of forming the linear bundles of microfilaments that are attached to the focal contacts, the role in movement of which has already been described. When a fibroblast is moving on fibers of a fraction of a micrometer in diameter as within the substance of a gel of fibrin or collagen (17), it no longer has the possibility of forming extended areas of adhesion, as it can beneath the leading lamella when it is on a plane surface. The shape of the cell becomes very different, with a number of narrow elongated branches that may taper to fine tips. Locomotion and adhesions have not been studied yet under these circumstances, obviously important since they are commonly presented to cells *in vivo*.

THE HYPOTHESES RELATING LOCOMOTION AND SHAPE

It may finally be useful to put together our tentative picture of the process of locomotion and its relation to shape in fibroblasts cultured on a plane surface. In such a moving fibroblast, there is an anterior region of protrusive activity, fed by a forward flow of material. It is possible that this material includes a large amount of surface components, which are being exteriorized at the leading edge. There is usually only one major region of protrusive activity because of some competitive relation between them, which is related to the microtubule system. The protrusive activity results in a slow extension (at best about 100 μm/hr) of the front end across the substratum. This advance is consolidated by the formation near the front end of adhesions to the substratum, of which the most important seem to be focal contacts. Protrusion, therefore, produces adhesion, and protrusion may at times itself be induced or encouraged by an adjacent area of adhesion. When a fibroblast is moving on small diameter fibrils, as in a gel, only a narrow region of adhesion is possible, and protrusion is on a narrow front.

The formation of the focal contacts is associated with the development of bundles of aligned microfilaments, usually running backward from the contacts; their orientation seems to correspond with the flow of material involved in protrusion. These bundles, possibly assisted by less organized actomyosin, contract against the resistance offered by more posterior adhesions to the substratum, and to other cells if present; they drag the body of the cell forward, toward their own focal contacts. The mode of formation of the posterior adhesions has not been followed yet. They do not, it seems, represent the focal contacts originally formed at the front of the cell. These fade out, by an unknown mechanism, after the first 10 or 20 μm of the cell has passed over them.

These processes of protrusion and contraction seem to be balanced to produce, on a given substratum, a roughly standard length of cell. A rounded-up cell will, therefore, soon expand again to that length. By suitable centrifugation in the presence of cytochalasin B a fibroblast can be drawn out into a fine thread more than a millimeter long, up to 15 times its normal length; returned to normal medium, it will draw itself together and regain its standard length (B. R. O'Brien, *personal communication*). With the standard length goes the standard shape of a free fibroblast, roughly triangular in plan, determined by adhesions to substratum, tensions between adhesions, and protrusive activity of varying extent at one region of its periphery. An internal rigid skeleton has not yet been demonstrated, and its existence seems to depend on the preservation of the adhesions. This standard shape is modified by contractile adhesions to other cells, involving contact inhibition, and by apparently spontaneous changes of direction.

This is our tentative scheme, at least for a chick heart fibroblast. There are, of course, many recognizably different kinds of fibroblasts, variable in shape according to substratum and liquid medium. We suggest, however, that with quantitative changes of emphasis of the various component parts of the mechanism, they can be fitted into the same framework.

ACKNOWLEDGMENT

The work of the authors is supported by a grant from the Medical Research Council of Great Britain.

REFERENCES

1. Abercrombie, M., and Dunn, G. A. (1975): Adhesions of fibroblasts to substratum during contact inhibition observed by interference reflection microscopy. *Exp. Cell Res.,* 92: 57–62.
2. Abercrombie, M., Heaysman, J. E. M., and Pegrum, S. M. (1970): The locomotion of fibroblasts in culture. I. Movements of the leading edge. *Exp. Cell Res.,* 59:393–398.
3. Abercrombie, M., Heaysman, J. E. M., and Pegrum, S. M. (1970): The locomotion of fibroblasts in culture. II. "Ruffling." *Exp. Cell Res.,* 60:437–444.
4. Abercrombie, M., Heaysman, J. E. M., and Pegrum, S. M. (1970): The locomotion of

fibroblasts in culture. III. Movements of particles on the dorsal surface of the leading lamella. *Exp. Cell Res.,* 62:389–398.

5. Abercrombie, M., Heaysman, J. E. M., and Pegrum, S. M. (1971): The locomotion of fibroblasts in culture. IV. Electron microscopy of the leading lamella. *Exp. Cell Res.,* 67: 359–367.

6. Ashton, F. T., Somlyo, A. V., and Somlyo, A. P. (1975): The contractile apparatus of vascular smooth muscle: Intermediate high voltage stereo electron microscopy. *J. Mol. Biol.,* 98:17–30.

7. Bragina, E. E., Vasiliev, J. M., and Gelfand, I. M. (1976): Formation of bundles of microfilaments during spreading of fibroblasts on the substrate. *Exp. Cell Res.,* 97:241–248.

8. Bretscher, M. S. (1976): Directed lipid flow in cell membranes. *Nature,* 260:21–23.

9. Brunk, V., Ericsson, J. L. E., Ponten, J., and Westermark, B. (1971): Specialization of cell surfaces in contact-inhibited human glia-like cells *in vitro. Exp. Cell Res.,* 67:407–415.

10. Buckley, I. K., and Porter, K. R. (1967): Cytoplasmic fibrils in living cultured cells. A light and electron microscope study. *Protoplasma,* 64:349–380.

11. Cooke, P. (1976): A filamentous cytoskeleton in vertebrate smooth muscle cells. *J. Cell Biol.,* 68:539–556.

12. Cornell, R. (1969): Cell-substrate adhesion during cell culture. *Exp. Cell Res.,* 58:289–295.

13. Curtis, A. S. G. (1964): The mechanism of adhesion of cells to glass. A study by interference reflection microscopy. *J. Cell Biol.,* 20:199–215.

14. DiPasquale, A. (1975): Locomotory activity of epithelial cells in culture. *Exp. Cell Res.,* 94:191–215.

15. DiPasquale, A. (1975): Locomotion of epithelial cells. Factors involved in extension of the leading edge. *Exp. Cell Res.,* 95:425–439.

16. Dunn, G. A., and Heath, J. P. (1977): A new hypothesis of contact guidance. *Exp. Cell Res.,* 101:1–14.

17. Elsdale, T., and Bard, J. (1972): Collagen substrate for studies on cell behaviour. *J. Cell Biol.,* 54:626–637.

18. Erickson, C. A., and Trinkaus, J. P. (1976): Microvilli and blebs as sources of reserve surface membrane during cell spreading. *Exp. Cell Res.,* 99:375–384.

19. Fenn, W. O. (1922): The adhesiveness of leucocytes to solid surfaces. *J. Gen. Physiol.,* 5:143–167.

20. Flaxman, B. A., Revel, J. P., and Hay, E. D. (1969): Tight junctions between contact inhibited cells in vitro. *Exp. Cell Res.,* 58:438–443.

21. Fleischer, M., and Wohlfarth-Bottermann, K. E. (1975): Correlation between tension force generation, fibrillogenesis and ultrastructure of cytoplasmic actomyosin during isometric and isotonic contractions of protoplasmic strands. *Cytobiologie,* 10:339–365.

22. Goldman, R. D. (1971): The role of three cytoplasmic fibres in BHK-21 cell motility. *J. Cell Biol.,* 51:752–762.

23. Goldman, R. D., and Knipe, D. (1972): Functions of cytoplasmic fibres in non-muscle cell motility. *Cold Spring Harbor Symp. Quant. Biol.,* 37:523–534.

24. Goldman, R. D., Lazarides, E., Pollack, R., and Weber, K. (1975): The distribution of actin in non-muscle cells. *Exp. Cell Res.,* 90:333–344.

25. Harris, A. K. (1973): Cell surface movements related to cell locomotion. In: *Locomotion of Tissue Cells, Ciba Foundation Symposium,* 14:3–20.

26. Harris, A. (1973): Location of cellular adhesions to solid substrata. *Dev. Biol.,* 35:97–114.

27. Harris, A. (1973): Behaviour of cultured cells on substrata of variable adhesiveness. *Exp. Cell Res.,* 77:285–297.

28. Harris, A. (1974): Contact inhibition of cell locomotion. In: *Cell Communication,* edited by R. P. Cox, pp. 147–185. Wiley, New York.

29. Heaysman, J. E. M., and Pegrum, S. M. (1973): Early contacts between fibroblasts. An ultrastructural study. *Exp. Cell Res.,* 78:71–78.

30. Isenberg, G., Rathke, P. C., Hulsmann, N., Franke, W. W., and Wohlfarth-Bottermann, K. E. (1976): Cytoplasmic actomyosin fibrils in tissue culture cells. Direct proof of contractility by visualization of ATP-induced contraction in fibrils isolated by laser microbeam dissection. *Cell Tissue Res.,* 166:427–443.

31. Ishikawa, H., Bischoff, R., and Holtzer, H. (1969): Formation of arrowhead complexes with heavy meromyosin in a variety of cell types. *J. Cell Biol.,* 43:312–328.

32. Izzard, C. S., and Lochner, L. R. (1976): Cell-to-substrate contacts in living fibroblasts: An interference reflexion study with an evaluation of the technique. *J. Cell Sci.,* 21:129–160.

33. James, D. W., and Taylor, J. F. (1969): The stress developed by sheets of chick fibroblasts *in vitro. Exp. Cell Res.,* 54:107–110.

34. Kelly, R. E., and Rice, R. V. (1969): Ultrastructural studies on the contractile mechanism of smooth muscle. *J. Cell Biol.,* 42:683–694.

35. Lazarides, E. (1975): Tropomyosin antibody: The specific localization of tropomyosin in non-muscle cells. *J. Cell Biol.,* 65:549–561.

36. Lazarides, E. (1976): Actin, α-actinin, and tropomyosin interaction in the structural organization of actin filaments in non-muscle cells. *J. Cell Biol.,* 68:202–219.

37. Lazarides, E., and Burridge, K. (1975): α-Actinin: Immunofluorescent localization of a muscle structural protein in non-muscle cells. *Cell,* 6:289–298.

38. Lesseps, R. J. (1963): Cell surface projections: Their role in the aggregation of embryonic cells as revealed by electron microscopy. *J. Exp. Zool.,* 153:171–182.

39. Lochner, L., and Izzard, C. S. (1973): Dynamic aspects of cell-substrate contact in fibroblast motility. *J. Cell Biol.,* 59:1991.

40. Luduena, M. A., and Wessells, N. K. (1973): Cell locomotion, nerve elongation and microfilaments. *Dev. Biol.,* 30:427–440.

41. Maroudas, N. G. (1973): Chemical and mechanical requirements for fibroblast adhesion. *Nature,* 244:353–355.

42. Middleton, C. A. (1976): Contact-induced spreading is a new phenomenon depending on cell-cell contact. *Nature,* 259:311–313.

43. Panner, B. J., and Honig, C. R. (1967): Filament ultrastructure and organization in vertebrate smooth muscle. Contraction hypothesis based on localization of actin and myosin. *J. Cell Biol.,* 35:303–321.

44. Revel, J. P., and Wolken, K. (1973): Electron microscope investigations of the underside of cells in culture. *Exp. Cell Res.,* 78:1–14.

45. Taylor, A. C. (1961): Attachment and spreading of cells in culture. *Exp. Cell Res. Suppl.,* 8:154–173.

46. Trinkaus, J. P., Betchaku, T., and Krulikowski, L. S. (1971): Local inhibition of ruffling during contact inhibition of cell movement. *Exp. Cell Res.,* 64:291–300.

47. Vasiliev, J. M., Gelfand, I. M., Domnina, L. V., Ivanova, O. Y., Komm, S. G., and Olshevskaja, L. V. (1970): Effect of colcemid on the locomotory behaviour of fibroblasts. *J. Embryol. Exp. Morphol.,* 24:625–640.

48. Weber, K., and Groeschel-Stewart, U. (1974): Antibody to myosin: The specific visualization of myosin-containing filaments in non-muscle cells. *Proc. Natl. Acad. Sci. U.S.A.,* 71:4561–4564.

49. Weber, K., Pollack, R., and Bibring, T. (1975): Antibody against tubulin: The specific visualization of cytoplasmic microtubules in tissue culture cells. *Proc. Natl. Acad. Sci. U.S.A.,* 72:459–463.

50. Weiss, P., and Garber, B. (1952): Shape and movement of mesenchyme cells as functions of the physical structure of the medium. *Proc. Natl. Acad. Sci. U.S.A.,* 38:264–280.

51. Witkowski, J. A., and Brighton, W. D. (1971): Stages of spreading of human diploid cells on glass surfaces. *Exp. Cell Res.,* 68:372–380.

Cell and Tissue Interactions, edited by
J. W. Lash and M. M. Burger. Raven
Press, New York, 1977.

The Collagens of the Extracellular Matrix

Edward J. Miller

*Department of Biochemistry and Institute of Dental Research, University of
Alabama Medical Center, University Station, Birmingham, Alabama 35294*

This chapter will be devoted to a summary of current information on the biosynthesis, biochemical properties, and possible physiological roles of the genetically distinct collagens in vertebrate organisms with special reference to the relatively few extant studies that suggest specific collagen-cell interactions. No attempt will be made to provide a comprehensive survey of current knowledge, since several reviews dealing with various aspects of these topics have been compiled within the past 4 years (30,39,40,42,52). Instead, emphasis is placed on the most recent relevant contributions as well as on areas in which considerable progress may be expected in the near future.

THE KNOWN COLLAGENS

The collagens to be discussed most extensively at this time are the relatively well characterized Types I, II, and III collagens. As indicated in Table 1, the Type I collagen molecule may be characterized as a hybrid molecule, since its three polypeptide chains include two $\alpha 1(I)$ chains plus an $\alpha 2$ chain. Fibers derived from Type I molecules are prevalent in virtually all of the major connective tissues. In contrast, the Type II collagen molecule is composed of three apparently identical $\alpha 1(II)$ chains, and fibers formed from Type II molecules are found chiefly in hyaline cartilages. The Type III collagen molecule is likewise composed of three identical chains, designated $\alpha 1(III)$, and fibers containing Type III molecules are prevalent in selected connective tissues, such as dermis, major vessels, and uterine wall, where they coexist with fibers derived from Type I collagen molecules. Each of the constituent α chains of these collagens exhibit a molecular weight of approximately 95,000 daltons and contains slightly more than 1,000 amino acid residues.

Since fibers derived from Types I, II, and III molecules are apparently always located between the cellular components of a given tissue or organ, these collagens are often referred to collectively as the interstitial collagens. Although these collagens undoubtedly account for the majority of the interstitial collagen in vertebrate organisms, there exist reports strongly sug-

TABLE 1. The known collagens

	Chain composition	Distribution
Type I	$[\alpha1(I)]_2\alpha2$	Major connective tissues
Type II	$[\alpha1(II)]_3$	Hyaline cartilages
Type III	$[\alpha1(III)]_3$	Selected connective tissues
"Type IV"		Basement membranes

gesting that additional types of interstitial collagen might be prevalent in certain eye tissues (53) as well as cardiac muscle (38). It may be assumed, then, that our information with respect to the total number of interstitial collagens is as yet incomplete.

Aside from considerations pertaining to the interstitial collagens, there appear to be additional collagen-like molecules localized in the basement membranes of vertebrate organisms. Based on the results of initial studies designed to characterize the collagen in several basement membranes (31) basement membrane collagen has, in certain instances, been designated as Type IV collagen. Nevertheless, as of this writing, there is considerable disagreement concerning the number of distinct collagens in basement membrane structures as well as the nature of their constituent chains (13,31,49). The present data, however, support the notion that the collagenous sequences in basement membranes occur in polypeptide chains which may be somewhat larger (13,31) or considerably smaller (49) than the constituent α chains of the interstitial collagens. Recent work in our own laboratory has, in general, confirmed several of these observations and further demonstrated that the nature of the collagenous components in basement membranes is likely to be dependent on the location and presumably the cellular origin of the basement membrane (10).

PREPARATION AND CHAIN COMPOSITION

The chain composition of Types I, II, and III collagens have been determined following extraction and purification of the respective collagens accompanied by careful analyses of their elution patterns when chromatographed in denatured form on carboxymethyl cellulose (40). In general, the dermis and some tendons of relatively young organisms represent good sources of monomeric Type I collagen molecules, which are readily extracted by employing neutral salt or dilute acid solvents. On denaturation, collagen prepared in this fashion chromatographs on carboxymethyl cellulose as a series of components representing $\alpha1(I)$ chains, $\alpha2$ chains, and dimers of these chains resulting from the formation of intramolecular cross links. Nevertheless, the $\alpha1(I)$ and $\alpha2$ chains, including those present in dimer form, are present in a 2:1 ratio, which has been shown to be indicative of the actual stoichiometry of the Type I collagen molecule.

Hyaline cartilages provide the most appropriate starting materials for the preparation of Type II collagen. Due to the prevalence of stable ketoamine cross links in cartilage collagen fibers as well as the tendency of these cross links to be intermolecular in nature, a substantial pool of readily extractable collagen does not accumulate in these tissues. The collagen may thus be extracted in monomeric form only from the cartilaginous tissues of lathyritic animals in which cross linking has been inhibited by the administration of a lathyrogen. Alternatively, the cartilage collagen may be rendered soluble by employing limited proteolysis with enzymes such as pepsin or papain, which degrade the nonhelical extremities of the molecules where cross links originate, thus allowing the dissolution of the fibers as somewhat truncated, but monomeric, molecules. Depending on the cartilage used to prepare the collagen, some Type I molecules may also be present. The Type I and II collagens can, however, be effectively separated in native form by selective precipitation at different ionic strengths from neutral salt solvents. When denatured and chromatographed as carboxymethyl cellulose, preparations of Type II collagen are eluted as a single somewhat heterogenous peak in a region closely corresponding to the elution position of $\alpha 1(I)$ chains, suggesting that the three $\alpha 1(II)$ chains of the Type II molecule are identical. The heterogeneity observed in the $\alpha 1(II)$ peak when chromatographed on carboxymethyl cellulose has been ascribed to a number of factors such as the relatively high content of hydroxylysine-linked carbohydrate and nonspecific proteolysis of the chains during preparation of the collagen. It is now apparent, however, that there may be a certain degree of genetic heterogeneity in $\alpha 1(II)$ chains (see below). Thus, the latter factor may ultimately be responsible for the observed chromatographic heterogeneity of these chains.

Type III collagen, similar to Type I collagen, is abundant in a variety of tissues but resembles Type II collagen in that it is quite resistant to extraction in native form. This accounts for the observations that preparations of Type I collagen obtained by employing neutral salt or dilute acid solvents normally contain very little, if any, Type III collagen. Although current information is somewhat fragmentary and in certain respects contradictory (4,19) it would appear that the relative resistance to extraction on the part of Type III collagen may be ascribed to a cross-linking pattern similar to that outlined above for Type II collagen.

Largely due to its resistance to extraction, Type III collagen was initially isolated from several tissues following solubilization by means of limited proteolysis with pepsin and separation from Type I collagen by differential salt precipitation (9). This approach remains the method of choice in the preparation of relatively large quantities of Type III collagen, although it was subsequently shown that some Type III collagen as well as a form of Type III procollagen can be extracted in neutral salt solvents from the skin of very young animals (8,34,51). On denaturation, Type III collagen is recovered largely as γ-components (molecular weight, 285,000 daltons) as the result of interchain disulfide bonding between the $\alpha 1$(III) chains. On reduction and alkylation, the high molecular weight components are converted to $\alpha 1$(III) chains, which chromatograph on carboxymethyl cellulose as a single homogenous peak near the point at which the $\alpha 2$ chain of Type I collagen is eluted.

Basement membrane collagen resembles Type II and III collagens in its resistance to extraction. However, it can likewise be effectively solubilized during limited proteolysis with pepsin (13,31). In our recent studies on tissues such as blood vessels and skin (10), we chose this approach to solubilize the tissue collagens and initiated the studies under the assumption that collagens of basement membrane origin could be separated from interstitial collagens by differential salt precipitation under appropriate conditions. When whole skin as well as selected regions of major vessels were investigated in this fashion, the results revealed that substantial quantities of collagen-like protein remained in solution under conditions in which Types I, II, and III collagens normally precipitate. Characterization of the more soluble collagens led to the isolation of three unique collagenous components, each of which exhibit compositional features indicative of their origin in basement membranes. One of the components has an apparent molecular weight of 55,000 daltons and is recovered following denaturation and reduction of the more soluble collagen in tissue lying adjacent to the lumen of aortas as well as of moderately sized veins. We have suggested, therefore, that a collagen-like molecule comprised of these chains occurs in basement membranes of endothelial origin. The second component exhibits a molecular weight of about 110,000 daltons. It is readily recovered following denaturation of the more soluble collagen found in pepsin digests of the medial layers of aortas and veins suggesting the origin of this collagen in basement membranes surrounding smooth muscle cells. The third component also has a molecular weight of approximately 110,000 daltons and is most readily retrieved on denaturation of the more soluble collagen of skin preparations, suggesting that the latter collagen is derived from epithelial basement membranes.

It is reasonable to presume that further work will eventually lead to a more complete biochemical characterization of these chains and more definitive information concerning the anatomical location of the respective collagens from which they are derived.

Furthermore, the basic technique used to isolate these collagens, i.e., selective fractionation of the collagens from the total pepsin-solubilized collagen of a given tissue is highly advantageous, since it does not require prior isolation of an anatomically defined basement membrane structure. This approach should be equally applicable to other tissues and organs containing basement membranes and considerably facilitate studies designed to characterize the collagen molecules of basement membranes and elucidate the molecular organization of these most important structures.

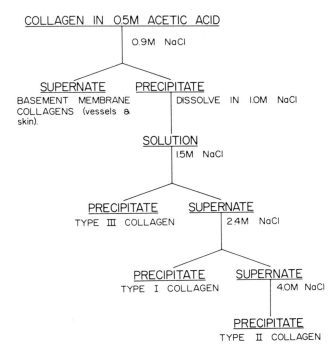

FIG. 1. The separation of various collagens by differential salt precipitation.

As noted often above, the various interstitial collagens as well as some of the collagens presumably originating in basement membranes exhibit widely different solubility properties once they have been extracted from the fibers in various tissues and dissolved in appropriate solvents. Exploitation of these properties to achieve separation of the collagens in native form has proven to be an extremely useful technique in the isolation of these proteins, and current methodology is summarized in Fig. 1. For the purposes of this illustration, a hypothetical mixture of basement membrane collagens from vessel wall and skin plus Types I, II, and III collagens are present in 0.5 M acetic acid. The addition of sufficient NaCl to bring the acetic acid solution to 0.9 M results in the precipitation of Types I, II, and III collagens, while the indicated basement membrane collagens remain in solution. The interstitial collagens may then be redissolved in neutral salt buffer and selectively precipitated at the indicated concentrations of NaCl. At the moment, the applicability of a similar approach to the separation of possible mixtures of basement membrane collagens remains to be investigated.

PRIMARY STRUCTURES

The biochemical properties and physiological roles of the different collagens are ultimately dependent on the primary structure of their constituent chains. Accordingly, investigations on the amino acid sequence of the $\alpha1(I)$, $\alpha2$, $\alpha1(II)$, and $\alpha1(III)$ chains are currently being pursued in several laboratories. Although information on the primary structure of the $\alpha2$, $\alpha1(II)$, and $\alpha1(III)$ chains remains fragmentary, the available comparative sequence data conclusively supports the concept that synthesis of each of the α chains is controlled by a distinct genetic locus. Thus, the bovine $\alpha1(II)$ chain exhibits largely conservative amino acid substitutions relative to the bovine $\alpha1(I)$ chain at approximately 20% of the positions along the chain (6,7). An even higher proportion of variant positions, on the order of 30%, has been observed in comparing homologous sequences of bovine $\alpha2$ and $\alpha1(I)$ chains (18) as well as of bovine $\alpha1(III)$ and $\alpha1(I)$ chains (17). When one considers that the α chains are composed of repetitive Gly-X-Y triplets throughout most of their length, and that glycyl residues must occupy every third position to allow formation of the native triple-helical structure, the level of sequence variation between the chains becomes even more striking, since substitutions are permitted only in two-thirds of the amino acid residues, i.e., in X and Y positions.

It should be pointed out that the figures cited above for the number of sequence differences between the chains represent average values and that the primary structure of the respective chains is likely to be highly conserved in certain specific regions. One such region appears to be the site at which the chains are cleaved by preparations of mammalian collagenase (41). As information on the primary structure of the chains is completed, it may be presumed that additional relatively invariant regions will be recognized and correlated with certain functional parameters of the protein such as cross linking and fiber formation.

One additional observation derived from primary structure studies warrants discussion at this point. Very recent studies on the $\alpha1(II)$ sequence indicate that preparations of these chains from bovine nasal cartilage most likely represent a mixture of two forms of the chain (Butler, W. T., Finch, J. E., and Miller, E. J., *manuscript in preparation*). Based on the relative amounts of the chains found in pooled cartilages from several animals as well as in the cartilage of single animals, the chains have been designated $\alpha1(II)$ Major and $\alpha1(II)$ Minor. Although the data are as yet quite incomplete, it appears that $\alpha1(II)$ Major and $\alpha1(II)$ Minor differ in primary structure at only a few positions, at approximately 1% of their respective amino acid residues. These results suggest that additional forms of the $\alpha1(I)$, $\alpha2$, and $\alpha1(III)$ chains might also exist, although no data in this regard are currently available.

IMMUNOCHEMICAL PROPERTIES

Interest in collagen as an antigen has been revived in recent years and it now appears clear that collagen is an immunogen, albeit not a potent one.

Numerous studies have been performed indicating the location and nature of the antigenic determinants on collagen and procollagen molecules as well as characterizing the response of various species to the administration of the antigens (20).

Of particular interest in the context of the present discussion are studies leading to the production and purification of specific antibodies to the various interstitial collagens and their respective biosynthetic precursors, the procollagens (44,45). The specific antibodies have proven to be of great value in visualizing the precise distribution of the antigens within a given tissue by employing the conventional techniques of indirect immunofluorescence (54,55). The differences in primary structure for the constituent α chains of Types I, II, and III collagens as well as chemical differences in their respective procollagen extension peptides undoubtedly account for the observed antigenic specificity. Furthermore, the purified antibodies show similar reactions with a given type of collagen or procollagen from a variety of species irrespective of the species of origin of the antigen. Thus, use of the

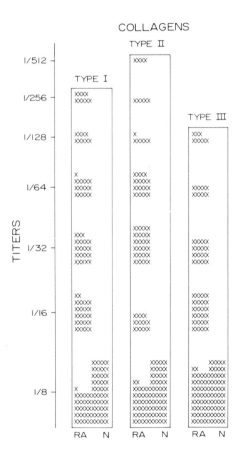

FIG. 2. Antibody titers to Types I, II, and III collagens in the sera of patients with rheumatoid arthritis **(RA)** and normal individuals **(N).**

specific antibodies promises to be a powerful tool in studies designed to investigate the distribution of the various collagens in tissues and organs containing more than one type of collagen as well as in studies designed to assess changing patterns of collagen synthesis and deposition during development or in pathological states.

Of additional interest in this regard are a number of recent studies indicating that patients with classical rheumatoid arthritis commonly exhibit moderately high titers to native human Types I, II, and III collagens and their component chains in both their sera (2) and joint fluids (1). The data for serum antibodies to the native collagens are summarized in Fig. 2, indicating that significant titers (> $\frac{1}{8}$) were observed against each collagen in approximately one-half of the patients under investigation. In contrast, similar studies on the sera from 50 normal individuals failed to reveal the presence of significant titers to any of these collagens. Moreover, hemagglutination inhibition experiments subsequently showed that patients with titers to the native collagens and their component chains had apparently produced antibodies specific for each type of collagen as well as for the component chains of each collagen (3).

These data extend current information indicating that rheumatoid arthritis exhibits several features of an autoimmune disease and suggest that the production of antibodies against the various collagens might at least serve to perpetuate the disease process. They further suggest that the sera of such patients might serve as a readily available source of monospecific antibodies against the human collagens and their constituent chains.

FIBER FORMATION

It may be assumed that the physiological roles of collagen are, for the most part, ultimately achieved following deposition of the protein in fibrous form. Fiber formation is, then, a critical event in the eventual function of the protein. Although the general molecular architecture of collagen fibers and the precision with which molecules are aligned within the fiber have been known for several years, the molecular interactions involved in fiber formation have been critically evaluated only recently. Such studies have been made possible by the availability of amino acid sequence data for the entire $\alpha 1(I)$ chain. Using the latter data, initial studies have suggested that both ionic and hydrophobic interactions are important determinants in specifying the axial displacement between molecules (29) and that these interactions on the contact edges of neighboring molecules are maximized if the pitch of the superhelix approximates a value of 30° (11). These studies have been informative with respect to the potential mechanisms involved in fiber formation as well as the structure of the native collagen molecule. The results must be considered as preliminary, however, since they are heavily dependent on subjective evaluations of the relative strengths of various interactions as a function of distance between potentially interacting sites. Further, the results are perhaps biased to a certain degree, since the contributions of the $\alpha2$ chain, which contains considerably more hydrophobic residues than $\alpha1(I)$, are not included in the analyses.

With respect to the latter point, it has recently been proposed that the amino acid sequence of the $\alpha2$ chain in the Type I collagen molecule might serve to specify the direction and magnitude of the axial displacement between adjacent molecules within the fiber (28). As a corollary of this proposal, it was further suggested that the axial displacement between molecules in fibers formed from Type II or Type III collagen molecules would be much less regular. These are intriguing proposals. If correct, they might account at least in part for the observations that fibers reconstituted *in vitro* from the latter collagens often appear in the electron microscope to be smaller and somewhat less well structured than fibers derived from Type I collagen molecules (50,56). These observations are also generally consistent with what one observes for fibers derived from the various collagens *in situ,* although a systematic study of the morphological characteristics of fibers derived from the various collagens in different tissues has not as yet been performed.

It should be noted that even though collagen-collagen interactions are most certainly operative in fiber formation, it seems equally plausible that the actual fiber size as well as the precise molecular architecture achieved *in vivo* may be dependent on several additional factors such as the environment in which fiber formation occurs and the state of the collagen molecules themselves. With regard to the latter factor, the conversion of procollagen to collagen may be inefficient for certain collagens in certain tissues leaving significant portions of the procollagen extension peptides still attached to the molecules as they enter the process of fiber formation. The presence of remnants of the extension peptides, then, would be expected to have some influence on the manner in which the molecules align themselves within the fibers.

TISSUE DISTRIBUTION

Some insight into the possible physiological roles of the different collagens has been obtained in studies detailing their occurrence in a variety of vertebrate tissues. These studies have been accomplished, for the most part, by utilizing unique chemical features of each type of collagen for identification and, in some instances, quantitation of the relative proportions of the collagens in a given tissue or organ. The chemical features most commonly employed are the solubility and chromatographic properties of extracted collagens as well as the cyanogen bromide peptide patterns observed for both extracted collagens and preparations of insoluble collagen.

The results of these studies have been extensively reviewed (40) and are

	TYPE I	TYPE II	TYPE III
BONE	+		
DENTIN	+		
TENDON	+		
DERMIS	+		+
HEART VALVES	+		+
INTESTINE	+		+
KIDNEY	+		+
LIVER	+		+
LUNG	+		+
MUSCLE	+		+
PERIODONTAL LIGAMENT	+		+
SPLEEN	+		+
SYNOVIUM	+		+
UTERUS	+		+
VESSELS	+		+
CARTILAGE			
HYALINE		+	
ELASTIC		+	
FIBROUS			
Meniscus	+		
Annulus	+	+	
Avian Articular	+	+	

FIG. 3. The distribution of Types I, II, and III collagens in various tissues and organs.

summarized along with more recent data (14–16,36) in Fig. 3. As noted, the distribution of the different collagens in the various tissues and organs allows division of the latter into three general categories. First, the more rigid or nonelastic connective tissues as exemplified by bone, dentin, and tendon contain predominantly Type I collagen with very little, if any, of the other interstitial collagens. In contrast, the more distensible connective tissues such as dermis, vessel walls, uterine wall, and intestinal wall largely contain a mixture of both Types I and III collagens. This situation also appears to prevail in the connective tissue stroma of a variety of organs such as the liver, lung, kidney, and spleen. And thirdly, there are the cartilaginous tissues that may contain only Type II collagen (hyaline and elastic cartilages) or a mixture of Types I and II collagens (fibrocartilages). There appears to be one important exception to the latter generalization, however, and that is the meniscus, a classic fibrocartilage, which apparently contains only Type I collagen.

It is perhaps hazardous to draw definitive conclusions from such data. As a first approximation, however, it seems reasonable to suggest that the Type I collagen molecules and fibers derived therefrom represent the most versatile collagen as the distribution of this collagen is virtually ubiquitous. In this regard, the Type I molecule might justifiably be termed the most advanced and complex collagen molecule, since it contains two $\alpha 1(I)$ chains and an $\alpha 2$ chain and therefore contains information derived from the activity of two structural genes. Further, it may be indicated that the simpler Type III collagen molecule is synthesized for specific functional roles in the more pliable soft tissues, where the smaller fibers derived from the collagen might serve to provide considerable support but also allow a great deal of distensibility. Similar considerations apply to the Type II collagen molecule, the small fibers of which appear to be particularly well suited for maximum dispersion throughout cartilaginous matrices in order to achieve, in conjunction with the other matrix components, a tissue of intermediate deformability and elasticity.

BIOSYNTHESIS

Although some uncertainty exists concerning the precise nature of the initial polypeptide chains, it now appears clear that the biosynthesis of all collagens involves the formation of a relatively large precursor molecule (procollagen) comprised of pro α chains, which contain rather lengthy noncollagenous sequences at both the NH_2- and COOH-terminal ends of their repetitive triplet sequences (40). Intracellular steps in the formation of the various procollagens likewise appear to be at least qualitatively identical and include assembly of the primary structure, hydroxylation of appropriate prolyl and lysyl residues in the nascent polypeptide chains, glycosylation of certain hydroxylysyl residues, chain association accompanied by interchain disulfide bonding, and triple helix formation.

Current evidence indicates that the hydroxylation reactions (12,24) as well as the glycosylation reactions (25,46) are initiated prior to release of the nascent chains from ribosomes, but that chain association leading to the formation of triple helix is delayed until completion and release of the chains from ribosomes (35). Since collagen molecules in native triple-helical conformation are not particularly effective substrates for the hydroxylating (5,48) and glycosylating (43,47) enzymes, these results suggest that the rapidity with which chain association and helix formation occurs is an important factor in regulating the extent of these posttranslocational modifications. In this regard, chain association and helix formation are apparently delayed for a considerable interval in cells synthesizing Type II (26) and basement membrane (23) collagens, the chains of which characteristically exhibit relatively high levels of hydroxylysine and hydroxylysine-bound carbohydrate. The basic mechanism responsible for the slow rate of chain association in the latter systems is not as yet clear. It is possible, however, that association of the chains is impaired due to the presence of bound enzymes at numerous sites on the newly synthesized chains. This implies either that the enzymes in question have a greater affinity for $\alpha 1(II)$ chains and the chains of basement membrane collagen or that the hydroxylating and glycosylating enzymes are cell and tissue specific. At the moment, there appears to be insufficient data for definitive conclusions in these matters.

The conversion of procollagen to collagen apparently occurs extracellularly and requires the activity of at least two specific proteases (40). The precise number of cleavages which occur, however, and the order in which they occur has not as yet been elucidated. The importance of these reactions is underscored by studies on the connective tissues of dermatosparactic cattle and individuals with Type VII Ehlers-Danlos syndrome, genetic disorders in which there is reduced capacity to remove the NH_2-terminal procollagen extension peptides. It appears likely, though, that even under normal circumstances significant portions of the procollagen extension peptides may remain in extracellular fibrous collagen as evidenced by the staining reaction of certain tissues with specific antibodies to Type I procollagen (55).

Of additional interest in the present discussion are considerations pertaining to the cells involved in the synthesis of the different collagens. The data reviewed above and summarized in Fig. 3 suggest, as a first approximation, that connective tissue cells might be classified on the basis of the collagen they normally synthesize. Thus, osteoblasts, odontoblasts, and tendon fibroblasts apparently produce only Type I collagen. Similarly, the chondroblasts of hyaline and elastic cartilages appear to synthesize only Type II collagen. On the other hand, fibroblasts of such tissues as dermis and heart valves as well as the smooth muscle cells of uterus and the major vessels seem to be capable of synthesizing both Types I and III collagens. Such information certainly suggests that there may normally exist two distinct populations of fibroblasts or smooth muscle cells in the indicated tissues. This does not, however, appear to be the case for dermal fibroblasts grown in culture, as recent studies employing immunohistological techniques have demonstrated that a single fibroblast may contain both Types I and III procollagens (21). Similar studies have not as yet been performed with smooth muscle cells or cells derived from fibrocartilages. Such studies will be of considerable interest with respect to the cellular origins of the various collagens.

COLLAGEN-CELL INTERACTIONS

Studies on collagen-cell interactions are abundant in the present-day literature. There are, for example, numerous studies devoted to the adherence, release phenomena, and aggregation of platelets in the presence of collagen fibers. Of more immediate relevance to the present discussion is the host of studies suggesting that collagen-cell interactions are important determinants in developmental processes (see reviews by Drs. Hay and Slavkin, *this volume*). For the most part, the evidence obtained in the latter studies has been indirect and emphasis is placed here on the relatively recent investigations in which the influence of collagen has been assayed directly.

Perhaps the greatest impetus for the view that collagen is involved in developmental processes was provided by the studies demonstrating that a collagenous substratum is required for *in vitro* differentiation of myoblasts to myotubes (27). This phenomenon has been reinvestigated with results indicating that Types I, II, and III collagens as well as an unspecified preparation of basement membrane collagen are equally effective as *in vitro* substrates for myoblast development (32). These results, then, confirmed the efficacy of a collagenous substratum in myoblast development but did not indicate a preferential effect on the part of any particular type of collagen. On the other hand, Type II collagen substrates have been shown to be somewhat more effective than Type I collagen substrates in enhancing *in vitro* somite chondrogenesis, as assayed by collagen and glycosaminoglycan synthesis by the cultured somites (33). Experiments of this nature, however, where the assay system depends heavily on the synthesis of matrix molecules, may reflect enhancement and/or acceleration of a function on the part of previously committed cells as opposed to a stimulus for true differentiation. Nevertheless, the results of such studies are intriguing and certainly suggest that matrix molecules such as collagen may serve to stabilize and maintain a differentiated state.

Regarding the latter point, it has been shown that the chondrocyte phenotype with respect to collagen synthesis is exquisitely sensitive to the environment in which the cells are grown and maintained *in vitro*. For example, when grown in the presence of chick embryo extract, chick embryo chondrocytes cease the production of Type II collagen and initiate synthesis of Type I collagen plus Type I-trimer (chain composition, $[\alpha 1(I)]_3$) in a matter of days (37). Human articular chondrocytes likewise appear to synthesize Type I collagen at the focal points of degenerative lesions, as determined by immunohistochemical techniques (22) and it was suggested that this alteration in collagen synthesis might be attributed to the loss of matrix surrounding the cells which accompanies the onset of osteoarthritis.

It seems clear, then, that collagen does play a role in certain processes of differentiation. It appears equally plausible that, once elaborated in a given system, collagen also functions in the maintenance of the differentiated state. Whether or not any particular type of collagen plays a unique or more efficacious role in certain processes of differentiation and development remains to be established, and will most certainly be the subject of future investigations.

REFERENCES

1. Andriopoulos, N. A., Mestecky, J., Miller, E. J., and Bennett, J. C. (1976): Antibodies to human native and denatured collagens in synovial fluids of patients with rheumatoid arthritis. *Clin. Immunol. Immunopathol.,* 6:209–212.
2. Andriopoulos, N. A., Mestecky, J., Miller, E. J., and Bradley, E. L. (1976): Antibodies to native and denatured collagens in sera of patients with rheumatoid arthritis. *Arthritis Rheum.,* 19:613–617.
3. Andriopoulos, N. A., Mestecky, J., Wright, G. P., and Miller, E. J. (1976): Characterization of antibodies to the native human collagens and their component α chains in the sera and joint fluids of patients with rheumatoid arthritis. *Immunochemistry,* 13:709–712.

4. Bailey, A. J., and Sims, T. J. (1976): Chemistry of the collagen cross-links: Nature of the cross-links in the polymorphic forms of dermal collagen during development. *Biochem. J.,* 153:211–215.
5. Berg, R. A., and Prockop, D. J. (1973): Purification of [^{14}C] protocollagen and its hydroxylation by prolyl-hydroxylase. *Biochemistry,* 12:3395–3401.
6. Butler, W. T., Miller, E. J., and Finch, J. E. (1976): The covalent structure of collagen. Amino acid sequence of the NH_2-terminal helical portion of the $\alpha 1(II)$ chain. *Biochemistry,* 15:3000–3006.
7. Butler, W. T., Miller, E. J., Finch, J. E., and Inagami, T. (1974): Homologous regions of collagen $\alpha 1(I)$ and $\alpha 1(II)$ chains. Apparent clustering of variable and invariant amino acid residues. *Biochem. Biophys. Res. Commun.,* 57:190–195.
8. Byers, P. H., McKenney, K. H., Lichtenstein, J. R., and Martin, G. R. (1974): Preparation of Type III procollagen and collagen from rat skin. *Biochemistry,* 13:5243–5248.
9. Chung, E., and Miller, E. J. (1974): Collagen polymorphism: Characterization of molecules with the chain composition $[\alpha 1(III)]_3$ in human tissues. *Science,* 183:1200–1201.
10. Chung, E., Rhodes, R. K., and Miller, E. J. (1976): Isolation of three collagenous components of probable basement membrane origin from several tissues. *Biochem. Biophys. Res. Commun.* 71:1167–1174.
11. Cunningham, L. W., Davies, H. A., and Hammonds, R. G. (1976): An analysis of the association of collagen based on structural models. *Biopolymers,* 15:483–502.
12. Cutroneo, K. R., Guzman, N. A., and Sharawy, M. M. (1974): Evidence for a subcellular vesicular site of collagen prolyl hydroxylation. *J. Biol. Chem.,* 249:5989–5994.
13. Daniels, J. R., and Chu, G. H. (1975): Basement membrane collagen of renal glomerulus. *J. Biol. Chem.,* 250:3531–3537.
14. Epstein, E. H., Jr., and Munderloh, N. H. (1975): Isolation and characterization of CNBr peptides of human $[\alpha 1(III)]_3$ collagen and tissue distribution of $[\alpha 1(I)]_2\alpha 2$ and $[\alpha 1(III)]_3$ collagens. *J. Biol. Chem.,* 250:9304–9312.
15. Eyre, D. R., and Muir, H. (1975): The distribution of different molecular species of collagen in fibrous, elastic and hyaline cartilages of the pig. *Biochem. J.,* 151:595–602.
16. Eyre, D. R., and Muir, H. (1975): Type III collagen: A major constituent of rheumatoid and normal synovial membrane. *Connect. Tissue Res.,* 4:11–16.
17. Fietzek, P. P., and Rauterberg, J. (1975): Cyanogen bromide peptides of Type III collagen: First sequence analysis demonstrates homology with Type I collagen. *FEBS Lett.,* 49:365–368.
18. Fietzek, P. P., and Rexrodt, F. W. (1975): The covalent structure of collagen: The amino acid sequence of $\alpha 2$-CB4 from calf skin collagen. *Eur. J. Biochem.,* 59:113–118.
19. Fujii, K., Tanzer, M. L., Nusgens, B. V., and Lapiere, C. M. (1976): Aldehyde content and cross-linking of Type III collagen. *Biochem. Biophys. Res. Commun.,* 69:128–134.
20. Furthmayr, H., and Timpl, R. (1976): Immunochemistry of collagens and procollagens. In: *International Review of Connective Tissue Research, Vol. 7,* edited by D. A. Hall and D. S. Jackson, pp. 61–99. Academic Press, New York.
21. Gay, S., Martin, G. R., Müller, P. K., Timpl, R., and Kühn, K. (1976): Simultaneous synthesis of Types I and III collagen by fibroblasts in culture. *Proc. Natl. Acad. Sci., U.S.A.,* 73:4037–4040.
22. Gay, S., Müller, P. K., Lemmen, C., Remberger, K., Matzen, K., and Kühn, K. (1976): Immunohistological study on collagen in cartilage-bone metamorphosis and degenerative osteoarthritis. *Klin. Wochenschr.,* 54:969–976.
23. Grant, M. E., Schofield, D. J., Kefalides, N. A., and Prockop, D. J. (1973): The biosynthesis of basement membrane collagen in embryonic chick lens. III. Intracellular formation of the triple helix and the formation of aggregates through disulfide bonds. *J. Biol. Chem.,* 248:7432–7437.
24. Guzman, N. A., Rojas, F. J., and Cutroneo, K. R. (1976): Collagen lysyl hydroxylation occurs within the cisternae of the rough endoplasmic reticulum. *Arch. Biochem. Biophys.,* 172:449–454.
25. Harwood, R., Grant, M. E., and Jackson, D. S. (1975): Studies on the glycosylation of hydroxylysine residues during collagen biosynthesis and the subcellular localization of collagen galactosyltransferase and collagen glucosyltransferase in tendon and cartilage cells. *Biochem. J.,* 152:291–302.
26. Harwood, R., Bhalla, A. K., Grant, M. E., and Jackson, D. S. (1975): The synthesis and secretion of cartilage procollagen. *Biochem. J.,* 148:129–138.

27. Hauschka, S. D., and Konigsberg, I. R. (1966): The influence of collagen on the development of muscle clones. *Proc. Natl. Acad. Sci., U.S.A.,* 55:119–126.
28. Hukins, D. W. L., and Woodhead-Galloway, J. (1976): Proposed role for the $\alpha2$ chain in resolving an ambiguity in self-assembly of the collagen fibril. *Biochem. Biophys. Res. Commun.,* 70:413–417.
29. Hulmes, D. J. S., Miller, A., Parry, D. A. D., Piez, K. A., and Woodhead-Galloway, J. (1973): Analysis of the primary structure of collagen for the origins of molecular packing. *J. Mol. Biol,* 79:137–148.
30. Kefalides, N. A. (1975): Basement membranes: Structural and biosynthetic considerations. *J. Invest. Dermatol.,* 65:85–92.
31. Kefalides, N. A. (1971): Isolation of a collagen from basement membranes containing three identical α chains. *Biochem. Biophys. Res. Commun.,* 45:226–234.
32. Ketley, J. N., Orkin, R. W., and Martin, G. R. (1976): Collagen in developing chick muscle *in vivo* and *in vitro. Exp. Cell Res.,* 99:261–268.
33. Kosher, R. A., and Church, R. L. (1975): Stimulation of *in vitro* somite chondrogenesis by procollagen and collagen. *Nature,* 258:327–330.
34. Lenaers, A., and Lapiere, C. M. (1975): Type III procollagen and collagen in skin. *Biochim. Biophys. Acta,* 400:121–131.
35. Lukens, L. N. (1976): Time of occurrence of disulfide linking between procollagen chains. *J. Biol. Chem.,* 251:3530–3538.
36. Mannschott, P., Herbage, D., Weiss, M., and Buffevant, C. (1976): Collagen heterogeneity in pig heart valves. *Biochim. Biophys. Acta,* 434:177–183.
37. Mayne, R., Vail, M. S., and Miller, E. J. (1977): The effect of embryo extract on the types of collagen synthesized by cultured chick chondrocytes. *Dev. Biol.,* 54:230–240.
38. McClain, P. E. (1974): Characterization of cardiac muscle collagen. *J. Biol. Chem.,* 249: 2303–2311.
39. Miller, E. J. (1973): A review of biochemical studies on the genetically distinct collagens of the skeletal system. *Clin. Orthop.,* 92:260–280.
40. Miller, E. J. (1976): Biochemical characteristics and biological significance of the genetically distinct collagens. *Mol. Cell. Biochem.* 13:165–192.
41. Miller, E. J., Harris, E. D., Chung, E., Finch, J. E., McCroskery, P. A., and Butler, W. T. (1976): Cleavage of Type II and III collagens with mammalian collagenase: Site of cleavage and primary structure at the NH_2-terminal portion of the smaller fragment released from both collagens. *Biochemistry,* 15:787–792.
42. Miller, E. J., and Matukas, V. J. (1974): The biosynthesis of collagen. *Fed. Proc.,* 33: 1197–1204.
43. Myllylä, R., Risteli, L., and Kivirikko, K. I. (1975): Glucosylation of galactosylhydroxylysyl residues in collagen *in vitro* by collagen glucosyltransferase. *Eur. J. Biochem.,* 58: 517–521.
44. Nowack, H., Gay, S., Wick, G., Becker, U., and Timpl, R. (1976): Preparation and use in immunohistology of antibodies specific for Type I and Type III collagen and procollagen. *J. Immunol. Meth.,* 12:117–124.
45. Nowack, H., Hahn, E., and Timpl, R. (1975): Specificity of the antibody response in inbred mice to bovine Type I and Type II collagen. *Immunology,* 29:621–628.
46. Oikarinen, A., Anttinen, H., and Kivirikko, K. I. (1976): Hydroxylation of lysine and glycosylation of hydroxylysine during collagen biosynthesis in isolated chick-embryo cartilage cells. *Biochem. J.,* 156:545–551.
47. Risteli, L., Myllylä, R., and Kivirikko, K. I. (1976): Partial purification and characterization of collagen galactosyltransferase from chick embryos. *Biochem. J.,* 155:145–153.
48. Ryhänen, L., and Kivirikko, K. I. (1974): Hydroxylation of lysyl residues in native and denatured protocollagen by protocollagen lysyl hydroxylase *in vitro. Biochim. Biophys. Acta,* 343:129–137.
49. Spiro, R. G. (1973): Biochemistry of the renal glomerular basement membrane and its alterations in diabetes mellitus. *N. Engl. J. Med.,* 288:1337–1342.
50. Stark, M., Miller, E. J., and Kühn, K. (1972): Comparative electron-microscope studies on the collagens extracted from cartilage, bone, and skin. *Eur. J. Biochem.,* 27:192–196.
51. Timpl, R., Glanville, R. W., Nowack, H., Wiedemann, H., Fietzek, P. P., and Kühn, K. (1975): Isolation, chemical and electron microscopical characterization of neutral-salt-soluble Type III collagen and procollagen from fetal bovine skin. *Hoppe-Seyler's Z. Physiol. Chem.,* 356:1783–1792.

52. Trelstad, R. L. (1973): The developmental biology of vertebrate collagens. *J. Histochem. Cytochem.*, 21:521–528.
53. Trelstad, R. L., and Kang, A. H. (1974): Collagen heterogeneity in the avian eye: lens, vitreous body, cornea, and sclera. *Exp. Eye Res.*, 18:359–406.
54. von der Mark, H., von der Mark, K., and Gay, S. (1976): Study of differential collagen synthesis during development of the chick embryo by immunofluorescence. *Dev. Biol.*, 48:237–249.
55. Wick, G., Nowack, H., Hahn, E., Timpl, R., and Miller, E. J. (1976): Visualization of Type I and II collagens in tissue sections by immunohistological techniques. *J. Immunol.*, 117:298–303.
56. Wiedemann, H., Chung, E., Fujii, T., Miller, E. J., and Kühn, K. (1975): Comparative electron microscope studies on Type III and Type I collagens. *Eur. J. Biochem.*, 51:363–368.

Cell and Tissue Interactions, edited by
J. W. Lash and M. M. Burger. Raven
Press, New York, 1977.

Structure and Function of Proteoglycans of Cartilage and Cell-Matrix Interactions

Helen Muir

*Biochemistry Division, Kennedy Institute of Rheumatology, Bute Gardens,
London W6 7DW, England*

PHYSICAL CHEMISTRY AND FUNCTION OF CARTILAGE

Cartilage is a highly specialized form of connective tissue that can withstand compressive forces to a remarkable degree. Being stiff but not rigid, it is able to distribute stress under impact loading (9). It has been calculated that even during walking the load borne transiently by small areas of articular cartilage is several times body weight (51). Although stiff and rubbery, human articular cartilage consists of about 70% water (32). When a load is applied, fluid pressure within the cartilage rises immediately, but water is driven out only slowly because the proteoglycans in the cartilage impede the flow of interstitial water. The proteoglycans, however, remain within the cartilage (9,29). Proteoglycans are large, highly hydrated compounds that are entrapped in the collagen network. The resilience or compressive stiffness of cartilage is therefore directly correlated with the proteoglycan content measured as glycosaminoglycan (partial correlation coefficient 0.854, $p < 0.001$) (26), whereas tensile stiffness and strength of cartilage depend on the collagen content (25).

In adult cartilage, there are relatively few cells distributed sparsely in a stiff matrix (Fig. 1) and cellular interactions are likely to be minimal. In embryonic cartilage, however, the cells are in relatively close proximity so that cellular interactions, which are necessary for development, can take place easily. The much greater cellularity and proximity of cells in embryonic cartilage must be borne in mind when conclusions drawn from embryonic cartilage are extended to adult tissue. Moreover, adult articular cartilage is almost impermeable to proteins, even to those of relatively small size, whereas the diffusion of small solutes such as glucose approaches that in water (31), so that the cartilage behaves as a molecular sieve.

Cartilage collagen is Type II (see Miller, *this volume*), a type found only in cartilage (5) and intervertebral discs (6). Proteoglycans of cartilage are also distinct from those of other connective tissues in several respects and, hence, both the collagen and proteoglycan are phenotypic of cartilage.

FIG. 1. Sagittal section of normal tibial cartilage, stained with safranin O and hematoxylin and eosin. ×25.

PROTEOGLYCAN STRUCTURE AND AGGREGATION

Proteoglycans consist of a protein core to which glycosaminoglycan chains are attached laterally. Those of cartilage contain chondroitin sulfate and keratan sulfate in variable proportions and their molecular weights range from 1×10^6 to 5.8×10^6 (7,30,42). They are generally rather polydisperse in molecular size and chemical composition, but in any given cartilage the population of molecules shows a characteristic range in these variables, which vary with the type of cartilage and which change with age.

A unique feature of cartilage proteoglycans is their ability to form multimolecular aggregates of very high molecular weight of the order of 50 to 100 million (reviewed in ref. 37). Aggregates have been obtained from many kinds of adult cartilage and also from rat chondrosarcoma (39) and cultures of embryonic chick chondrocytes (21). Dissociative procedures to extract proteoglycans from cartilage without disruptive homogenization of the tissue were first introduced by Sajdera and Hascall (45). They were also the first to apply equilibrium density gradient centrifugation in cesium chloride to the purification of proteoglycans and showed that aggregates were dissociated in 4 M guanidinium chloride (20). These techniques have enabled major advances to be made in the understanding of the structure of these complex macromolecules.

Proteoglycans are generally extracted with 4 M guanidinium chloride,

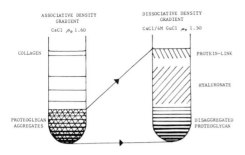

FIG. 2. Equilibrium density gradient centrifugation in cesium chloride. **A:** Associative conditions. Starting density usually 1.6 g/ml. Centrifuged at 100,000 × g for 48 hr at 20°C. **B:** Dissociative conditions. Lower fraction from associative gradient mixed with an equal volume of 7.5 M guanidinium chloride at pH 5.8. Starting density adjusted to 1.5 g/ml with cesium chloride. Centrifuged at 100,000 × g for 48 hr at 20°C.

i.e., under conditions that dissociate aggregates. The dissociated components then reassociate on dialysis to low ionic strength. The proteoglycans are then purified by density gradient centrifugation in cesium chloride under associative conditions (Fig. 2). Proteoglycans separate at the bottom of the gradient from collagen and other contaminants, which float to the top. The proteoglycan fraction, which contains both aggregated and nonaggregated species, is then subjected to a second density gradient centrifugation under dissociative conditions in the presence of 4 M guanidinium chloride, which dissociates aggregates when the constituents of the aggregate separate at different buoyant densities (Fig. 2). These procedures are now in general use in the field, starting densities being changed according to the buoyant densities of the proteoglycans under investigation, which vary according to the proportion of carbohydrate to protein. Aggregated and dissociated proteoglycans can be distinguished by gel-permeation chromatography on large-pore gels such as Sepharose 2B (Fig. 3).

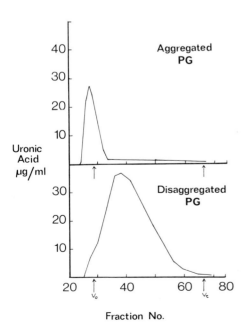

FIG. 3. Gel-permeation chromatography on Sepharose 2B of aggregated and dissociated proteoglycans prepared as shown in Fig. 2. V_o denotes the void volume of the column. The uronic acid content of eluant fractions was measured.

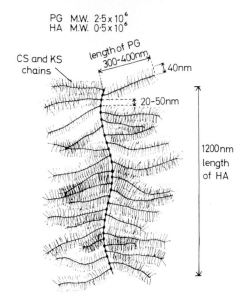

PG M.W. 2.5×10^6
HA M.W. 0.5×10^6

CS and KS chains

length of PG 300-400nm

40nm

20-50nm

1200nm length of HA

FIG. 4. The proteoglycan-hyaluronic acid complex. The dimensions of the model were deduced from the stoichiometry of the interaction (14). **PG,** proteoglycan; **HA,** hyaluronic acid; **CS,** chondroitin sulfate; **KS,** keratan sulfate. (From ref. 17a.)

In 1972, Hardingham and Muir (13) found that dissociated proteoglycans interacted with hyaluronic acid in a unique manner, in which many proteoglycan molecules became bound to a single chain of hyaluronate, which acts as a thread linking proteoglycan molecules together. Subsequently, when proteoglycan aggregates were dissociated, hyaluronic acid was identified in the middle of the second dissociative density gradient (16). It represented rather less than 1% of the total uronic acid in the aggregate. That aggregation depends on the interaction of proteoglycans with hyaluronate is now generally recognized. The interaction has been examined in detail by Hardingham and Muir (13) using viscometry and gel chromatography, since interaction leads to a large increase in molecular size. From such studies it was concluded that many molecules of proteoglycan interacted with a single chain of hyaluronic acid and that each proteoglycan molecule possessed a single binding site for hyaluronate, because a gel did not form at higher proportions of hyaluronate. Using published molecular weights for proteoglycan and hyaluronic acid, it was possible to deduce a model (Fig. 4) for the complex whose dimensions agreed reasonably well with those calculated from electron micrographs of proteoglycan aggregates (42–44). Although the interaction is not covalent, it is extremely specific to hyaluronate, and does not occur with any other closely related glycosaminoglycan, even chondroitin (desulfated chondroitin sulfate) (19), which differs from hyaluronate only in the configuration of the hydroxyl group on C_4 of the hexosamine residues.

The chondroitin sulfate chains are not necessary for binding, since binding occurs after they have been largely removed by chondroitinase

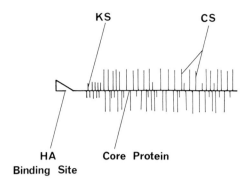

KS   CS

FIG. 5. Diagram of cartilage proteoglycan molecule.

HA   Core Protein
Binding Site

digestion of proteoglycan (19). The minimum length of hyaluronate that interacts strongly is ten sugar residues, since oligosaccharides of hyaluronate compete strongly with hyaluronic acid and inhibit interaction between proteoglycan and hyaluronate, whereas octasaccharides and smaller oligosaccharides have little effect (15,19). The hyaluronate binding region of the proteoglycan, however, is quite large. Sequential chrondroitinase and trypsin digestion of the proteoglycan-hyaluronate complex indicate that this region has a molecular weight of about 90,000 (24).

Proteoglycan aggregation is prevented by reduction of cystine residues (20). Although reduction and alkylation of proteoglycan monomers abolished hyaluronate interaction, there was no change in molecular size or loss of protein, and it is therefore concluded that the tertiary structure of the hyaluronate binding region depends on five to seven intramolecular disulfide bridges (11). The structure of the binding site is fairly sensitive to chemical modification. Amino groups are very important in the functional structure of the binding region and may take part in direct subsite interactions with the carboxyl groups of hyaluronate (11). Many cooperative subsite interactions are involved in the proteoglycan-hyaluronate interaction, which in addition to arginine and ϵ-amino groups of lysine depends on intact tryptophan residues, although fluorescence measurements suggest that tryptophan may not be directly involved in subsite interactions but in maintaining the conformation of the binding site (11). An analogy for the hyaluronate-proteoglycan interaction is provided by lysozyme, which has a binding site for the hexasaccharide $(GlcNAc-MurNAc)_3$ that involves at least 15 subsite interactions (2).

Proteoglycans of cartilage consist of a population of molecules that vary in molecular size, chemical composition, and buoyant density in cesium chloride. Since different fractions are all able to interact with hyaluronate, it is concluded that all contain the same binding site and that the molecule, therefore, consists of an invariant region comprising the hyaluronate binding site devoid of carbohydrate and a region bearing glycosaminoglycan chains (Fig. 5) (11,24) that is variable in length. This conclusion is consistent

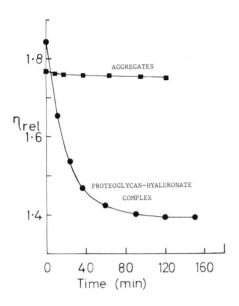

FIG. 6. The effect of hyaluronate oligo-saccharides on the viscosity of proteo-glycan aggregates and of the proteo-glycan-hyaluronate complex in 0.5 M guanidinium chloride, pH 5.8 at 30°C (Hardingham, *unpublished results*).

with the results of electron microscopic studies of proteoglycan monomers (48) and with chemical results obtained with proteoglycans of different buoyant density (11,22,23) where molecular size, buoyant density, and carbohydrate content decreased together. Buoyant density in cesium chloride depends on the relative proportion of carbohydrate to protein.

Aggregates also contain a third constituent known as "protein-link" (20), which separates at the top of the dissociative density gradient and which was at first thought to induce aggregation. It has yet to be fully characterized, but it appears to function in stabilizing the proteoglycan-hyaluronate complex, which is in equilibrium with its dissociation products. Although the equilibrium is far on the side of interaction in the complex, the aggregate does not exhibit an equilibrium, presumably because the link-protein stops dissociation entirely (17). Oligosaccharides of hyaluronate that interact with proteoglycan (15,19) dissociate the proteoglycan-hyaluronate complex, which results in a decrease in relative viscosity. Aggregates which contain protein-link, however, are unaffected by the oligosaccharides (19) (Fig. 6). Protein-link binds to hyaluronate (18) and in bovine nasal cartilage there appear to be two with molecular weights of about 45,000 and 65,000 (18), but in chondrosarcoma (39) and embryonic chick cartilage there is only one of about 45,000 molecular weight (21). The interrelationship of the three components of the aggregate are shown diagrammatically in Fig. 7.

The function of aggregation is not known. It presumably helps to immobilize proteoglycans in the collagen network, when an external load is applied and water is expressed (see ref. 31). Another purpose of aggregation may be to protect proteoglycans to some degree from the effects of protein-

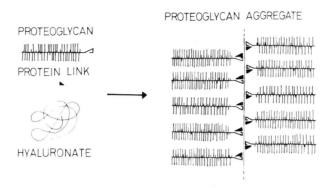

FIG. 7. Diagram of aggregation.

ases, since controlled partial degradation was only possible with proteoglycan aggregates, whereas monomers were rapidly degraded (18,24).

In addition to the proteoglycans that are able to interact with hyaluronate, there are lesser amounts of proteoglycans that do not interact and are therefore not present as aggregates. These may preferentially be extracted in 0.15 M NaCl. No hyaluronic acid can be separated from them when they are subjected to the dissociative procedure. In several respects, they appear to represent a different population of proteoglycan (16). From the amino acid composition, it appears that the hyaluronate binding region is incomplete or absent (11). Such proteoglycans are of relatively small molecular size and contain less protein and very little keratan sulfate (50) compared with aggregating proteoglycans. The incorporation of ^{35}S-sulfate into proteoglycans of different size suggested that the smaller nonaggregating proteoglycans are neither precursors nor degradation products of those that are able to aggregate and interact with hyaluronate (14).

THE INFLUENCE OF CELLULAR ENVIRONMENT

Cartilage cells are extremely sensitive to changes in the condition of the matrix that surrounds them. For example, depletion of the matrix of chick embryonic cartilage in organ culture stimulated the synthesis of proteoglycans (8) and collagen (1) to such an extent that the loss was more than made good within 4 days. After removal of about 80% of the chondroitin sulfate from the matrix with hyaluronidase, the total synthesis of proteoglycan was five times greater than in the controls (12). Since a chemically defined medium alone was used, somatomedin was not involved in this response. The greater part of the newly formed proteoglycan was found in the medium, particularly during the early phase of recovery from the effects of the enzyme, when the matrix was still permeable to large molecules. Chondrocytes in cell culture have far less matrix surrounding them than in whole cartilage

and they may therefore be in a chronic state of repair, attempting to sur-
round themselves with a normal matrix. It is probable, therefore, that under
cell culture conditions, the synthesis of matrix constituents is greater than
it would otherwise be either *in vivo* or in tissue slices where the cells remain
within their original matrix. In keeping with this suggestion is the finding that
the synthesis of proteoglycans by cultures of chick embryonic chondrocytes
began to diminish when cartilage nodules had developed and metachromatic
material was appearing round the cells (21).

THE INFLUENCE OF HYALURONIC ACID
ON PROTEOGLYCAN SYNTHESIS

The possibility that one or more constituents of cartilage matrix may influence the synthesis of proteoglycans has been investigated by several
laboratories. The presence of chondroitin sulfate cartilage proteoglycan in
the medium was found to stimulate proteoglycan synthesis by embryonic
chondrocytes (38,49) and those isolated from adult cartilage (53). This
effect on chondrocytes, however, was not specific to cartilage constituents
but was shown by other polyanions such as dextran sulfate. Undifferentiated
chondroblasts, however, were stimulated only by cartilage proteoglycans
and not by other polyanions (27). On the other hand, hyaluronic acid, which
is present in small amounts in cartilage (16,18) when present in the medium
in low concentrations, had an inhibitory effect on proteoglycan synthesis by
suspensions of chondrocytes from adult cartilage (53) and by cultures of
embryonic chondrocytes (47,49), although it had no effect on chondroblasts
(27). The effect, at least with cells from adult tissue, was restricted to chondrocytes, since hyaluronate had no effect on dermal fibroblasts or synovial
cells. It was entirely specific, since other closely similar compounds such
as chondroitin were without effect. Moreover, larger oligosaccharides of
hyaluronate also had an inhibitory effect (52). Inhibition of proteoglycan
synthesis involved the attachment of hyaluronate to the cell surface, as
shown by the use of labeled hyaluronate (52). The inhibitory effect of hyaluronate involved an interaction that resembled in several respects the interaction of hyaluronate with proteoglycans. To be effective, the hyaluronate
had to be free and not already bound to proteoglycan as in the complex or in
aggregates (52). Hyaluronate had no effect, however, on total protein synthesis (10) nor on the level of activity of the glycosyl transferases involved
in chondroitin sulfate chain synthesis (10).

With the use of β-D-xylosides, it is possible to divorce chondroitin sulfate
synthesis from the synthesis of the core protein, because the xylosides compete with the core protein as acceptors for the sugar transferases involved
in building up chondroitin sulfate chains (41). Hyaluronic acid had no effect
on chondroitin sulfate chain synthesis initiated by β-D-xylosides. Handley
and Lowther (10) suggest that either hyaluronate reduces the synthesis of

core protein or it reduces the activity of the xylosyl transferase that attaches xylose to serine residues of the core protein, a reaction which is the first step in chondroitin sulfate chain synthesis.

OSTEOARTHROSIS: A CELL-TISSUE RESPONSE

Osteoarthrosis is a progressive destructive disease of joints of varying severity that is widespread in man and other species. It is characterized by progressive deterioration and localized erosion of articular cartilage and by remodeling of bone at the joint margins (28,46) and is accompanied by a variety of biochemical changes in cartilage (33). Since the disease is asymptomatic in the initial phases, it is detectable only later when radiological changes are apparent. In order to study the changes that initiate the disease, experimental models of osteoarthrosis have been developed. In particular, osteoarthrosis has been produced surgically by cutting the anterior cruciate ligament of the knee (40). This can also happen naturally and the consequent joint laxity leads to osteoarthrosis of the knee (Fig. 8) (34–36).

Using this experimental model, it has been possible to examine the earliest biochemical changes occurring in the cartilage of the knee before lesions can be seen by conventional histology. Since lesions develop in the same

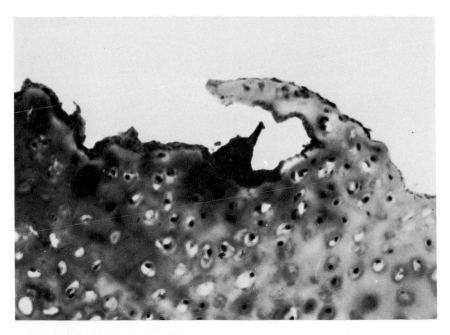

FIG. 8. Sagittal section of cartilage from equivalent area of the tibial condyle as in Fig. 1 16 weeks after resection of the anterior cruciate ligament. Staining and magnification as in Fig. 1.

area of the tibial condyle in each animal, this area can be sampled before lesions appear. One of the earliest changes seen within 2 weeks was an increase in water content of 2 to 3% in the entire cartilage of the joint. As a consequence, the proteoglycans became more extractable and the tissue softer and more permeable, since permeability is related to water content (31). This change in the state of hydration of the tissue led to profound metabolic changes throughout the joint cartilage. Proteoglycans of a different chemical composition were produced that contained much more chondroitin sulfate than those originally present, which appeared to have been largely replaced. Although the newly formed proteoglycans were of similar molecular size as those of control cartilage and were equally able to interact with hyaluronate, preliminary results nevertheless suggest that the proportion of aggregated proteoglycans was less even 7 weeks after surgery (McDevitt and Muir, *unpublished results*).

Collagen synthesis was also stimulated, but the newly formed collagen was entirely of the normal Type II (4). Protein synthesis became largely diverted to collagen synthesis, since in controls collagen synthesis represented 15 to 20% of total protein synthesis compared with 80 to 88% in osteoarthritic cartilage. Collagen and proteoglycan synthesis are not closely integrated processes, however, since in cartilage slices *in vitro* each can proceed normally without the other for at least 20 hr (3).

As the disease progressed, the DNA content of the cartilage increased and the tissue appeared more cellular with marked changes in the morphological appearance of the cells (Fig. 8). Although it took several weeks for lesions to develop, with increasing damage to the articular surfaces, all the biochemical changes began before lesions appeared and, moreover, occurred in regions of the joint cartilage that did not develop lesions. This suggests that in the initial stages of osteoarthrosis there is a profound change in metabolism of the entire cartilage of the joint, which is set in train by a change in the state of hydration of the tissue.

The changes described here are an example of a response to an altered cellular environment *in vivo* where conditions would differ from normal far less than they would even under optimal conditions *in vitro*. That the changes were brought about by an increase in hydration of only a few percent emphasizes how sensitive chondrocytes, even those in adult cartilage, are to changes in the state of the matrix. Cells from immature developing tissue are likely to be much more responsive to their environment, and, hence, when *in vitro* systems are used, culture conditions must be critically examined if conclusions of general application to whole organisms are to be valid.

REFERENCES

1. Bosmann, H. B. (1968): Cellular control of macromolecular synthesis: Rates of synthesis of extracellular macromolecules during and after depletion by papain. *Proc. R. Soc. Lond.* [*Biol.*], 169:399–425.

2. Chipman, D. M., and Sharon, N. (1969): Mechanism of lysozyme action. Lysozyme is the first enzyme for which the relation between structure and function has become clear. *Science,* 165:454–465.
3. Dondi, P., and Muir, H. (1976): Collagen synthesis and deposition in cartilage during disrupted proteoglycan production. *Biochem. J.,* 160:117–121.
4. Eyre, D. R., McDevitt, C. A., and Muir, H. (1975): Experimentally induced osteoarthrosis in the dog. *Ann. Rheum. Dis.,* 34 (Suppl. 2):137–140.
5. Eyre, D. R., and Muir, H. (1975): The distribution of different molecular species of collagen in fibrous, elastic and hyaline cartilages of the pig. *Biochem. J.,* 151:595–602.
6. Eyre, D. R., and Muir, H. (1976): Types I and II collagens in intervertebral disc. Interchanging radial distributions in annulus fibrosus. *Biochem. J.,* 157:267–270.
7. Eyring, E. J., and Yang, J. T. (1968): Conformation of protein polysaccharide complex from bovine nasal septum. *J. Biol. Chem.,* 243:1306–1311.
8. Fitton-Jackson, S. (1970): Environmental control of macromolecular synthesis in cartilage and bone: Morphogenetic response to hyaluronidase. *Proc. R. Soc. Lond. [Biol.],* 175:405–453.
9. Freeman, M. A. R., and Kempson, G. E. (1973): Load carriage. In: *Adult Articular Cartilage,* edited by M. A. R. Freeman, pp. 228–246. Pitman Medical, London.
10. Handley, C. J., and Lowther, D. A. (1976): Inhibition of proteoglycan biosynthesis by hyaluronic acid in chondrocytes in cell culture. *Biochim. Biophys. Acta,* 444:69–74.
11. Hardingham, T. E., Ewins, R. J. F., and Muir, H. (1976): Cartilage proteoglycans. Structure and heterogeneity of the protein core and the effects of specific protein modifications on the binding to hyaluronate. *Biochem. J.,* 157:127–143.
12. Hardingham, T. E., Fitton-Jackson, S., and Muir, H. (1972): Replacement of proteoglycans in embryonic chick cartilage in organ culture after treatment with testicular hyaluronidase. *Biochem. J.,* 129:101–112.
13. Hardingham, T. E., and Muir, H. (1972): The specific interaction of hyaluronic acid with cartilage proteoglycans. *Biochim. Biophys. Acta,* 279:401–405.
14. Hardingham, T. E., and Muir, H. (1972): Biosynthesis of proteoglycans in cartilage slices. Fractionation by gel chromatography and equilibrium density-gradient centrifugation. *Biochem. J.,* 126:791–803.
15. Hardingham, T. E., and Muir, H. (1973): Binding of oligosaccharides of hyaluronic acid to proteoglycans. *Biochem. J.,* 135:905–908.
16. Hardingham, T. E., and Muir, H. (1974): Hyaluronic acid in cartilage and proteoglycan aggregation. *Biochem. J.,* 139:565–581.
17. Hardingham, T. E., and Muir, H. (1975): Structure and stability of proteoglycan aggregates. *Ann. Rheum. Dis.,* 34 (Suppl. 2):26–28.
17a. Hardingham, T. E., and Muir, H. (1974): The function of hyaluronic acid in proteoglycan aggregation. In: *Normal and Osteoarthrotic Articular Cartilage,* edited by S. Y. Ali, M. W. Elves, and D. H. Leaback, pp. 51–58. Institute of Orthopaedics (University of London), London.
18. Hascall, V. C., and Heinegard, D. (1974): Aggregation of proteoglycans. I. The role of hyaluronic acid. *J. Biol. Chem.,* 249:4232–4241.
19. Hascall, V. C., and Heinegard, D. (1974): Aggregation of cartilage proteoglycans. II. Oligosaccharide competitors of the proteoglycan-hyaluronic acid interaction. *J. Biol. Chem.,* 249:4242–4249.
20. Hascall, V. C., and Sajdera, S. W. (1969): Proteinpolysaccharide complex from bovine nasal cartilage. The functions of glycoprotein in the formation of aggregates. *J. Biol. Chem.,* 244:2384–2396.
21. Hascall, V. C., Oegema, T. R., and Brown, M. (1976): Isolation and characterisation of proteoglycans from chick limb bud chondrocytes grown *in vitro. J. Biol. Chem.,* 251: 3511–3519.
22. Heinegard, D. (1975): Structure of cartilage proteoglycans. *Ann. Rheum. Dis.,* 34 (Suppl. 2):29–31.
23. Heinegard, D. (1977): Polydispersity of cartilage proteoglycans: Structural variations with size and buoyant density of the molecules. *J. Biol. Chem.,* 252:1980–1989.
24. Heinegard, D., and Hascall, V. C. (1974): Aggregation of proteoglycan. III. Characteristics of the proteins isolated from trypsin digests of aggregates. *J. Biol. Chem.,* 249:4250–4256.
25. Kempson, G. E., Muir, H., Pollard, C., and Tuke, M. (1973): The tensile properties of

cartilage of human femoral condyles related to the content of collagen and glycosaminoglycans. *Biochim. Biophys. Acta,* 297:456–472.

26. Kempson, G. E., Muir, H., Swanson, S. A. V., and Freeman, M. A. R. (1970): Correlations between the compressive stiffness and chemical constituents of human articular cartilage. *Biochim. Biophys. Acta,* 215:70–77.

27. Kosher, R. A., Lash, J. W., and Minor, R. R. (1973): Environmental enhancement of *in vitro* chondrogenesis. IV. Stimulation of somite chondrogenesis by exogenous chondromucoprotein. *Dev. Biol.,* 35:210–220.

28. Lee, P., Rooney, P. J., Starrock, R. D., Kennedy, A. C., and Dick, W. C. (1974): The etiology and pathogenesis of osteoarthrosis: A review. *Semin. Arthritis Rheum.,* 3:189–218.

29. Linn, F. C., and Sokoloff, L. (1965): Movement and composition of interstitial fluid of cartilage. *Arthritis Rheum.,* 8(4):481–494.

30. Luscombe, M., and Phelps, C. F. (1967): The composition and physicochemical properties of bovine nasal septa protein-polysaccharide complex. *Biochem. J.,* 102:110–119.

31. Maroudas, A. (1973): Physico-chemical properties of articular cartilage. In: *Adult Articular Cartilage,* edited by M. A. R. Freeman, pp. 131–170. Pitman Medical, London.

32. Maroudas, A., Muir, H., and Wingham, J. (1969): The correlation of fixed negative charge with glycosaminoglycan content of human articular cartilage. *Biochim. Biophys. Acta,* 177:492–500.

33. McDevitt, C. A. (1973): Biochemistry of articular cartilage. Nature of proteoglycans and collagen of articular cartilage and their role in aging and osteoarthrosis. *Ann. Rheum. Dis.,* 32:364–378.

34. McDevitt, C. A., Gilbertson, E., and Muir, H. (1977): An experimental model of osteoarthritis: Early morphological and biochemical changes. *J. Bone Joint Surg.,* 59B:24–35.

35. McDevitt, C. A., and Muir, H. (1976): Biochemical changes in the cartilage of the knee in experimental and natural osteoarthritis in the dog. *J. Bone Joint Surg.,* 58-B:94–101.

36. McDevitt, C. A., Muir, H., and Pond, M. J. (1974): Biochemical events in early osteoarthrosis. In: *Normal and Osteoarthrotic Articular Cartilage,* edited by S. Y. Ali, M. W. Elves, and D. H. Leaback, pp. 207–217. Institute of Orthopaedics, London.

37. Muir, H., and Hardingham, T. E. (1975): Structure of proteoglycans. In: *Biochemistry of Carbohydrates. MTP International Review of Science,* edited by W. J. Whelan, pp. 153–222. Butterworths, London, and University Park Press, Baltimore.

38. Nevo, Z., and Dorfman, A. (1972): Stimulation of chondromucoprotein synthesis in chondrocytes by extracellular chondromucoprotein. *Proc. Natl. Acad. Sci. U.S.A.,* 69:2069–2072.

39. Oegema, T. R., Hascall, V. C., and Dziewiatkowski, D. D. (1975): Isolation and characterisation of proteoglycans from the Swarm rat chondrosarcoma. *J. Biol. Chem.,* 250:6151–6159.

40. Pond, M. J., and Nuki, G. (1973): Experimentally induced osteoarthritis in the dog. *Ann. Rheum. Dis.,* 32:387–388.

41. Robinson, H. C., Brett, M. J., Tralaggan, P. J., Lowther, D. A., and Okayama, M. (1975): The effect of D-xylose, β-D-xylosides and β-D-galactosides on chondroitin sulphate biosynthesis in embryonic chicken cartilage. *Biochem. J.,* 148:25–34.

42. Rosenberg, L., Hellman, W., and Kleinschmidt, A. K. (1970): Macromolecular models of protein polysaccharide from bovine nasal cartilage based on electron microscopic studies. *J. Biol. Chem.,* 245:4123–4130.

43. Rosenberg, L., Hellman, W., and Kleinschmidt, A. K. (1975): Electron microscopic studies of proteoglycan aggregates from bovine articular cartilage. *J. Biol. Chem.,* 250:1877–1883.

44. Rosenberg, L., Pal, S., Beale, R., and Schubert, M. (1970): A comparison of protein-polysaccharides of bovine nasal cartilage isolated and fractionated by different methods. *J. Biol. Chem.,* 245:4112–4122.

45. Sajdera, S. W., and Hascall, V. C. (1969): Proteinpolysaccharide complex from bovine nasal cartilage. A comparison of low and high shear extraction procedures. *J. Biol. Chem.,* 244:77–87.

46. Sokoloff, L. (1969): *The Biology of Degenerative Joint Disease.* University of Chicago Press, Chicago.

47. Solursh, M., Vaerewyck, S. A., and Reiter, R. S. (1974): Depression by hyaluronic acid of glycosaminoglycan synthesis by cultured chick embryo chondrocytes. *Dev. Biol.,* 41:233–244.
48. Thyberg, J., Lohmander, S., and Heinegard, D. (1975): Proteoglycans of hyaline cartilage. Electron microscopic studies of isolated molecules. *Biochem. J.,* 151:157–166.
49. Toole, B. P. (1973): Hyaluronate and hyaluronidase in morphogenesis and differentiation. *Am. Zool.,* 13:1061–1065.
50. Tsiganos, C. P., and Muir, H. (1969): Studies on protein-polysaccharides from pig laryngeal cartilage. Heterogeneity, fractionation and characterization. *Biochem. J.,* 113:885–894.
51. Walker, P., Dowson, D., Longfield, M., and Wright, V. (1969): A joint simulator. In: *Lubrication and Wear in Joints,* edited by V. Wright, pp. 104–109. Sector Publishing Ltd., London.
52. Wiebkin, O. W., Hardingham, T. E., and Muir, H. (1975): The interaction of proteoglycans and hyaluronic acid and the effect of hyaluronic acid on proteoglycan synthesis by chondrocytes of adult cartilage. In: *Dynamics of Connective Tissue Macromolecules,* edited by P. M. C. Burleigh and A. R. Poole, pp. 81–104. North Holland Publishing Co., Amsterdam.
53. Wiebkin, O. W., and Muir, H. (1973): The inhibition of sulphate incorporation in isolated adult chondrocytes by hyaluronic acid. *FEBS Lett.,* 37:42–46.

Cell and Tissue Interactions, edited by
J. W. Lash and M. M. Burger. Raven
Press, New York, 1977.

Tissue Interactions and Extracellular Matrix Components

J. W. Lash and N. S. Vasan

Department of Anatomy, School of Medicine/G3, University of Pennsylvania,
Philadelphia, Pennsylvania 19104

If integral development is defined as a discrete event (as integers are discrete from one another) in the differentiation of an organism, or of its components, then integral development is not the rule in biological systems. The fertilized egg, in its uterus or shell, approximates an integer, but the genetic information contained within the egg regulates and controls the development of the embryo. Infinite interactions must occur between the developing cells and tissues for the presumptive embryo to be realized as a fulfillment of its genetic potential. Most of these interactions still escape our understanding, but there are some areas that are more clearly understood. The papers in this volume represent some of the recent advances in our understanding of cell and tissue interactions at the extracellular level, at the cell surface, and within the cell. The extracellular matrix components are an example of one area that is becoming better understood in relation to cell and tissue interactions, largely due to the advances in collagen and pro-teoglycan research. These two extracellular matrix components are known to play an important role in the control and regulation of some developmental processes (6,15–17,29,31,36,38), and the evidence is increasing that the effects of the extracellular matrix components may be mediated by the cell surface membrane and its components (2,22,35).

This report will deal with recent evidence that the proteoglycans and collagens of the extracellular matrix play an important role in the regulation of chondrogenesis. In the development of vertebral cartilage, it has long been known that the embryonic spinal cord and notochord interact with the somitic mesenchyme, and that this interaction is necessary for the differentiation of somitic cartilage (4,11,18–20,37).

The exact nature of the interaction between the inducing tissues (spinal cord and notochord) and the responding tissue (somitic sclerotome) is not fully understood, and there is some difference of opinion between the few laboratories pursuing this problem. It has been proposed that the interaction results in the somitic cells undergoing a "quantal" mitosis, which transforms the precartilaginous cells into cells programmed for synthesizing cartilage matrix (12). Most of the evidence, however, suggests that it is the microenvironment of the precartilaginous cells (i.e., the extracellular matrix products of the notochord), which stimulates the already programmed cells to synthesize and accumulate cartilage matrix (15–17,20,30,38). These conflicting concepts will be examined further at the end of this report. We will review here only the sequence of events in the stimulation of somitic sclerotomal cells to synthesize and accumulate cartilage matrix.

The earliest somites are epithelial vesicles, containing a few core cells (42). The medial portion of the somite becomes mesenchymal, and these cells (sclerotome cells) migrate toward the notochord. The sclerotome cells are at this stage of development synthesizing glycosaminoglycans, proteoglycans, and collagen. As the cells are migrating toward the notochord, the notochord is also synthesizing glycosaminoglycans, proteoglycans, and collagen. Whereas the glycosaminoglycans have been well characterized in both of these tissues (1,15,22), the proteoglycans have been described, but not well characterized (12). The collagen of the notochord has been characterized as Type II (26), but the collagen synthesized by the somites has not yet been definitely characterized (see ref. 41, and the statements in ref. 34).

As the consequence of this medial migration of the sclerotome cells, they become surrounded by the halo of proteoglycans and collagen synthesized by the notochord (3,30,36). The mechanism is still unclear, but as the result of this juxtaposition of extracellular matrix products and the sclerotome cells, the cells increase their synthetic activity and begin to accumulate cartilage matrix.

In spite of the negative statements reported by Okayama et. al. (34), it is now clearly established that isolated proteoglycans and collagen can effectively substitute for the notochord in stimulating somites to undergo chondrogenic differentiation (15,17). The assays in these reports were histological and the biochemical detection of glycosaminoglycans. In this report, we extend these observations to analyze the proteoglycans synthesized by the somites in response to exogenous proteoglycans and collagen.

METHODOLOGY

Details of the methodology will be found elsewhere (40), but a brief summary will be presented here.

Culture Techniques

Somites were obtained from stage 17 white Leghorn chick embryos (staging series of Hamburger and Hamilton, ref. 9). The somite explants, consisting of a cluster of 10 to 12 somites, were cultured in various ways.

Nuclepore Filters

The explants were placed on the surface of a Nuclepore filter (0.8-μm pore size, Nuclepore Corp., Pleasanton, Calif.). The filter was placed on a Nitex grid (656-μm mesh, Tetko, Inc., Elmsford, N.Y.), which supported the filters at the surface of the liquid nutrient medium. The medium used was either F12X (8), or Eagle's MEM (13). The media were obtained from Gibco, Grand Island, N.Y. Proteoglycans (200 μg/ml) were prepared and added to the medium as described by Kosher et. al., (15). Collagen (Types I, II, and III, kindly supplied by E. Miller) was dissolved in 0.5 M acetic acid (5) and dialyzed against sterile Simms' balanced salt solution (SBSS) for 24 hr at 4°C. The collagen solution was then added to the surface of the Nuclepore filter and allowed to gel at 37°C. The concentration of the collagen on the filter ranged between 2 and 12 μg/mm^2. In some instances an artificial matrix consisting of a collagen/proteoglycan lattice was prepared (32), and this was added to the surface of the filter. The concentration of proteoglycan was 0.2 to 0.4 μg/mm^2, and that of collagen was 12 to 37 μg/mm^2.

Nutrient Agar

Somites were placed directly on nutrient agar (8), or on collagenous substrates, which rested on the agar. The collagenous substrates were Type I collagen lattice on a piece of Millipore filter (kindly supplied by R. Kosher), or a sheet of Ethicon collagen (Type I) (kindly supplied by E. Katz).

Biochemical Analyses

Proteoglycan synthesis was monitored with the use of radioactive sulfate (5 or 10 μCi/ml of nutrient medium). The proteoglycans were sequentially extracted using 0.4 M, then 4.0 M guanidinium hydrochloride (GuHCl). The GuHCl was buffered and contained proteolytic inhibitors, as described by Oegema et. al. (33). The 0.4 M extract contained small molecular weight proteoglycans (proteoglycan monomers), and the subsequent 4.0 M extract contained the dissociated monomers of the larger proteoglycan aggregates. Sucrose density gradient (5 to 20%) centrifugation was performed on the extracts containing the proteoglycans obtained under "associative" conditions (0.4 M GuHCl) and "dissociative" conditions (4.0 M GuHCl). Complete methodology will be found in Vasan and Lash (40).

Glycosaminoglycan synthesis was measured according to the procedures of Kosher and Lash (16).

Scanning electron micrographs were obtained using a JSM-U-3 Joel and a AMR 1000A scanning electron microscope.

RESULTS

Glycosaminoglycan Synthesis

In all instances, exogenous proteoglycans and/or collagen stimulated glycosaminoglycan synthesis and increased matrix production. Table 1 presents some representative results. Whereas the magnitude of the response differed, with collagen usually being a more effective stimulator, the combined collagen-proteoglycan lattice did not give an additive response.

TABLE 1. *Stimulation of GAG synthesis in somite explants*

Explant	Age (days)	GAG stimulation[a]
Control	3	1.0
"Ethicon"	3	7.13
PGA	3	1.60
Collagen Type I	3	2.91
Collagen Type II	3	5.11
PGA/Collagen Type II	3	1.8

Ethicon is commercial Type I collagen. PGA is proteoglycan aggregate prepared from 14-day embryonic chick sterna. PGA/Coll-Type II represents somites grown on an artificial matrix.

[a] Glycosaminoglycan (GAG) stimulation was measured according to methods previously published (16). Since the control values varied according to the method of culture, all values are expressed as the degree of stimulation over the controls. Each experiment has been repeated at least three times, and representative values are given above. More complete data will be given by Lash et al. (23).

Figures 1 to 4 show the surface structure of some of the substrates used, as seen in the scanning electron microscope. It is interesting to note that the PGA-collagen lattice bears a close resemblance to cartilage matrix, and that the lattice made with Type III collagen appears different from that made with Type II collagen. The obvious orientation of the artificial matrix made with Type II collagen is unexplained. The orientation in the Ethicon (Type I) collagen is the result of the methods of preparation.

Proteoglycan Analyses

Proteoglycan synthesis was assayed in somites cultured with or without notochord (Figs. 5 to 7). Figure 5 shows the amount of proteoglycan synthesized after 6 days by somite explants. The smaller (monomeric) proteoglycans are heterogeneous with respect to size, and outnumber the aggregate form (4.0 M). The stimulation in proteoglycan synthesis by the notochord

FIG. 1. Scanning electron micrograph of Ethicon collagen filter (Type I). × 40.

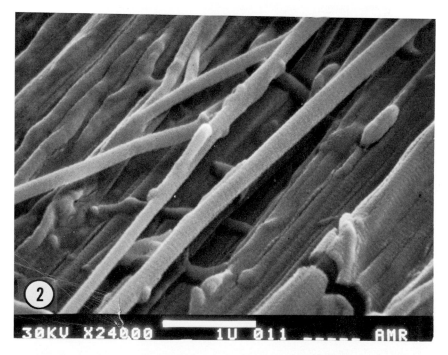

30KV X24000 1U 011 AMR

FIG. 2. Scanning electron micrograph of Ethicon collagen filter. × 24,000.

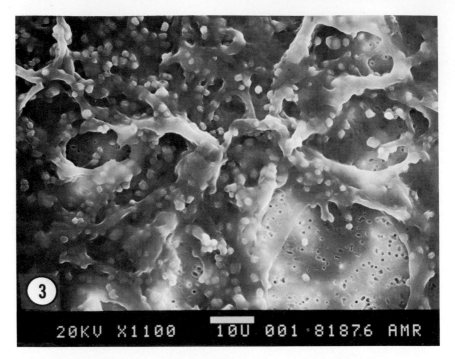

FIG. 3. Collagen Type II/proteoglycan lattice, on a Nuclepore filter (0.8-μm pore size). Note the orientation within the matrix. × 1,700.

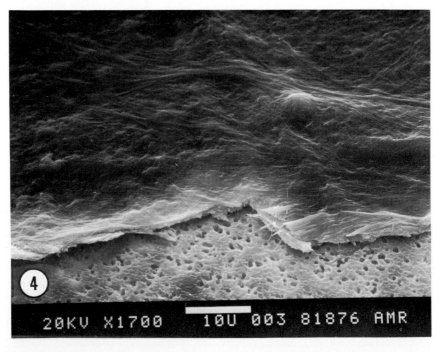

FIG. 4. Collagen Type III/proteoglycan lattice, on a Nuclepore filter (0.8-μm pore size). The amorphous matrix is very similar to the matrix synthesized by embryonic somites. × 1,100.

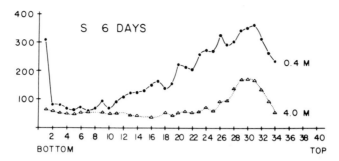

FIG. 5. Sucrose density gradient centrifugation of labeled proteoglycan sequentially extracted (0.4 M and 4.0 M GuHCl) from 6-day somite cultures. The profile shows the amount of proteoglycan monomer (0.4 M) and proteoglycan aggregate (4.0 M). The abscissa is DPM radioactive sulfate and the ordinate is the fraction (0.5 ml) number.

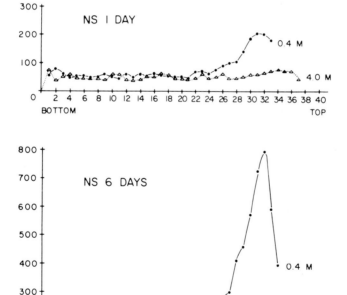

FIGS. 6 and 7. Sucrose density gradient centrifugation of labeled proteoglycan sequentially extracted from notochord-somite cultures, 1 day and 6 days old. There is relatively little increase in the aggregate form of proteoglycans (4.0 M), but a significant increase in the amount of proteoglycan monomers (0.4 M). Abscissa and ordinate as in Fig. 5.

can be seen in Figs. 6 and 7. After only 1 day in culture, there is an increase in proteoglycan synthesis, primarily in the monomeric form (0.4 M). After 6 days in culture, there is a marked increase in the monomeric form, and the proteoglycans are more homogeneous in size. Figure 8 shows the proteoglycans synthesized by notochord-somite explants after 14 days in culture. With age, there is not only an increase in the size of the monomeric form of the proteoglycans, but other size classes also appear. It should be noted that, in all figures except Fig. 8, the proteoglycan monomer as well as the aggregate form is being assayed. The 0.4 M GuHCl peak represents the nonaggregated, or monomeric forms, whereas the 4.0 M GuHCl peak represents the dissociated (monomeric) form of the proteoglycan aggregate. Extraction using only 4.0 M GuHCl pulls out all of the existing proteoglycans as the (monomeric) dissociation form, and does not permit a distinction between the types of proteoglycans present in the tissue (see ref. 34).

Figures 9 and 10 show the results after stimulation by the proteoglycan/collagen Type II lattice. After only 3 days in culture, there is a marked stimulation in the synthesis of proteoglycan aggregates. In some instances the smaller proteoglycan monomer increases more than the aggregate (see Figs. 5 to 7), whereas Fig. 10 shows that the aggregate form increases more than the monomeric form. We have no explanation yet for these differences, and this is currently under intensive study. There is also no explanation yet for the apparent difference in size between the proteoglycans we have extracted from somites and the size published by others (12,34). In spite of these points, which need clarification, the increase in the synthesis of glycosaminoglycans, and the increase in histologically detectable matrix is

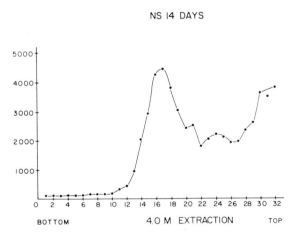

FIG. 8. Sucrose density gradient centrifugation of labeled proteoglycan extracted from 14-day-old notochord-somite cultures, showing not only an increase in the size of the monomeric form of the proteoglycans, but the appearance of other size classes. Abscissa and ordinate as in Fig. 5.

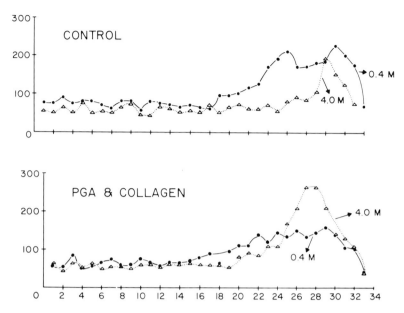

FIGS. 9 and 10. Density gradient centrifugation as in Figs. 5 to 8. Control somite explants and somites explanted on the surface of a proteoglycan/collagen Type II lattice (see Fig. 3). The explants were sequentially extracted after 3 days of culture. There is an increase in the 4.0 M extract, which indicates that there is an increase in the aggregated form of the proteoglycans as the result of stimulation. Abscissa and ordinate as in Fig. 5.

correlated with an increased synthesis of proteoglycans. It also appears that with the maturation of cartilage, there is an increase in the aggregated form of proteoglycans.

CONCLUSIONS AND DISCUSSION

With the recent advances in connective tissue biochemistry, it is becoming possible to define more precisely what is meant by chondrogenic differentiation. Mature cartilage matrix is now known to contain at least three forms of proteoglycans. They exist as relatively small proteoglycan monomers, which are capable of aggregating with hyaluronic acid by means of link glycoproteins to form the very large proteoglycan aggregate. Another form is the nonaggregating proteoglycan, which does not appear to be able to aggregate (10). In addition, cartilage is the only adult tissue in which collagen Type II has been found. Collagen Type II has been reported from two noncartilaginous embryonic tissues, chick notochord (26), and corneal epithelium (27). Thus, the differentiation of cartilage can be considered as the acquisition of these particular biochemical attributes. A note of caution is needed, however, in that differentiating cartilage need not necessarily have the exact same molecular profile as mature cartilage. It has been reported that the relative proportion of chondroitin-4-sulfate, chondroitin-6-sulfate, and heparan sulfate change during chondrogenic differentiation (28). It is assumed that the presence of proteoglycan aggregates is one of the biochemical definitions for cartilage matrix. The work reported here, however, suggests that cartilage matrix can form if the predominate proteoglycan is the small monomeric form. In fact, somite cells infected with RSVts24 virus will form matrix with only small monomers and no detectable aggregates (24).

The evidence is clear now that the extracellular matrix components, proteoglycans and collagen, play a regulatory role in cell and tissue interactions (6,15–17,29,31,36,38). In the instance of somite chondrogenesis, the evidence is becoming incontrovertible that exogenous collagen and/or proteoglycans can effectively substitute for the notochord in stimulating chondrogenesis (15,17). Strudel has independently arrived at similar conclusions, using a slightly different methodology (39). In spite of the evidence, there is the recurring attempt to explain somite chondrogenesis according to a scheme couched in terms of cell lineage and quantal cell cycles (12). This simplistic view is put forth with meager and incomplete evidence, and may indeed seem to explain results obtained under adverse conditions where most of the precartilaginous cells die in culture (see ref. 8). It is, however, inadequate to explain somite chondrogenesis under conditions that are not so hostile. Killing precartilaginous cells tells little about chondrogenic differentiation except that, for its occurrence, living cells are necessary.

Thus, there are cells in the embryonic somites which do possess a chondrogenic bias (see ref. 21). They are synthesizing proteoglycans and collagen, although whether these are identical to the molecules found in mature cartilage is not known. There is some evidence, however, that indicates that the precartilaginous limb buds have the same type of proteoglycans as the latter differentiating cartilage (7). This view is disputed by others (34). With the constant refinement in biochemical techniques, the concept of microheterogeneity is becoming more important. How different must proteoglycan molecules be to be classified as different types of molecules? The proliferation of types and subtypes of collagen is a good example of heterogeneity within a class of molecules. Although their mobility would indicate similarity, the proteoglycans of the precartilaginous somite undoubtedly differ in some respects from those in mature cartilage. We know that the glycosaminoglycan profiles change, and it may be that the core protein or the link proteins also change with age.

As for collagen synthesis in somites, much less is known. Although it was erroneously misquoted in Okayama et al. (34) that embryonic somites synthesize Type I collagen, this fact has not yet been established. It would not be surprising to find that Type III collagen is synthesized by the chondrogenic sclerotomal cells of the somites. Any report on collagen synthesis by embryonic somites would have to be interpreted with caution, since the somites are a heterogenous tissue. The dermatome and myotome may be synthesizing Type I collagen (see ref. 41), whereas the sclerotome may be producing another type. Whether the synthesis of Type II collagen by the notochord is related to its role in cartilage induction is not known, it could either be a fortuitous or a meaningful coincidence.

In conclusion, it can be stated that in the interaction between the embryonic notochord and adjacent somites, which results in somite chondrogenesis, intercellular matrix components play a major role. At the time of interaction, the notochord is producing collagen and proteoglycans, and the evidence is strong that these molecules act upon the somites and stimulate chondrogenic differentiation (6,15–17,29,31,36,38). At the time of interaction, it has been shown that the somite cells have a definite chondrogenic bias (21). The interaction can thus be looked upon as a permissive one, not requiring the invocation of unprovable theories such as quantal cell cycles (see ref. 12; see also ref. 14). It is known that the collagen and proteoglycan synthesized by the notochord are effective stimulators of somite chondrogenesis (16), and it is also known that other collagens and proteoglycans are effective (16,17). There does, however, appear to be more specificity with respect to the source of the proteoglycans than the source of collagen (23). If there are differences in the response to the different collagens and proteoglycans, our current methods of assay do not permit the detection of such differences. If we assume that the extracellular matrix components will be especially adapted to the local microenvironment, then microheterogeneities in the inducer matrix components may be capable of evoking differential responses by the chondrogenic somites.

ACKNOWLEDGMENTS

Supported by USPHS Research Grant HD-00380. We are grateful for the assistance of Gladys Treon and Mark Tyson. We also thank Elizabeth Belsky for her generous assistance with the scanning electron microscopy.

REFERENCES

1. Abrahamsohn, P. A., Lash, J. W., Kosher, R. A., and Minor, R. R. (1975): The ubiquitous occurrence of chondroitin sulfates in chick embryos. *J. Exp. Zool.*, 194:511–518.
2. Burger, M., and Jumblatt J. (1977): Membrane involvement in cell-cell interactions: A two component model system for cellular recognition that does not require live cells. In: *Cell and Tissue Interactions*, edited by J. Lash, and M. M. Burger. (*This volume.*)
3. Ebendal, T. (1976): Cellular locomotion in extracellular fibrous matrices *in vivo* and *in vitro*. *J. Gen. Physiol.*, 68:4a.
4. Ellison, M. L., and Lash, J. W. (1971): Environmental enhancement of *in vitro* chondrogenesis. *Dev. Biol.*, 26:486–496.
5. Elsdale, T., and Bard, J. (1972): Collagen substrata for studies on cell behavior. *J. Cell Biol.*, 54:626–637.
6. Fitton Jackson, S. (1975): The influence of tissue interactions and extracellular macromolecules on the control of phenotypic expression and synthetic capacity of bone and cartilage. In: *Extracellular Matrix Influences on Gene Expression*, edited by H. C. Slavkin, and R. C. Greulich, pp. 489–495. Academic Press, New York.

7. Goetinck, P. F., and Royal, P. D. (1976): The *in vitro* synthesis of sulfated proteoglycans by skin fibroblasts from normal and nanomelic embryos. *J. Gen. Physiol.,* 68:6a.
8. Gordon, J. S., and Lash, J. W. (1974): *In vitro* chondrogenesis and differential cell viability. *Dev. Biol.,* 36:88–104.
9. Hamburger, V., and Hamilton, H. L. (1951): A series of normal stages in the development of the chick embryo. *J. Morphol.,* 88:49–92.
10. Hardingham, T. E., and Muir, H. (1974): Hyaluronic acid in cartilage and proteoglycan aggregation. *Biochem. J.,* 139:565–581.
11. Holtzer, H., and Detwiler, S. R. (1953): An experimental analysis of the development of the spinal column. III. Induction of skeletogenous Cells. *J. Exp. Zool.,* 123:335–366.
12. Holtzer, H., Rubinstein, N., Fellini, S., Yeoh, G., Chi, J., Birnbaum, J., and Okayama, M. (1975): Lineages, quantal cell cycles and the generation of cell diversity. *Q. Rev. Biophys.,* 8:523–557.
13. Konigsberg, I. R. (1963): Clonal analysis of myogenesis. *Science,* 40:1273–1284.
14. Konigsberg, I. R., and Buckley, P. A. (1974): Regulation of the cell cycle and myogenesis by cell-medium interaction. In: *Concepts of Development,* edited by J. Lash, and J. R. Whittaker, pp. 179–193. Sinauer Associates, Inc., Sunderland, Mass.
15. Kosher, R. A., Lash, J. W., and Minor, R. R. (1973): Environmental enhancement of *in vitro* chondrogenesis. IV. Stimulation of somite chondrogenesis by exogenous chondromucoprotein. *Dev. Biol.,* 35:210–220.
16. Kosher, R. A., and Lash, J. W. (1975): Notochordal stimulation of *in vitro* somite chondrogenesis before and after enzymatic removal of perinotochordal materials. *Dev. Biol.,* 42:362–378.
17. Kosher, R. A., and Church, R. L. (1975): Stimulation of *in vitro* somite chondrogenesis by procollagen and collagen. *Nature,* 258:327–330.
18. Lash, J. W. (1963): Tissue interaction and specific metabolic responses: Chondrogenic induction and differentiation. In: *Cytodifferentiation and Macromolecular Synthesis,* edited by M. Locke, pp. 235–260. Academic Press, New York.
19. Lash, J. W. (1967): Differential behavior of anterior and posterior embryonic chick somites *in vitro. J. Exp. Zool.,* 165:47–56.
20. Lash, J. W. (1968): Chondrogenesis: Genotypic and phenotypic expression. *J. Cell Physiol.,* 72 (Suppl. 1):35–46.
21. Lash, J. W. (1968): Somitic mesenchyme and its response to cartilage induction. In: *Epithelial-Mesenchymal Interaction,* edited by R. Fleischmajer, pp. 165–172. William & Wilkins, Baltimore.
22. Lash, J. W., Rosene, K., Minor, R. R., Daniel, J. C., and Kosher, R. A. (1973): Environmental enhancement of *in vitro* chondrogenesis. III. The influence of external potassium ions and chondrogenic differentiation. *Dev. Biol.,* 35:370–375.
23. Lash, J. W., and Vasan, N. S. (1977): *In preparation.*
24. Lash, J. W., Vasan, N. S., and Kaji, A. (1977): *In preparation.*
25. Linsenmayer, T. F., Toole, B. P., and Trelstad, R. L. (1973): Temporal and spatial transitions in collagen types during embryonic chick limb development. *Dev. Biol.,* 35:232–239.
26. Linsenmayer, T. F., Trelstad, R. L., and Gross, J. (1973): The collagen of chick embryonic notochord. *Biochem. Biophys. Res. Comm.,* 53:39–45.
27. Linsenmayer, T. F., Smith, G. N., Jr., and Hay, E. D. (1977): *In vitro* synthesis of two collagen types of embryonic chick corneal epithelium. *Proc. Natl. Acad. Sci. U.S.A. (In press).*
28. Mathews, M. B. (1975): Polyanionic glycans in development and aging of vertebrate cartilage. In: *Molecular Biology, Biochemistry and Biophysics: Connective Tissue,* Vol. 19, edited by A. Kleinzeller, G. F. Springer, and H. G. Wittman, pp. 156–171. Springer-Verlag, New York.
29. Meier, S., and Hay, E. D. (1975): Control of corneal differentiation *in vitro* by extracellular matrix. In: *Extracellular Matrix Influences on Gene Expression,* edited by H. C. Slavkin, and R. C. Greulich, pp. 185–196. Academic Press, New York.
30. Minor, R. R. (1973): Somite chondrogenesis. *J. Cell Biol.,* 56:27–50.
31. Nevo, Z., and Dorfman, A. (1972): Stimulation of chondromucoprotein synthesis in chondrocytes by extracellular chondromucoprotein. *Proc. Natl. Acad. Sci. U.S.A.,* 69:2069–2071.

32. Oegema, T. R., Laidlow, J., Hascall, V. C., and Dziewiatkowski, D. D. (1975): The effect of proteoglycans on the formation of fibrils from collagen solution. *Arch. Biochim. Biophys.* 170:698–709.
33. Oegema, T. R., Hascall, V. C., and Dziewiatkowski, D. D. (1975): Isolation and characterization of proteoglycans from the swarm rat chondrosarcoma. *J. Biol. Chem.,* 250:6151–6159.
34. Okayama, M., Pacifici, M., and Holtzer, H. (1976): Differences among sulfated proteoglycans synthesized in non-chondrogenic cells, presumptive chondroblasts and chondroblasts. *Proc. Natl. Acad. Sci. U.S.A.,* 73:3224–3228.
35. Roth, S., Shurr, B. O., and Durr, R. (1977): A possible enzymatic basis for some cell recognition and migration phenomena in early embryogenesis. In: *Cell and Tissue Interactions,* edited by J. Lash, and M. M. Burger. (*This volume*).
36. Ruggeri, A. (1972): Ultrastructural, histochemical, and autoradiographic studies on the developing chick notochord. *Z. Anat. Entwicklungsgesch.,* 138:20–33.
37. Strudel, G. (1953): L'Influence morphogene de tube nerveux sur differentiation de la colonne vertebrale. *C. R. Soc. Biol.,* 147:132–133.
38. Strudel, G. (1971) Matériel extracellulaire et chondrogéneses vertebrale. *C. R. Acad. Sci. Paris,* 272:473–476.
39. Strudel, G. (1973): Relationship between the chick periaxial metachromatic extracellular material and vertebral chondrogenesis. In: *Biology of the Fibroblast,* edited by E. Kulonen, and J. Pikkarainen, pp. 93–102. Academic Press, New York.
40. Vasan, N. S., and Lash, J. W. (1977): Proteoglycan heterogeneity in articular and epiphyseal cartilages of embryonic chicks. *In preparation.*
41. von der Mark, H., von der Mark, K., and Gay, S. (1976): Study of differential collagen synthesis during development of the chick embryo by immunoflourescence. *Dev. Biol.,* 48:237–249.
42. Williams, L. W. (1910): The somites of the chick. *Am. J. Anat.,* 11:55–99.

Cell and Tissue Interactions, edited by
J. W. Lash and M. M. Burger. Raven
Press, New York, 1977.

Interaction Between the Cell Surface and Extracellular Matrix in Corneal Development

Elizabeth D. Hay

Department of Anatomy, Harvard Medical School, Boston, Massachusetts 02115

In this volume, Doctors Miller and Muir have described the molecular structure of two of the most important components of the extracellular matrix (ECM): collagen and proteoglycan. I would like to begin by illustrating the relation of ECM to the cell surface in electron micrographs of two embryonic tissues that have been exploited in recent studies of cell-matrix interaction *in vitro:* embryonic cornea and the developing somite. Dr. Lash has reviewed in this volume the evidence that collagen and proteoglycans play a role in stabilizing chondrogenesis in somite mesenchyme. The evidence that collagen and glycosaminoglycans (GAG) enhance collagen and GAG production by developing corneal epithelial cells will be summarized in this chapter. In addition, data that the cell-ECM interaction is located at the cell surface and is autocatalytic in nature will be presented here. Finally, a tentative model of the possible sites of interaction between ECM and the cell surface will be proposed.

MORPHOLOGY OF CELL SURFACE-ECM INTERACTION

Ruthenium red is a polycation that both fixes[1] and stains GAG (22,34) and that beautifully preserves the small, proteinaceous filaments of embryonic and adult ECM (8,21,34). In the vicinity of the embryonic notochord and neural tube, these filaments appear to be aggregating into larger fibrils (Fig. 1), some of which are undoubtedly collagenous in nature (20). The proteoglycan particles revealed by the ruthenium red method (7,34) seem to coat and to link the small filaments together and may play a role in their aggregation, both in embryonic ECM (Figs. 1 and 2) and in cartilage (Fig. 3). The particles are composed of enzyme-sensitive GAG, probably largely chondroitin sulfate (8,32,34), which may be linked to protein and, in turn, to hyaluronate as described elsewhere by Muir (*this volume*). What is visual-

---

[1] Ruthenium red fixation consists of treating tissues in glutaraldehyde-formaldehyde containing ruthenium red followed by osmium containing ruthenium red (22,34). For enzyme studies, the tissue is lightly fixed in formaldehyde, digested with enzyme *en bloc,* and then fixed in the ruthenium red mixtures (8).

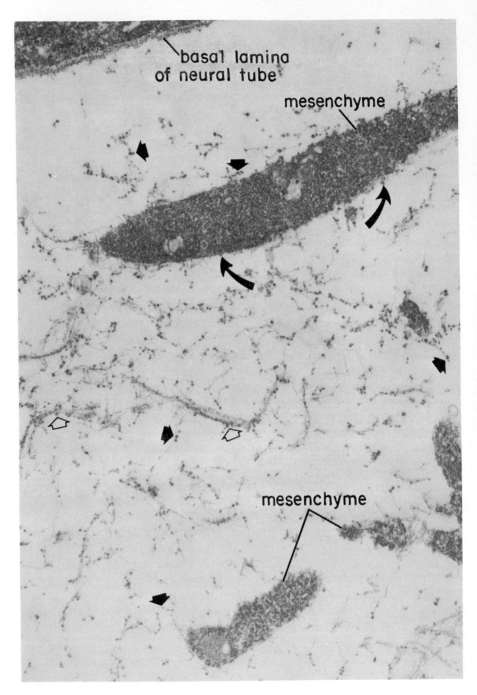

FIG. 1. Electron micrograph of a section of ruthenium red-fixed (see footnote 1 in text) trunk of a 3-day-old chick embryo. The neural tube lies above the picture and the notochord below and to the left. The mesenchymal cells shown here derived from sclerotome and are migrating toward the left. The matrix through which they are migrating consists of proteoglycan particles **(short, solid arrows)** attached to small filaments 50 Å in diameter. The filaments appear to be aggregating into fibrils, presumably collagenous in nature, with which anion-rich particles are also associated **(open arrows).** Filaments and particles **(curved arrows)** also appear to be attached to the cell surface. × 21,000.

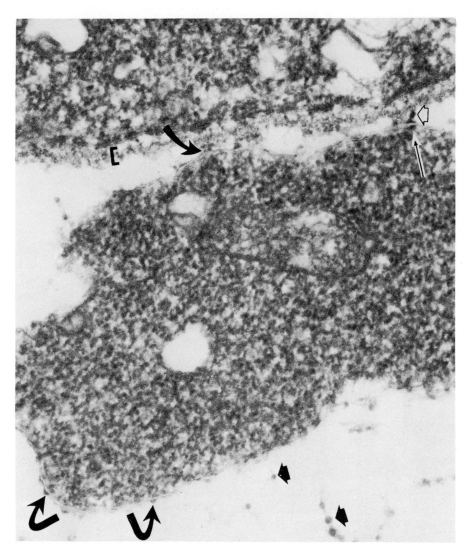

FIG. 2. Another electron micrograph of the same section depicted in Fig. 1. The area shown is located to the upper right of the area shown in Fig. 1. The mesenchymal cell in the center of the picture is migrating to the left and is in close contact with the basal lamina **(bracket)** of the neural tube. At this magnification, the tripartite nature of the plasma membrane can be appreciated. Numerous anion-rich sites reside on the outer leaflet of the membrane **(curved arrows).** In one region, a particle 300 Å in diameter **(open arrow)** associated with the basal lamina seems attached to the plasma membrane; a corresponding density appears in the adjacent cytoplasm **(thin arrow).** Proteoglycan particles, 200 to 300 Å in diameter, are associated with the filaments of the extracellular matrix **(short, solid arrows).** × 72,250.

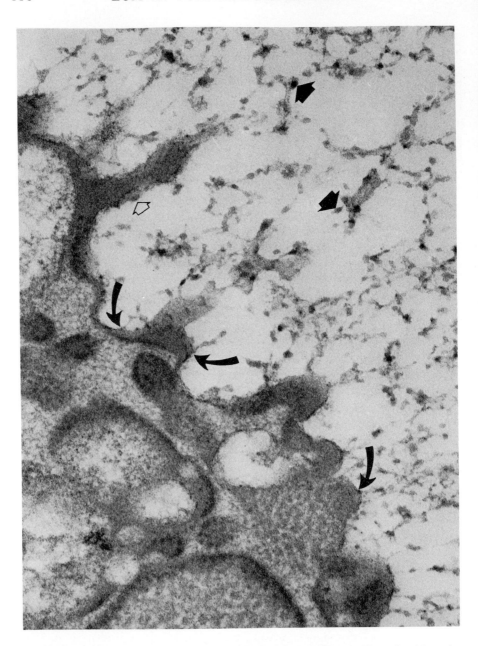

FIG. 3. Electron micrograph of a section of adult mouse hyaline cartilage fixed in ruthenium red (7). Proteoglycan particles **(short, solid arrows)** in the cartilage matrix are attached to collagenous filaments and small fibrils (24). At the open arrow, such a particle is seen to be closely associated with the plasma membrane of a chondrocyte. In other areas **(curved arrows),** small filaments of the extracellular matrix seem to attach to anion-rich sites on the plasma membrane. The cytoplasm of the cell is not well preserved by the ruthenium red method (7). × 90,000.

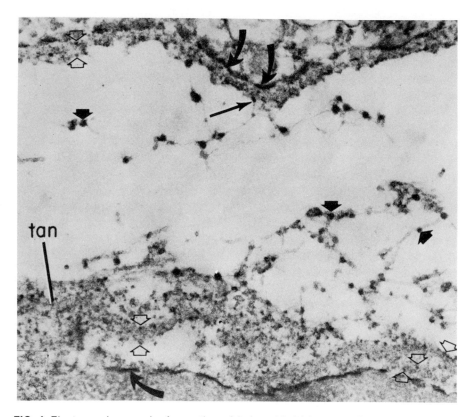

FIG. 4. Electron micrograph of a section of 3-day-old chick cornea fixed with ruthenium red prior to the formation of the orthogonal stroma (8). The base of the corneal epithelium appears at the top and the lens at the bottom. Small proteoglycan particles ($\sim$ 100 Å in diameter) lie in the inner and outer layers of the epithelial basal lamina **(open arrows, top left)**. In tangential **(tan)** view of the basal lamina or capsule of the lens, proteoglycan particles may be seen spaced throughout the lamina **(open arrows, bottom)**; they are approximately 600 Å apart and they are composed mainly of chondroitin sulfate (32). The chondroitin sulfate-rich particles **(short, solid arrows)** of the matrix are 200 to 400Å in diameter. $\times$ 60,000. (From ref. 8.)

ized on electron micrographs is a ruthenium red-OsO_4 precipitate (22) of what may have been a more expanded proteoglycan macromolecule in the living state.

In the developing cornea (Figs. 4 to 8), the small particle-rich filaments are connected to proteoglycan particles (7,8,32) in the basal lamina (basement membrane) of the corneal epithelium (top half, Fig. 4) and of the lens (Fig. 4, bottom half). The first fibrils recognizable as collagen by their striation appear along the lens capsule (the basal lamina of the lens) and seem to derive from the GAG-rich filamentous precursors (7,8). Collagen fibrils characteristic of the definitive corneal stroma are 250 Å in diameter

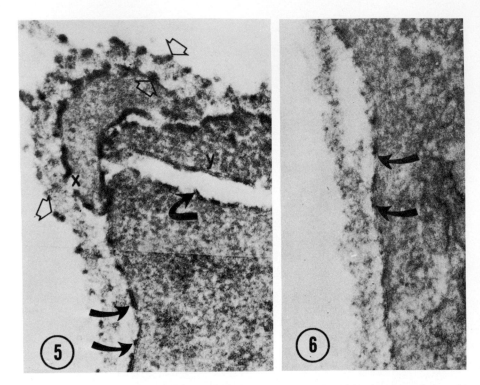

FIGS. 5 and 6. Electron micrographs of the anterior lens basal lamina from 3-day-old chick embryo untreated (Fig. 5) and treated with testicular hyaluronidase (Fig. 6) before fixation in ruthenium red (see footnote 1 in text). In Fig. 5, the proteoglycan granules **(open arrows)** of the basal lamina are shown at higher magnification than in Fig. 4. Ruthenium red stains an anion-rich material on the outer leaflet of the plasma membrane **(curved arrows, Figs. 4 and 5),** which is resistant to testicular hyaluronidase treatment of the tissue **(Fig. 6).** Testicular hyaluronidase digests chondroitin sulfate, chondroitin, and hyaluronate (8). In between these anion-rich areas, the plasma membrane is often poorly preserved on both the basal **(x, Fig. 5)** and lateral **(y, Fig. 5)** surfaces of the epithelial cell. × 92,000. (From ref. 8.)

and are coated with proteoglycan particles of smaller diameter (32). The role of the particles in fibril polymerization and the possible collagenous nature of the filamentous precursors are discussed elsewhere (7).

What is relevant to the theme of this symposium is the interaction of the proteoglycan particles, the proteinaceous filaments, and the collagen fibrils of the ECM with the plasmalemma of the developing chondrocytes and corneal epithelial cells. The mesenchymal cells of the sclerotome migrate through the ECM produced by neural tube and notochord (Figs. 1 and 2) before they differentiate into the overt cartilage (axial skeleton) that encloses the neural tube and notochord. The plasmalemma of the migrating mesenchymal cell is studded with ruthenium-red-staining particles, presumably proteoglycan, which connect the small filaments to the cell surface (Fig. 2).

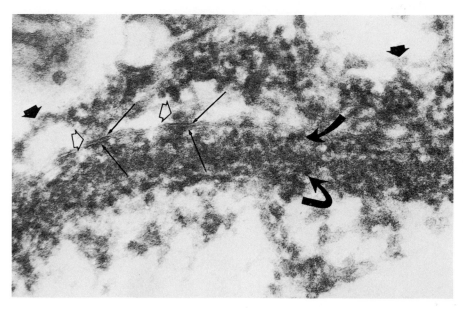

FIG. 7. Electron micrograph of the basal lamina and epithelial basal cytoplasm of a 9-day-old chick cornea fixed in ruthenium red (7). The inner and outer leaflets of the epithelial plasma membrane appear densely stained **(thin arrows)** in certain areas. The plasma membrane in between these areas is not well preserved by ruthenium red. In the specimen depicted here, the ground cytoplasm was poorly preserved and has washed out to a large extent. The proteinaceous cytoplasmic filaments that remain **(short, solid arrows)** appear to attach to the plasmalemma **(open arrows)** at the dense membrane plaques. A proteoglycan particle in the inner domain of the basal lamina is labeled by the curved arrow, while the U-shaped arrow indicates a granule in the outer domain of the lamina. × 115,000. (From ref. 7.)

After the mesenchymal cell has differentiated into cartilage, proteoglycan particles and small filaments, presumably collagenous in nature (24), are intimately associated with the plasmalemma (Fig. 3). In the definitive cornea, mesenchymal cells may attach to fully formed collagen fibrils (Fig. 8).

The corneal epithelial cells reside on a basal lamina rich in chondroitin sulfate (8,32). The proteoglycan particles of the inner layer of the lamina are attached to the basal plasmalemma of the epithelial cells (Fig. 5). After enzyme digestion to remove chondroitin sulfate and hyaluronate, it can be seen that the filamentous component of the lamina (presumably collagen) is also attached to the plasmalemma (Fig. 6). The plasmalemma in the areas of attachment is coated with enzyme-resistant anions that bind stain (Fig. 6) and may be heparan sulfate (8) or an unidentified glycoprotein lacking sialic acid (7). Filamentous cytoskeletal elements are closely associated with the cytoplasmic side of the ECM-rich areas of plasmalemma and may be attached there (Fig. 7).

FIG. 8. Electron micrograph of a fibroblast cell process in a 12-day-old cornea fixed in ruthenium red (7). A bundle of striated collagen fibrils traverses the field from left to right. The fibroblast plasmalemma is in intimate association with a collagen fibril (X). As in the case of the epithelial cell (Figs. 4 to 7), ruthenium red preserves anion-rich areas of the plasma membrane **(curved arrows)**, which alternate with poorly preserved areas of membrane. The densely staining regions of the plasma membrane seem to be associated with the same band **(open arrows)** of the collagen fibril. This appears to be the band to which proteoglycan particles are attached in other fibrils **(thin arrows).** × 140,000.

Thus, in both developing cartilage and cornea the morphologically identifiable components of the ECM are closely associated with the plasmalemma of the component cells. Proteinaceous filaments that may be composed of collagen (7,24) seem to be attached to anion-rich areas of the membrane, whereas proteoglycan particles seem to reside in or on the membrane. The trypsin and collagenase treatments used to isolate somite cells, chondrocytes, and corneal epithelium (5,14,17,29) remove the ECM or sever its attachment to the cell, leaving the plasmalemma blebbing and highly contorted (5). That the removal of the ECM also has a deleterious effect on collagen and GAG production by the cells is shown by the studies of Dorfman, Kosher, Lash, and others on chondrogenesis reported by Dr. Lash in this volume and by studies on the cornea to be described now.

EFFECT OF COLLAGEN AND GAG
ON CORNEAL ECM PRODUCTION

The embryonic corneal epithelium synthesizes most, if not all, of the collagen and GAG comprising the primary corneal stroma of the chick embryo (1,2,6,25,32). The primary stroma is located between the lens and the corneal epithelium (Figs. 9 and 10) and at 4 days of incubation consists of about 25 orthogonally arranged layers of collagen fibrils (11). Subsequently, after the endothelium has formed (Fig. 9), the stroma is invaded by corneal fibroblasts of neural crest origin, which add collagen to its interstices and convert it to its adult form.

If the corneal epithelium is enzymatically isolated at 5 days of incubation and combined with killed lens or isolated lens capsule (Fig. 11), its blebbing naked basal surface smooths out and makes close contact with the lens capsule ECM (2,5). Within 24 hr, the epithelium produces a new stroma

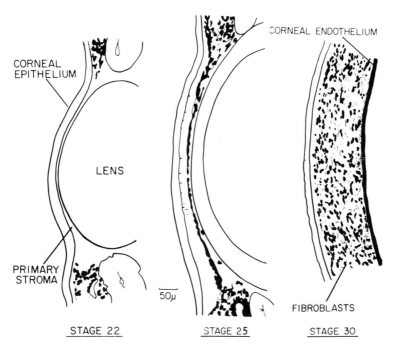

FIG. 9. Cameral lucida drawings of sections of the developing chick cornea. Between 3 and 4 days of incubation, the corneal epithelium lays down from the primary corneal stroma on the surface of the lens (stage 22, 4 days). Between 4 and 5 days, the corneal endothelium migrates into place under the stroma and separates it from the lens (stage 25, 5 days). At the end of stage 27 (5½ days), the stroma swells and is immediately invaded by mesenchymal cells destined to become fibroblasts. By stage 30 (7 days), over 10 layers of fibroblasts are present, all of which are contributing collagen to the stroma. Eventually, about 25 layers of fibroblasts will occupy the stroma. (From ref. 11.)

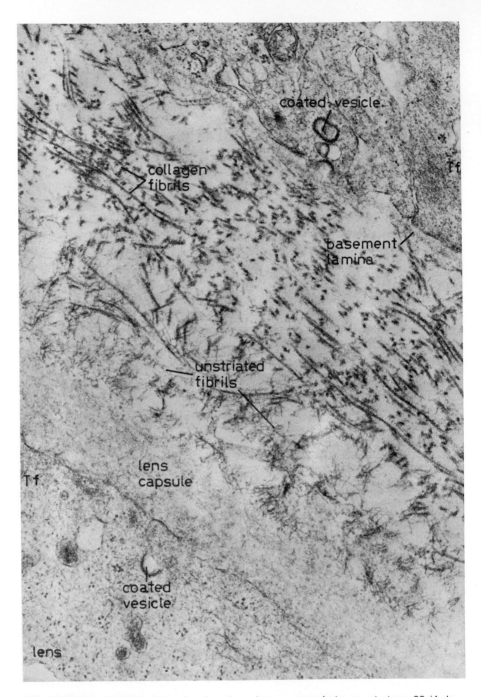

FIG. 10. Electron micrograph showing the primary corneal stroma at stage 22 (4 days; see Fig. 9). The collagen fibrils are arranged in layers at right angles to each other. Next to the corneal epithelium, the fibrils contact the basement lamina of the epithelium. At the junction between the stroma and the lens capsule, smaller unstriated fibrils occur. Tonofilaments (Tf) appear in the lens cells and also in the corneal epithelium. × 40,000. (From ref. 11.)

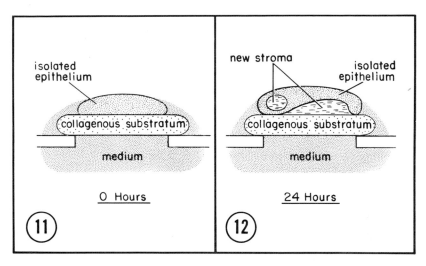

FIGS. 11 and 12. Diagrams of the culture method. Corneal epithelium isolated by trypsin-collagenase (26) is grown on a collagenous substratum, such as frozen-killed lens (1), purified lens capsule, or cartilage collagen (26). The epithelium makes close contact with the collagenous substratum **(Fig. 11)** and within 24 hr has secreted a facsimile of the removed corneal stroma **(Fig. 12).**

composed of orthogonally arranged collagen fibrils and GAG located either along the lens capsule or within the epithelial interstices (Fig. 12). Linsenmayer et al. (19) have shown that the stroma secreted in culture by corneal epithelium contains Type I and Type II collagen. Moreover, a small amount of new basal lamina is also produced by the corneal epithelium when it is cultured on a collagenous substratum (6).

Meier and Hay (26) measured collagen and GAG synthesis as incorporation of radioactive isotope during the initial 24 hr in culture and were able to show that over twice as much collagen and GAG was produced by epithelium grown on lens capsule as by epithelium on either Millipore filter or plastic, while the DNA content of the two types of cultures was essentially the same. If chondroitin sulfate or heparan sulfate are added to the medium, GAG synthesis is stimulated further, but collagen synthesis is not affected. Even epithelium grown on Millipore filter is stimulated to produce more GAG by these two GAGs. Whereas chondrocytes are inhibited by hyaluronate (29,31), hyaluronate has no detectable effect on production of GAG by corneal epithelium (27). Interestingly, the only GAGs the epithelium seems to be making at this time are chondroitin and heparan sulfates (27).

The effect of GAG on ECM production by corneal epithelium thus seems to be relatively specific. The effect of collagen, however, is not. As we have seen, not only collagen synthesis but also GAG production is en-

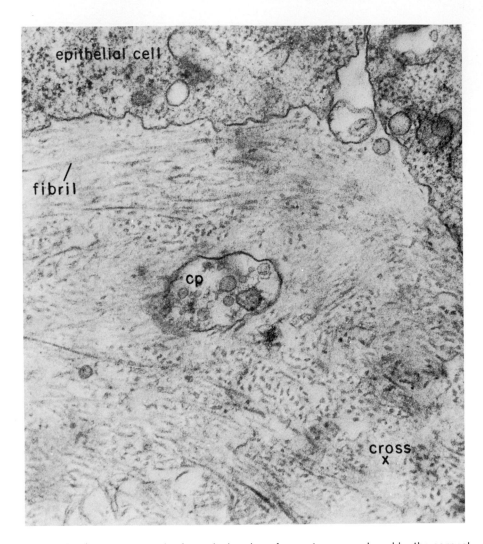

FIG. 13. Electron micrograph of a typical region of new stroma produced by the corneal epithelium grown, in this case, on purified Type II collagen. The new collagen fibrils in longitudinal section (fibril) are arranged at right angles to those in cross section **(cross x).** The fibrils have the same diameter as those of native corneal collagen (26). An epithelial cell process **(cp)** was trapped in the stroma shown here. The purified Type II collagen was a gift of Dr. George Martin. × 36,000. (From ref. 26.)

hanced by collagen substrates such as lens capsule. Moreover, gels of Type I (rat tendon) or Type II (cartilage) collagen are as effective as lens capsule (Type IV) collagen (26). Meier and Hay (26–28) provide the following evidence that the component of the substrate that stimulates the combined production of collagen and GAG is collagen and not a contaminant of the

ECM substrate: (a) known "contaminants," such as GAG, do not stimulate collagen synthesis; (b) all collagens so far tested have the same effect on collagen synthesis regardless of the degree and type of "contaminating proteins;" (c) substrates of other proteins, such as keratin and albumin, have no effect; (d) purification of lens capsule collagen by NaOH extraction does not destroy its effect; (e) highly purified Type II collagen prepared from rat chondrosarcoma has the same stimulating effect as lens capsule; in both cases incorporation of isotope into collagen and GAG is enhanced twofold (26), and visible stroma is produced (Fig. 13).

EXPERIMENTAL EVIDENCE FOR CELL SURFACE AND ECM INTERACTION

The normal location of collagen and proteoglycan is external to the cell surface (Figs. 1 to 8). Collagen is so insoluble under physiological conditions that it is impossible to get enough of it into solution to do a dose-response curve in our corneal *in vitro* system (26). Therefore, it seems very likely that the collagen-cell interaction is at the cell surface and depends on contact of the external plasmalemma surface with the collagenous substratum. To test this hypothesis, Meier and Hay (28) grew epithelium on Millipore filters (0.45 μm pore size) transfilter to lens capsule (Fig. 14).

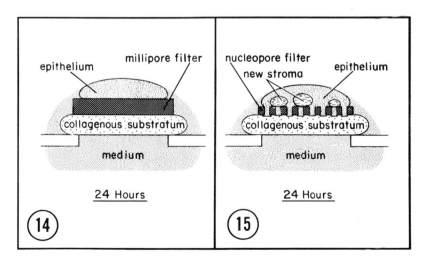

FIGS. 14 and 15. Diagrams of cultures employing filters to separate the epithelium from the collagenous substratum. The corneal epithelium sends shallow cell processes into Millipore filters, but in 24 hr these processes do not traverse the filters completely. The isolated epithelium, thus deprived of contact with collagen, fails to produce a new stroma **(Fig. 14)**. On the other hand, the epithelium can send cell processes quickly across a Nucleopore filter to contact an underlying collagenous substratum, and in this case it produces a substantial amount of new stroma **(Fig. 15)**.

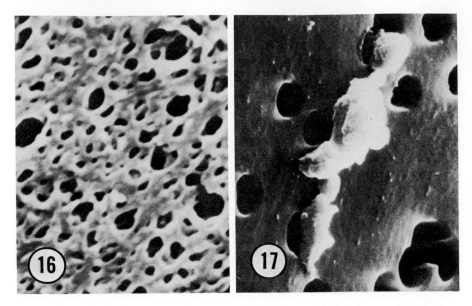

FIGS. 16 and 17. Scanning electron micrographs of the undersurface of a Millipore filter **(Fig. 16)** and a Nucleopore filter **(Fig. 17)**, each of which had corneal epithelium growing on the upper surface. Cell processes have traversed the Nucleopore filter but not the Millipore filter in 24 hr. × 100,000. (From ref. 28.)

Scanning electron microscopy of the undersurface of the Millipore filters showed that epithelial cell processes did not traverse the filters in the 24-hr period studied (Fig. 16). There was no stimulation of collagen synthesis by an underlying lens capsule that the epithelium could not contact (28). Thus, a freely diffusible factor emanating from the collagenous substratum is not involved in the enhancement of stroma production observed when the epithelium is in direct contact with the substratum (Fig. 12).

More direct evidence for a role of cell-ECM contact in the enhancement of corneal differentiation by ECM was obtained by Meier and Hay (28), by growing the epithelium on Nucleopore filters transfilter to lens capsule (Fig. 15). These filters, first used for studies of embryonic induction by Dr. Saxén (*this volume*) and co-workers (33), are very thin (10 μm) polycarbonate structures with straight pores that readily permit cell processes to traverse them (Fig. 17). Comparable Millipore filters are at least 25 μm thick and are composed of a complex cellulose acetate meshwork (Fig. 16). The corneal epithelium sent cell processes through the Nucleopore filters to contact the underlying lens capsule (Fig. 18), and the epithelium produced stroma on the epithelial side of the filter (Fig. 19). Nucleopore filters (0.8 μm pore size) did not significantly inhibit the effect of lens capsule or Type I or Type II collagen on corneal epithelial collagen synthesis (Table 1).

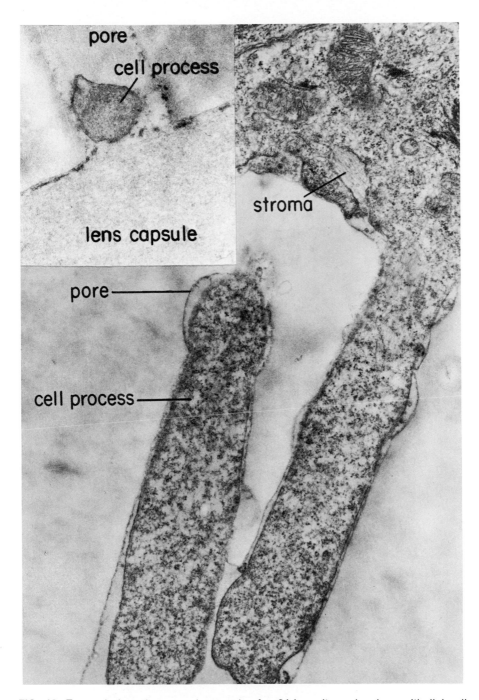

FIG. 18. Transmission electron micrograph of a 24-hr culture showing epithelial cell processes extending into the filter. They contact the killed lens capsule on the undersurface of the filter **(inset)**. Stroma accumulates on the upper surface of the filter under and within such epithelia provided the cell processes contact the underlying ECM Nucleopore filter. × 30,000. (From ref. 9.)

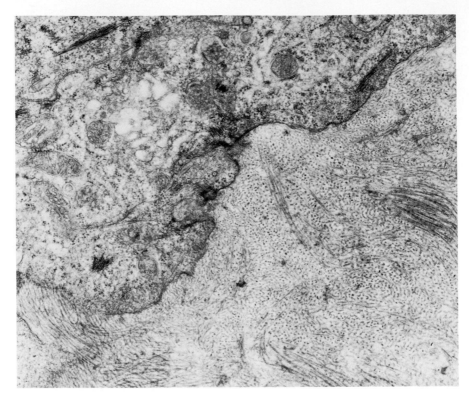

FIG. 19. Transmission electron micrograph of an area of newly synthesized corneal stroma within a lateral intercellular cleft of a corneal epithelium grown transfilter to lens capsule on Nucleopore filter. Fibrils in the stroma tend to be arranged at right angles to each other; some (dots) are cut in cross sections and others (orthogonal to the former) in long section. The epithelium appears in the upper left. × 24,000. (From ref. 9.)

TABLE 1. *The enhancement of collagen production by corneal epithelium cultured transfilter to collagenous substrata*[a]

Substratum	^{3}H-proline in collagen
Nucleopore filter alone	1,657 ± 122
Nucleopore filter over killed lens capsule	4,117 ± 204
Nucleopore filter over rat tail tendon gel	3,561 ± 177
Nucleopore filter over purified chondrosarcoma collagen	3,507 ± 182
Nucleopore filter over NaOH-extracted lens capsule	3,302 ± 120

[a] The results are expressed as the mean value of five determinations (eight epithelia each) plus or minus the standard deviation. Cultures were labeled with 5 μCi/ml proline-^{3}H for 24 hr. All Nucleopore filters were 0.8 μm pore size (10 μm thick). See Hay and Meier (9) for further description.

The epithelial cell surface in contact with the collagenous substratum can be measured in scanning electron micrographs of the undersurface of the Nucleopore filter after removing the substratum (28). Nucleopore filters of 0.8 μm pore size permit twice as many cell processes to cross as do those of 0.4 μm pore size. Smaller pores permit even fewer cell processes to cross in the 24-hr period under study, and the corresponding cell surface exposed to the collagenous substratum is smaller (Table 2). The amount of collagen produced, measured as incorporation of ³H-proline into hot TCA extractable protein, is directly proportional to the area of cell surface in contact with the substratum (Table 2). On 0.1 μm pore size filters, so few cell processes cross that the production of stroma by the epithelium is similar to the base line. The base line is defined as the amount of collagen and GAG produced in the same 24-hr period by epithelium grown on Nucleopore filter (0.8 μm pore size) without an underlying collagenous substratum.

TABLE 2. *The relation of pore size to exposed cell surface in transfilter stimulation of epithelial collagen synthesis by ECM*[a]

Pore size (μm)	Porosity (open space) (%)	Exposed surface area (μm²)	Collagen synthesis (cpm)
0.8	21.0 ± 1.4	8,922 ± 1,842	566 ± 42
0.4	17.1 ± 1.6	4,447 ± 1,122	452 ± 19
0.2	12.3 ± 1.4	1,436 ± 406	368 ± 24
0.1	8.5 ± 1.2	350 ± 108	312 ± 11

[a] Results are expressed per epithelium as the mean of five determinations plus or minus the standard deviation. Cultures were labeled for 24 hr with 5 μCi/ml proline-³H. See Meier and Hay (28) for further description.

Thus, the production of stroma by corneal epithelium *in vitro* is directly related to the area of transfilter cell contact with collagenous "inducer" ECM, and stroma production is not stimulated by freely diffusible molecules that cross filters. Neither conditioned medium nor serum is apparently involved (9). Hay and Meier (9) showed, moreover, that the corneal epithelium is not ingesting the collagenous substratum. Epithelium was grown for 24 hr transfilter (0.8 μm Nucleopore) to radioactive (³H-proline-labeled) collagen (NaOH-extracted lens capsule). Autoradiographs showed that no radioactivity appeared in the cell processes in the filter or in the overlying epithelium during the experiment (Figs. 20 and 21). Biochemical data confirmed the conclusion that no collagen from the substratum enters the epithelial cells (9). Thus, the interaction of collagen with corneal epithelium is most likely to be at the external cell surface.

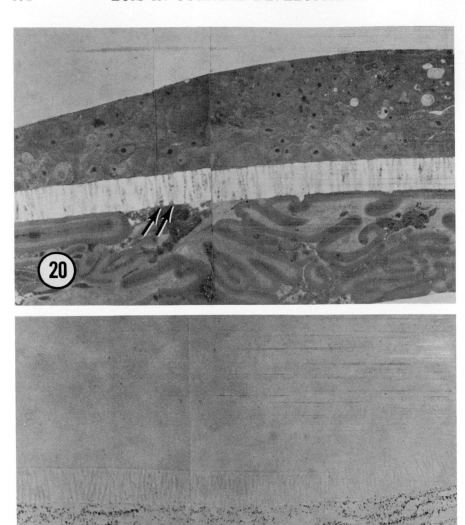

FIGS. 20 and 21. Light micrographs of sections of corneal epithelium grown transfilter to lens capsule on Nucleopore filter. Figure 20 is a stained section. Figure 21 in an autoradiograph of an unstained section. Cell processes are rarely transected their full length because they pass obliquely through the filter to contact the lens capsule **(arrows, Fig. 20).** In this experiment, the lens capsule was previously labeled with ^{3}H-proline. Radioactivity is retained in the lens capsule **(Fig. 21, bottom)** and is not taken up by overlying epithelium **(above).** × 450. (From ref. 9.)

AUTOCATALYTIC NATURE OF THE CELL-ECM INTERACTION

A further conclusion that can be derived from Nucleopore filter experiments is that the collagen-cell surface interaction is autocatalytic. Epithelium can be cultured transfilter (0.8 μm Nucleopore) to lens capsule for 24 hr and then transferred on the filter to another substratum. Epithelia transferred to plain rafts showed a 50% drop in synthetic activity compared to epithelia grown without interruption on lens capsule and a 30% drop compared to epithelia that were disrupted by remounting on new lens capsules, but in no case was the drop all the way back to the base-line level (9). In other words, the epithelium had benefitted in some way from transfilter contact with the lens capsule, and this benefit was transferred with it when it was moved to plain rafts.

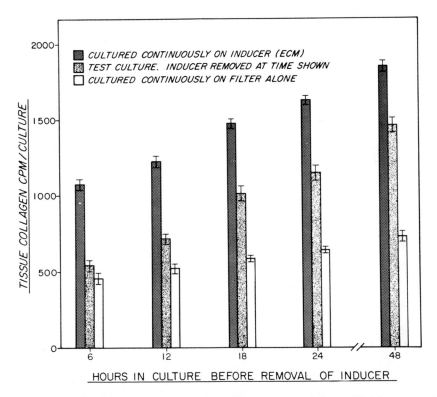

FIG. 22. Effect on epithelial collagen production of removing the transfilter lens capsule. Epithelium was cultured for 6, 12, 18, 24, or 48 hr transfilter to lens capsule on Nucleopore filter and then transferred with the filter to new lens capsule **(first bar of each set)** or to rayon raft **(second bar).** The base-line level **(third bar)** was determined on epithelium cultured for the same total period on Nucleopore filter alone. The test culture did increasingly better than the base-line culture with the increasing length of its initial exposure to "inducer." See Hay and Meier (9) for further description.

One of the obvious things that would be transferred with the epithelium and filter is the new ECM deposited by the epithelium in the initial 24 hr in culture (Fig. 15). In another experiment, epithelia were grown trans-filter to lens for shorter and for longer periods and then removed with their filters to plain rafts (Fig. 22). Epithelium grown transfilter to lens for only 6 hr did not produce significantly more collagen in the next 6-hr period than the base line (epithelium cultured on filter alone the whole period). With time in culture, however, epithelium grown on collagen is better and better able to continue to produce collagen after the underlying "inducer" is removed (Fig. 22). The most reasonable conclusion is that epithelial differentiation or productivity, if you will, is partly stabilized by the ECM the epithelium deposits on its own cell surface.

Such data as exist for chondrogenesis studied *in vitro* support a similar conclusion (3,13–17). Somites cultured in a confined space seem to produce enough ECM, thus sequestered, to enhance their own differentiation into cartilage, that is, to "self-differentiate" (3). Even with time, in culture corneal epithelium did not produce a visible stroma when grown on filter alone (2), but it is possible that conditions could be found that would enable the epithelium to "self-differentiate" without the crutch provided by the lens capsule or similar substratum.

CONCLUDING REMARKS

Experiments have been described here in which it was possible to measure *in vitro* the effect on corneal epithelial productivity of severing the epithelium's connection to its underlying ECM. The major conclusion that can be drawn is that a metabolically significant interaction takes place between the epithelial plasmalemma and the underlying ECM. The underlying ECM at 5 days of development consists of basal lamina and a stroma composed of several types of collagen, but any type of collagen is able to serve as an effective substratum, stimulating both collagen and GAG synthesis *in vitro* to levels that are probably comparable to those of the intact epithelium *in vivo*.

In addition to the effect of collagen on corneal epithelial differentiation, GAG of the kind made by the corneal epithelium has an enhancing effect on production of the same GAG by the epithelium. It is tempting to think that both collagen and GAG act at the level of the external surface of the cell membrane; experiments to date along these lines have shown that the collagen molecule seems to remain external to the developing cells with which it interacts, and it is predictable that GAG will be shown to have a similar location during chondrogenesis and corneal morphogenesis.

Assuming that a certain amount of speculation is desirable at a symposium such as this, let me state here the present condition of our working hypothesis. From the limited data that are available, it seems reasonable to believe that collagen and GAG do not interact with the plasmalemma in the same way or at the same site. Their end result is different in the cornea. The effect of either chondroitin or heparan sulfate is to stimulate synthesis of both chondroitin and heparan sulfates, the GAG normally made by the corneal epithelium. In a model of the cell membrane, we might place GAG in or on the external leaflet.

On the other hand, collagen of any type acts as an effective substratum for enhanced corneal epithelial production of both collagen and GAG. This nonspecificity of the "inducer" is reminiscent of platelet-collagen interaction, in which cell adhesion to collagen may be an important component (23). The small filaments described here which may be collagen seem to be attached to anionic sites on the membrane, and cytoplasmic filaments are adherent to the inner surface of the plasmalemma in the same region. A possible model is that collagen attaches to a membrane glycoprotein which serves to anchor cytoplasmic skeletal proteins that stabilize the secretory organelles of the cell. There is evidence that a plasma membrane glycoprotein might mediate adhesion of cells to collagen (30). Other models are possible, of course. Since collagen has no significant stretch of hydrophobic groups, it is unlikely to be a membrane protein (18). In the case of somite differentiation, it has been suggested that collagen and GAG act to lower intracellular cyclic AMP (13).

Further work is obviously needed to clarify the nature of cell-matrix interaction. It is tempting to think that cells in the adult remain intimately linked to the ECM in the manner described here. Yet it is quite clear that collagen and GAG synthesis are turned off or greatly reduced in adult corneal epithelium. The informational content of these macromolecules may be entirely a "permissive" one, as has been proposed for embryonic inducers in the past (4,10,12); their presence on the plasmalemma may be necessary for cell-matrix adhesion and active cell metabolism, but other factors probably determine which genes are read by any given cell at any given time.

REFERENCES

1. Dodson, J. W., and Hay, E. D. (1971): Secretion of collagenous stroma by isolated epithelium grown *in vitro. Exp. Cell Res.,* 65:215–220.
2. Dodson, J. W., and Hay, E. D. (1974): Secretion of collagen by corneal epithelium. II. Effect of the underlying substratum on secretion and polymerization of epithelial products. *J. Exp. Zool.,* 189:51–72.
3. Ellison, M. L., and Lash, J. W. (1971): Environmental enhancement of *in vitro* chondrogenesis. *Dev. Biol.,* 26:486–496.
4. Grobstein, C. (1967): Mechanisms of organogenetic tissue interaction. *Natl. Cancer Inst. Monogr.,* 26:279–299.
5. Hay, E. D. (1973): Origin and role of collagen in the embryo. *Am. Zool.,* 13:1085–1107.
6. Hay, E. D., and Dodson, J. W. (1973): Secretion of collagen by corneal epithelium. I. Morphology of the collagenous products produced by isolated epithelia grown on frozen-killed lens. *J. Cell Biol.,* 57:190–213.

7. Hay, E. D., Hasty, D. L., and Kiehnau, K. (1977): Fine structure of collagens and their relation to GAG. *First Munich Symposium on Biology of Connective Tissues* (*in press*).
8. Hay, E. D., and Meier, S. (1974): Glycosaminoglycan synthesis by embryonic inductors: Neural tube, notochord and lens. *J. Cell Biol.*, 62:889–898.
9. Hay, E. D., and Meier, S. (1976): Stimulation of corneal differentiation by interaction between cell surface and extracellular matrix. II. Further studies on the nature and site of transfilter "induction." *Dev. Biol.*, 52:141–157.
10. Hay, E. D., and Meier, S. (1977): Concept of embryonic induction; inductive mechanisms in orofacial morphogenesis. In: *Textbook of Oral Biology*, edited by J. Shaw and S. Meller. Saunders, Philadelphia (*in press*).
11. Hay, E. D., and Revel, J. P. (1969): Fine structure of the developing avian cornea. In: *Monographs in Developing Biology, Vol. 1*, edited by A. Wolsky and P. S. Chen, pp. 1–144. Karger, Basel.
12. Konigsberg, I. R., and Hauschka, S. D. (1965): Cell and tissue interactions in the reproduction of cell type. In: *Reproduction: Molecular, Subcellular and Cellular*, edited by M. Locke, pp. 243–290. Academic Press, New York.
13. Kosher, R. A. (1977): Inhibition of "spontaneous," notochord-induced, and collagen-induced *in vitro* somite chondrogenesis by cyclic AMP derivatives and theophylline. *Dev. Biol.*, 53:265–276.
14. Kosher, R. A., and Church, R. L. (1975): Stimulation of *in vitro* somite chondrogenesis by procollagen and collagen. *Nature*, 258:327–330.
15. Kosher, R. A., and Lash, J. W. (1975): Notochordal stimulation of *in vitro* somite chondrogenesis before and after enzymatic removal of perinotochordal material. *Dev. Biol.*, 42:362–378.
16. Kosher, R. A., Lash, J. W., and Miner, R. R. (1973): Environmental enhancement of *in vitro* chondrogenesis. IV. Stimulation of somite chondrogenesis by exogenous chondromucoprotein. *Dev. Biol.*, 35:210–220.
17. Lash, J. W. (1968): Somitic mesenchyme and its response to cartilage induction. In: *Epithelial-Mesenchymal Interactions*, edited by R. Fleischmajer & R. E. Billingham, pp. 165–172. Williams & Wilkins, Baltimore.
18. Lichtenstein, J. R., Bauer, E. A., Hoyt, R., and Wedner, H. J. (1976): Immunologic characterization of the membrane-bound collagen in normal human fibroblasts: Identification of a distinct membrane collagen. *J. Exp. Med.*, 144:145–154.
19. Linsenmayer, T. F., Smith, G. N., and Hay, E. D. (1977): *In vitro* synthesis of two collagen types by embryonic chick corneal epithelium. *Proc. Natl. Acad. Sci. U.S.A.*, 74:39–43.
20. Linsenmayer, T. F., Trelstad, R. L., and Gross, J. (1973): The collagen of chick embryo notochord. *Biochem. Biophys. Res. Commun.*, 53:39–45.
21. Low, F. N. (1968): Extracellular connective tissue fibrils in the chick embryo. *Anat. Rec.*, 160:93–108.
22. Luft, J. H. (1971): Ruthenium red and violet. II. Fine structural localization in animal tissues. *Anat. Rec.*, 171:369–415.
23. Marx, R. (ed.) (1977): *Collagen Platelet Interaction. First Munich Symposium on Biology of Connective Tissues* (*in press*).
24. Matukas, V. J., Panner, B. J., and Orbison, J. L. (1967): Studies on ultrastructural identification and distribution of protein-polysaccharide in cartilage matrix. *J. Cell Biol.*, 32:365–377.
25. Meier, S., and Hay, E. D. (1973): Synthesis of sulfated glycosaminoglycans by embryonic corneal epithelium. *Dev. Biol.*, 35:318–331.
26. Meier, S., and Hay, E. D. (1974): Control of corneal differentiation by extracellular materials. Collagen as a promoter and stabilizer of epithelial stroma production. *Dev. Biol.*, 38:249–270.
27. Meier, S., and Hay, E. D. (1974): Stimulation of extracellular matrix synthesis in the developing cornea by glycosaminoglycans. *Proc. Natl. Acad. Sci. U.S.A.*, 71:2310–2313.
28. Meier, S., and Hay, E. D. (1975): Stimulation of corneal differentiation by interaction between cell surface and extracellular matrix. I. Morphometric analysis of transfilter induction. *J. Cell Biol.*, 66:275–291.
29. Nevo, A., and Dorfman, A. (1972): Stimulation of chondromucoprotein synthesis in chondrocytes by extracellular chondromucoprotein. *Proc. Natl. Acad. Sci. U.S.A.*, 69:2069–2072.

30. Pearlstein, E. (1976): Plasma membrane glycoprotein which mediates adhesion of fibroblasts to collagen. *Nature,* 262:497–500.
31. Toole, B. P., Jackson, G., and Gross, J. (1972): Hyaluronate in morphogenesis: Inhibition of chondrogenesis *in vitro. Proc. Natl. Acad. Sci. U.S.A.,* 69:1384–1386.
32. Trelstad, R. L., Hayashi, K., and Toole, B. P. (1974): Epithelial collagens and glycosaminoglycans in the embryonic cornea. Macromolecular order and morphogenesis in the basement membrane. *J. Cell Biol.,* 62:815–830.
33. Wartiovaara, J., Nordling, S., Lehtonen, E., and Saxen, L. (1974): Transfilter induction of kidney tubules: Correlation with cytoplasmic penetration into Nucleopore filters. *J. Embryol. Exp. Morphol.,* 31:667–682.
34. Wight, T. N., and Ross, R. (1975): Proteoglycans in primate arteries. I. Ultrastructural localization and distribution in the intima. *J. Cell Biol.,* 67:660–674.

Cell and Tissue Interactions, edited by
J. W. Lash and M. M. Burger. Raven
Press, New York, 1977.

Developmental Roles of Hyaluronate and Chondroitin Sulfate Proteoglycans

Bryan P. Toole,*† Minoru Okayama,* Roslyn W. Orkin,*‡
M. Yoshimura,** M. Muto,** and A. Kaji**

*Developmental Biology Laboratory, Departments of Medicine and Anatomy,
Harvard Medical School at Massachusetts General Hospital, Boston, Massachusetts 02114, and **Department of Microbiology, School of Medicine, University of Pennsylvania, Philadelphia, Pennsylvania 19104*

Each stage in the morphogenesis and differentiation of a particular tissue type during embryogenesis or adult tissue remodeling is characterized not only by unique cell types but also by a unique extracellular matrix. The major macromolecular components of these matrices appear to be important both structurally and as environmental signals influencing cell behavior. We wish to discuss briefly examples of these two types of roles for hyaluronate and chondroitin sulfate proteoglycans in several developing situations. In doing this, we will introduce four major relationships: (a) hyaluronate synthesis with mesenchymal cell migration; (b) hyaluronate removal with cell differentiation; (c) proteoglycan-collagen interactions with structural matrix deposition; and (d) lipids with cell-associated hyaluronate.

CORRELATION OF HYALURONATE SYNTHESIS AND HYALURONIDASE ACTIVITY WITH MORPHOGENETIC EVENTS *IN VIVO*

The observation that hyaluronate is relatively enriched in embryonic or young tissues in comparison to adult tissues has been made by many investigators in the past. However, these studies have usually been addressed to general maturational or aging trends in tissue structure, e.g., in water content, rather than to morphogenetic sequences. By studying hyaluronate metabolism at discrete stages in early development, at times and in tissues where striking morphogenetic cell movements and differentiative changes are taking place, close correlations between hyaluronate synthesis and cell migration and between hyaluronidase activity and cell

---

† Established Investigator, American Heart Association.
‡ Fellow, Arthritis Foundation.

differentiation have now been revealed. These studies have been reviewed in detail recently (48), but a summary of some of them is given below.

A striking example of the correlation between cell migration and hyaluronate synthesis comes from our studies of chick embryo corneal development (54). The fibroblasts of the mature corneal stroma derive from the mesenchymal shelf at either edge of the acellular, epithelium-derived, corneal stroma and migrate into this stroma from the end of stage 27 of development (about $5\frac{1}{2}$ days of incubation) until stage 35, about 3 days later. The beginning of migration coincides with a marked swelling of the stroma, due to increased hydration, and its cessation coincides with deswelling (8) (see Fig. 1a). As shown in Fig. 1b, the major glycosaminoglycan component being synthesized by the cornea during this migratory stage is hyaluronate. The appearance of detectable hyaluronidase activity in the corneal tissue occurs when the cells have ceased their migration into the stroma and while the tissue dehydrates under the control of thyroid hormone to become transparent (5). At this stage, the mesenchymal cells differentiate to corneal fibroblasts (8), producing large amounts of collagen (55) and sulfated glycosaminoglycan (Fig. 1b). The change in major polysaccharide from hyaluronate early in development of the cornea to sulfated glycosaminoglycan in the mature stroma has been confirmed chemically (31).

It should be emphasized at this point that the onset of increased hydration of the corneal stroma correlates quite precisely with the onset of cell migration, but the onset of hyaluronate synthesis occurs earlier (Fig. 1). Hyaluronate has been shown to be responsible for high degrees of hydration in several other situations (6,19,42). The extent of this tissue hydration, however, is highly dependent on the negative charge of the glucuronate moieties which comprise approximately 50% of hyaluronate, a high molecular weight polymer of the disaccharide N-acetylglucosamine-glucuronate. In aqueous solutions, mutual repulsion between the charged carboxyl groups of the glucuronate moieties causes each molecule to occupy a large domain. The overlapping of these domains and the interaction between hyaluronate chains give rise to a molecular meshwork in solution, which at tissue concentrations can "trap" or restrict the free flow of water and exert an osmotic pressure (6,13,25,26). If, however, the charge of the carboxyl groups is blocked by a strongly interacting counterion, e.g., Ca^{++}, or a positively charged peptide, the molecular domain of the hyaluronate molecule will collapse to a much smaller size (10,34,35). It seems likely then that the sudden increase in hydration of the corneal stroma at stage 28 of development might be triggered by a change in configuration of the already present hyaluronate molecules from a collapsed to an extended form, e.g., in response to a change in Ca^{++} distribution. This would in turn result in separation of the endothelium and epithelium cell layers and also of the collagen layers within the stroma, thus facilitating migration of the outlying mesenchymal cells into the stroma.

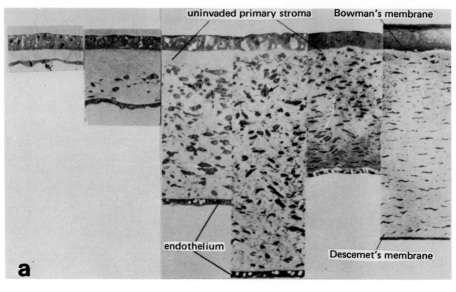

uninvaded primary stroma   Bowman's membrane

endothelium   Descemet's membrane

a

stage: **27**   **28**   **30**   **35**   **40**   **H**

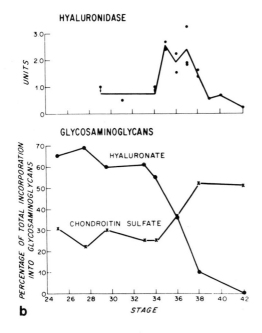

HYALURONIDASE

UNITS

3.0

2.0

1.0

0

GLYCOSAMINOGLYCANS

HYALURONATE

PERCENTAGE OF TOTAL INCORPORATION INTO GLYCOSAMINOGLYCANS

70

60

50

40

30

20

10

0

CHONDROITIN SULFATE

24 26 28 30 32 34 36 38 40 42

STAGE

b

FIG. 1. Comparison of morphology **(a)** and glycosaminoglycan metabolism **(b)** in chick embryo cornea. Hyaluronate synthesis is predominant during stromal swelling and the migration of mesenchymal cells into the stroma (stages 27 to 35). Hyaluronidase activity appears at the end of the migratory stage, concomitant with stromal deswelling (stages 35 to 40). (From ref. 48.)

Concurrence of hyaluronate synthesis with cell migration, and the very close correlation of increased extracellular matrix volume with the onset of migration of chick neural crest cells (30) and of primary mesenchyme at gastrulation in the chick (37) suggest that this phenomenon is widespread during embryogenesis.

Hyaluronate is also synthesized during the migration and proliferation of blastemal cells in newt limb regeneration (49), of chick embryo sclerotomal cells around the notochord (45), and of chick embryo limb mesoderm (45). In these three systems, a second correlation emerges, *viz.* between hyaluronate removal and cell differentiation. In these systems, as well as in the cornea (Fig. 1), hyaluronidase activity increases in the tissues at the end of the stage of active cell migration or proliferation. These observations have led us to propose that hyaluronate might suppress differentiation until the appropriate time of development of the particular tissue. Supporting this possibility is the finding that very small amounts of hyaluronate, added to high-density cultures of stage 26 chick embryo somite cells, inhibit the formation of cartilage-like nodules in these cultures (50). Hyaluronate also causes a reduction, by approximately 50%, in chondroitin sulfate-proteoglycan synthesis by these cells (46). A similar inhibition of proteoglycan synthesis by mature chondrocytes has also been described by other workers (21,38,57). In addition, we have obtained preliminary evidence that removal of hyaluronate, by hyaluronidase treatment, from the surface of nonfusing Rous sarcoma virus-transformed myoblasts allows many of the cells to fuse and differentiate (29).

Thus, our present working hypothesis is that a hyaluronate-rich extracellular matrix provides a beneficial milieu for mesenchymal cell migration and proliferation, and also prevents precocious differentiation.

PROTEOGLYCAN-COLLAGEN INTERACTION IN LIMB DEVELOPMENT

At the time of transition from the early morphogenetic phase to the differentiative phase in chick embryo limb development, there is not only an increase in hyaluronidase activity but also increased synthesis of chondroitin sulfate proteoglycan and collagen (Fig. 2). Moreover, the major proteoglycan and the major collagen synthesized by the differentiated cartilage cells of the limb are different qualitatively to the major proteoglycan and collagen of the stage 24 limb mesoderm (7,16,17). It has been proposed by several investigators (14,18,23,24,52) that the interaction of cartilage proteoglycan and collagen may be an important controlling event in cartilage matrix deposition. Thus, we have asked the question whether there is a transition in the developing limb from a noninteracting population to an interacting population of collagen and proteoglycan molecules at the time that structural cartilage matrix deposition begins.

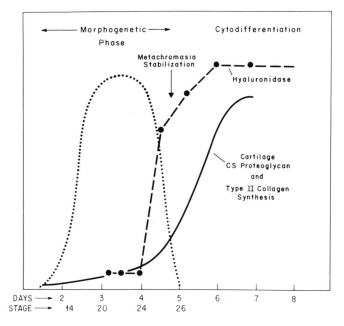

FIG. 2. Correlation of hyaluronidase activity in the developing chick embryo limb with the phase of cytodifferentiation in the scheme of Zwilling (58). (From ref. 46.)

Chondroitin sulfate proteoglycans from cartilages and dermatan sulfate proteoglycans from heart valves, skin, and tendons have been shown to react with soluble collagen *in vitro* at physiological ionic strength and pH to form insoluble aggregates that can be separated by centrifugation from remaining soluble material (52,53). An electron micrograph of such aggregates is shown in Fig. 3. They have been shown to contain both collagen and proteoglycan (Table 1) and can be formed with soluble collagens Type I, II, or III. Aggregate formation is dependent on both the protein and chondroitin sulfate portions of the proteoglycan (47). Newly formed fibrils precipitated *in vitro* by incubating at 37° in the presence of proteoglycan are more stable than in the absence of proteoglycan (44). *In vivo* the association of proteoglycan with collagen can be visualized in the electron microscope (Fig. 4) (22,36,55).

To study this phenomenon biochemically with embryonic tissue, we have devised a radioactive assay based on the ability of reactive proteoglycan to bind to and cause the precipitation of collagen as insoluble aggregates. In this assay (Fig. 5) (47), radioactively labeled proteoglycan is mixed with unlabeled Type I collagen in solution at ionic strength 0.16, pH 7.5, and 4°. The resultant fine precipitate is centrifuged down and washed. The precipitate and the combined supernatant plus washings are then analyzed separately for glycosaminoglycan content, thus giving a balance sheet of

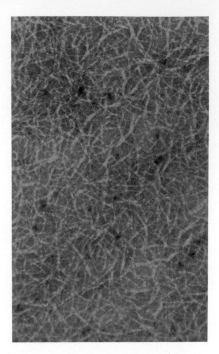

FIG. 3. Electron micrograph of aggregates formed by the interaction of cartilage chondroitin sulfate protein and Type I collagen *in vitro*. × 50,000.

reactive (in the precipitate) and nonreactive (in the supernatant plus washings) proteoglycan.

In order to examine proteoglycan reactivity in limb chondrogenesis, radioactively labeled proteoglycans were prepared from 4-day embryonic limb mesoderm and 8-day limb cartilage. The tissues were incubated with ^{35}S-sulfate, and then the labeled proteoglycans extracted with 4 M guanidine HCl and fractionated by the sucrose density gradient technique of Kimata et al. (11). Individual components from the gradients were then analyzed for their reactivity in the radioactive proteoglycan-collagen precipitation assay (Fig. 5).

TABLE 1. *Composition of proteoglycan-collagen aggregates[a]*

Proteoglycan preparation	Percent total collagen in aggregates	Percent total proteoglycan in aggregates	Ratio of collagen to proteoglycan in aggregates
Chondroitin sulfate proteoglycan	65	96	4.1
Dermatan sulfate proteoglycan	55	70	2.6

[a] The preparations used were Type I collagen from bovine fetal skin (1.2 mg/ml), chondroitin sulfate proteoglycan from bovine nasal cartilage (200 μg/ml), and dermatan sulfate proteoglycan from bovine heart valves (360 μg/ml). The results were taken from Toole and Lowther (52,53).

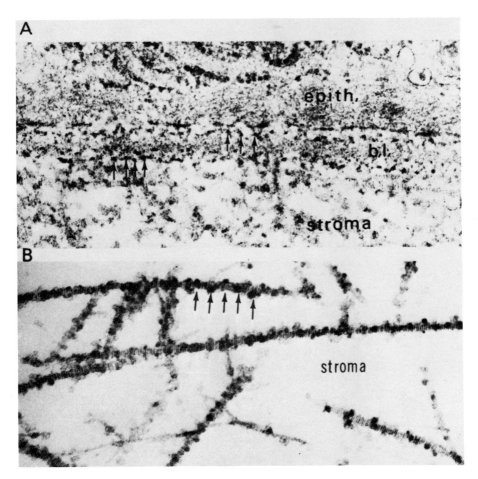

FIG. 4. Relationship of proteoglycan to basement membrane **(a)** (× 85,000 before reduction) and stromal **(b)** (× 52,000 before reduction) collagens in chick embryo cornea. In the basement membrane, ruthenium-red-stained chondroitin sulfate proteoglycan **(arrows)** is arranged in a regular pattern on either side of the electron-dense basal lamina **(bl)** (55). Proteoglycan is also associated with the stromal fibrils in a specific repeated location. (From ref. 48.)

Sucrose density gradient fractionation of 8-day limb cartilage proteoglycans gave rise to two peaks, the major one of which centrifuged to the bottom of the gradient and comprised greater than 90% of the total radioactively labeled glycosaminoglycan (Fig. 6). Ninety-eight percent of this major peak was reactive with collagen when tested in the radioactive precipitation assay, whereas only 39% of the minor, floating peak was reactive. A similar result was obtained with chondroitin sulfate-proteoglycan extracted from embryonic sternal cartilage. The proteoglycans of 4-day limb mesoderm, however, gave a different pattern in the sucrose density gradients

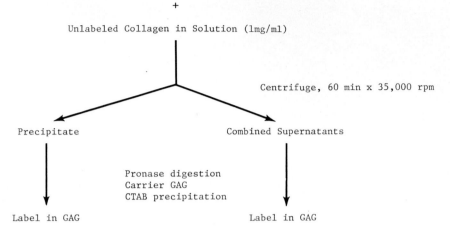

FIG. 5. Radioactive binding assay for reactivity of embryonic proteoglycans with collagen. **GAG,** glycosaminoglycan; **CTAB,** cetyl trimethylammonium bromide. (From ref. 51.)

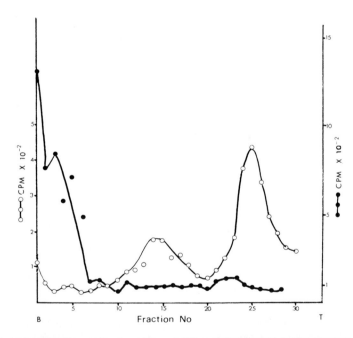

FIG. 6. Sucrose density gradient profiles of ^{35}S-sulfate-labeled proteoglycans extracted from 4-day chick embryo limb buds (———O———O———) and 8-day limb cartilages (———●———●———). **B,** bottom of gradient; **T,** top of gradient. The sedimenting peak at the bottom of the gradient for 8-day cartilage was fully reactive in the radioactive proteoglycan-collagen precipitation assay (Fig. 5), whereas the top peak for 4-day limb buds was only 20% reactive and the middle peak 55% reactive.

yielding two major floating peaks with virtually no proteoglycan at the bottom of the gradient (Fig. 6). This distribution of proteoglycan was similar to that obtained previously by Okayama and Holtzer (27). The smaller molecular weight component, from the top of the gradient, was virtually nonreactive with collagen in the radioactive precipitation assay. Approximately 50% of the second fraction, from the middle of the gradient, reacted with collagen, indicating that this is probably a mixture of non-reactive and reactive material. It is clear, however, that the major proportion of mature cartilage proteoglycan is reactive with collagen, whereas that in the 4-day limb mesoderm is mainly nonreactive. Normal and bromo-deoxyuridine-treated chondrocytes in culture were also compared. In each case, two ^{35}S-sulfate-labeled peaks were obtained in the sucrose gradients, one floating to the top and one sedimenting to the bottom. For the normal chondrocytes, the major peak (83% of the total proteoglycan) was at the bottom and was fully reactive with collagen. For the treated cells, the major peak (56% of the total proteoglycan) was at the top and was nonreactive.

We conclude from the above experiments that chondroitin sulfate proteo-glycans formed during the two phases of limb development (Fig. 2) are not only chemically different but have different functional properties. These properties suggest entirely different roles for each proteoglycan, which would be related specifically to the stage of tissue formation in which they are synthesized.

ASSOCIATION OF CELL-BOUND HYALURONATE WITH LIPID

We have already discussed the probable importance of the hydrodynamic properties of hyaluronate in the hydration of extracellular matrix, physical separation of cell and collagen layers, and migration of mesenchymal cells. The possible relationship of hyaluronate to suppression of differentiation, however, infers a direct action at the cell surface. Thus, we have sought evidence to support this postulate.

Many people have obtained indirect evidence that a proportion of the hyaluronate synthesized by fibroblasts in culture is associated with the cell surface (2–4,12,20,33,43). The total amount of hyaluronate synthesized and the proportion that is associated with the cell layer in culture varies greatly with growth rate, cell density, and viral transformation.

Recently Okayama et al. (28) described the increased synthesis of hya-luronate by chondrocytes in response to transformation by Rous sarcoma virus, and the association of a significant proportion of this hyaluronate with the cell layer (Table 2). The morphology of the transformed cells was also altered as a result of viral infection and this alteration is shown in Fig. 7. The morphological and biochemical changes were reversible if a tempera-ture-sensitive mutant of the viral strain was used for infection of the cells,

TABLE 2. *Increased hyaluronate synthesis in virus-transformed chondrocytes*

	Normal		Transformed	
	Cell layer	Medium	Cell layer	Medium
cpm in hyaluronate[a]				
per μg DNA	1,745	3,030	9,955	60,014
Percent of total glycosaminoglycan	5.5	7.7	51	76

[a] ¹⁴C-acetate-labeled glycosaminoglycans were treated with *Streptomyces* hyaluronidase and the products were analyzed by paper chromatography (32). Results were taken from Okayama et al. (28).

and the temperature raised to the nonpermissive temperature of 41° after demonstration of transformation at the permissive temperature of 36°.

We have initiated an investigation of (a) the differences that may exist in the chemical nature of the hyaluronate associated with the Rous-sarcoma-virus-transformed cell compared with that secreted into the medium and, (b) the possible existence of lipid-bound oligosaccharide intermediates in the biosynthesis of polymeric hyaluronate in this system. The basis for this investigation comes from the recent discovery of lipid-bound intermediates

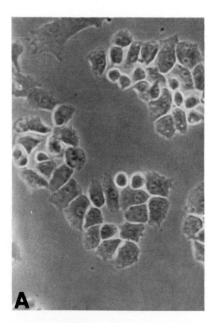 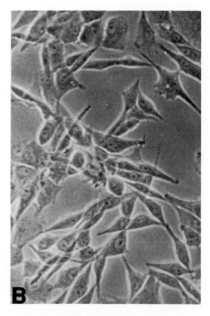

FIG. 7. Morphological change in chondrocytes transformed with Rous sarcoma virus. **a:** Normal chondrocytes from 11-day chick embryo vertebrae. **b:** Chondrocytes infected with wild-type Rous sarcoma virus.

in glycoprotein biosynthesis. In this pathway, the oligosaccharide moiety of the glycoprotein is assembled while attached to a phosphorylated poly-isoprenol acceptor. The oligosaccharide is then transferred *en bloc* from the lipid to the protein core of the glycoprotein being synthesized (see refs. 9,15).

To search for these potential forms of hyaluronate and its precursors, we have administered radioactively labeled precursors of hyaluronate, *viz.* ^{14}C-acetate and ^{14}C-glucose, to the Rous-sarcoma-virus-transformed chondrocyte, separated the cell layer and medium, and then dialyzed, lyophilized, and extracted each separately. Sequential extraction with chloroform: methanol (2:1) and chloroform:methanol:water (10:10:3) (containing 0.1 M ammonium acetate, pH 5.5) as described by Struck and Lennarz (40) followed by extraction with 4 M guanidine HCl, 0.05 M acetate, pH 5.5, was performed. Chloroform:methanol:water would be a suitable solvent for the extraction of lipid-bound forms of hyaluronate-related oligosaccharides. Whether free or bound to lipid or peptide, 4 M guanidine HCl might be expected to solubilize hyaluronate polymer. Aliquots of these extracts after concentration were then applied to paper and chromatographed in iso-butyric acid: 0.5 M NH_3 (5:3), in which solvent polyisoprenol-carbohydrate derivatives migrate to the front (1) and oligosaccharides or disaccharides resulting from digestion of hyaluronate with *Streptomyces* hyaluronidase or bacterial chondroitinases are separated (32).

As shown in Figs. 8 and 9, the labeled materials in the 4 M guanidine extracts of both cell layer and medium were confined mainly to the origin, as would be expected for macromolecular hyaluronate or proteoglycan.

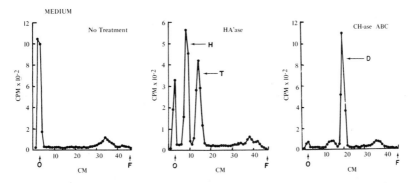

FIG. 8. Paper chromatography of ^{14}C-glucose-labeled materials in 4 M guanidine HCl extracts of lyophilized media from cultures of Rous-sarcoma-virus-transformed chondrocytes. An aliquot of untreated extract is compared with aliquots treated with *Streptomyces* hyaluronidase **(HA'ase)** and chondroitinase ABC **(CH-ase)**. *Streptomyces* hyaluronidase is specific for hyaluronate and gave rise to the characteristic unsaturated hexasaccharide **(H)** and tetrasaccharide **(T)** breakdown products of hyaluronate. Chondroitinase ABC gave rise to the unsaturated disaccharide **(D)** breakdown product of hyaluronate. **O**, origin; **F**, front.

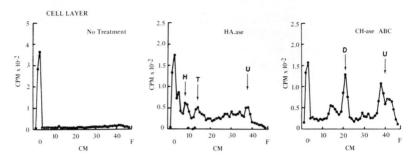

FIG. 9. Paper chromatography of ¹⁴C-glucose-labeled materials in 4 M guanidine HCl extracts of lyophilized cell layers from cultures of Rous-sarcoma-virus-transformed chondrocytes. An aliquot of untreated extract is compared with aliquots treated with *Streptomyces* hyaluronidase **(HA'ase)** and chondroitinase ABC **(CH-ase).** *Streptomyces* hyaluronidase gave rise to the unsaturated hexasaccharide **(H)** and tetrasaccharide **(T)** breakdown products of hyaluronate, but in addition yielded uncharacterized lipid-like material **(U).** Chondroitinase ABC gave rise to the unsaturated disaccharide **(D)** breakdown product of hyaluronate as well as uncharacterized lipid-like material **(U).** **O,** origin; **F,** front.

After digestion with *streptomyces* hyaluronidase or chondroitinase ABC, the unsaturated oligosaccharide or disaccharide breakdown products of hyaluronate migrated to the expected positions in the chromatogram. In the case of the medium, these breakdown products were the only significant migrating products (Fig. 8). In the case of the cell-associated material, however, lipid-like products appeared, which migrated to the front of the chromatogram (Fig. 9). This material was produced after digestion with either of the two enzymes. The simplest explanation of these results is that the lipid-like material is a lipid-oligosaccharide complex that has been released from the bulk of the hyaluronate polymer as a result of enzyme digestion. Prior to enzyme digestion, the attachment to polymeric hyaluronate would prevent its migration in the isobutyric acid/NH₃ solvent. Since a significant amount of labeled material, to which some of the hyaluronate could also have been attached, remained at the origin after enzyme digestion (Fig. 9), it is not yet possible to estimate the proportion of hyaluronate that was bound to the lipid-like material. Further characterization of the lipid-like product of hyaluronidase digestion and especially of residual-attached carbohydrate will clarify this issue.

When similar experiments to the above were performed with the chloroform:methanol:water extracts of the cell layer, the major component chromatographed with an R_f value of 0.26. After treatment with chondroitinase ABC, this component was dramatically reduced, and a new component appeared with an R_f value of 0.48. No such component was present in the extract of the lyophilized medium. In the case of the cell layer, a small amount of labeled material with the R_f value of 0.29 of the saturated disaccharide of hyaluronate was also produced by chondroitinase treatment.

A reasonable interpretation of these results would be that the major labeled component of the chloroform:methanol:water extract of the cell layer was a lipid-linked oligosaccharide terminating in glucuronate-N-acetylglucosamine-glucuronate at the nonreducing end, a likely precursor of lipid-bound polymeric hyaluronate. Again, complete characterization of this component as well as demonstration of a precursor-product relationship between the putative oligosaccharide and polymeric forms will be necessary to consolidate this claim.

Other important questions remaining in relation to these possible lipid-associated components are: (a) whether they are associated with the outer cell surface membrane; (b) whether they occur in nontransformed cells; (c) whether they are precursors of the secreted form of hyaluronate, which does not appear to be associated with lipid; and (d) whether there is transfer of lipid-bound polymeric hyaluronate to a peptide or protein core. In relation to the final question, we have found that hyaluronate synthesis by the Rous-sarcoma-virus-transformed chondrocytes is partially inhibited by treatment of the cells with puromycin as would be expected if transfer to peptide core occurs. In the case of hyaluronate synthesis by streptococci, however, puromycin has little effect (39), thus suggesting that the mechanism of synthesis in the two systems might be different. The difficulty encountered in detecting peptide material bound to hyaluronate extracted from connective tissue matrices (41,56) could be due either to an extremely small ratio of peptide to hyaluronate or to proteolytic scission of most of the peptide from the hyaluronate. If a lipid-bound form of hyaluronate existed as a structural entity at the cell surface, it would be expected to exert a considerable influence on the physical properties and the charge characteristics of the plasma membrane. This influence could conceivably be of importance in the regulation of cell behavior, e.g., in modifying or blocking cell recognition interactions involving cell surface glycoproteins. Alternatively, lipid-linked forms of hyaluronate might simply be transistory intermediates in the synthesis of the secreted form of hyaluronate.

In summary, the synthesis and degradation of hyaluronate and proteoglycans are stage specific in many developmental sequences. This specificity seems to relate in a functional manner to the morphogenetic events and the building of tissue structure that coincide with synthesis of the particular extracellular components.

ACKNOWLEDGMENTS

We wish to thank Geraldine Jackson, Patrick Cunningham, and Scott Luria for technical assistance, Dr. Robert Trelstad for assistance with electron microscopy, and Dr. Jerome Gross for his continued support. The projects described have been funded by American Heart Association Grant 73757 and USPHS Grants AM3564 CA19497, and DE04220. This is publication No. 727 of the Robert W. Lovett Memorial Group for the Study of Diseases Causing Deformities and contribution No. 5 of the series *Cell Transformation and Differentiation* from the laboratory of A. K.

REFERENCES

1. Anderson, J. S., Matsuhashi, M., Haskin, M. A., and Strominger, J. L. (1967): Biosynthesis of the peptidoglycan of bacterial cell walls. II. Phospholipid carriers in the reaction sequence. *J. Biol. Chem.,* 242:3180–3190.
2. Burger, M. M., and Martin, G. S. (1972): Agglutination of cells transformed by Rous Sarcoma virus by wheat germ agglutinin and concanavalin A. *Nature [New Biol.],* 237:9–12.
3. Clarris, B. J., and Fraser, J. R. (1968): On the pericellular zone of some mammalian cells *in vitro. Exp. Cell Res.,* 49:181–193.
4. Cohn, R. H., Cassiman, J. J., and Bernfield, M. R. (1976): Relationship of transformation, cell density and growth control to the cellular distribution of newly synthesized glycosaminoglycan. *J. Cell Biol.,* 71:280–294.
5. Coulombre, A. J., and Coulombre, J. L. (1964): Corneal development. III. The role of the thyroid in dehydration and the development of transparency. *Exp. Eye Res.,* 3:105–114.
6. Fessler, J. (1960): A structural function of mucopolysaccharide in connective tissue. *Biochem. J.,* 76:124–132.
7. Goetinck, P. F., Pennypacker, J. P., and Royal, P. D. (1974): Proteochondroitin sulfate synthesis and chondrogenic expression. *Exp. Cell Res.,* 87:241–248.
8. Hay, E. D., and Revel, J. P. (1969): Fine structure of the developing avian cornea. In: *Monographs in Developmental Biology, Vol. I,* edited by A. Wolsky, and P. S. Chen. Karger, Basel.
9. Hughes, R. C. (1976): *Membrane Glycoproteins.* Butterworths, London.
10. Katchalsky, A. (1964): Polyelectrolytes and their biological interactions. *Biophys. J.,* 4 (Suppl.): 9–41.
11. Kimata, K., Okayama, M., Oohira, A., and Suzuki, S. (1974): Heterogeneity of proteochondroitin sulfates produced by chondrocytes at different stages of cytodifferentiation. *J. Biol. Chem.,* 249:1646–1653.
12. Kojima, K., and Maekawa, A. (1972): A difference in the architecture of surface membrane between two cell types of rat ascites hepatomas. *Cancer Res.,* 32:847–852.
13. Laurent, T. C. (1966): *In vitro* studies on the transport of macromolecules through the connective tissue. *Fed. Proc.,* 25:1128–1134.
14. Lee-Own, V., and Anderson, J. C. (1976): Interaction between proteoglycan subunit and Type II collagen from bovine nasal cartilage, and the preferential binding of proteoglycan subunit to Type I collagen. *Biochem. J.,* 153:259–264.
15. Lennarz, W. J. (1975): Lipid-linked sugars in glycoprotein synthesis. *Science,* 188:986–991.
16. Levitt, D., and Dorfman, A. (1974): Concepts and mechanisms of cartilage differentiation. In: *Current Topics in Developmental Biology, Vol. 8,* edited by A. Moscona, pp. 103–149. Academic Press, New York.
17. Linsenmayer, T. F., Toole, B. P., and Trelstad, R. L. (1973): Temporal and spatial transitions in collagen types during embryonic chick limb development. *Dev. Biol.,* 35:232–239.
18. Mathews, M. B. (1965): The interaction of collagen and acid mucopolysaccharides. A model for connective tissue. *Biochem. J.,* 96:710–716.
19. McCabe, M. (1972): The diffusion coefficient of caffeine through agar gels containing a hyaluronic acid-protein complex. A model system for the study of the permeability of connective tissues. *Biochem. J.* 127:249–253.
20. Morris, C. C. (1960): Quantitative studies on the production of acid mucopolysaccharides by replicate cell cultures of rat fibroblasts. *Ann. N.Y. Acad. Sci.,* 86:878–894.
21. Muir, H. (1976): Structure and function of proteoglycans of cartilage and cell-matrix interactions. In: *Cell and Tissue Interactions,* edited by J. Lash and M. M. Burger. (*This volume.*)
22. Nakao, K., and Bashey, R. I. (1972): Fine structure of collagen fibrils as revealed by ruthenium red. *Exp. Mol. Pathol.,* 17:6–13.

23. Obrink, B. (1973): The influence of glycosaminoglycans on the formation of fibers from monomeric tropocollagen *in vitro*. *Eur. J. Biochem.*, 34:129–137.
24. Oegema, T. R., Laidlaw, J., Hascall, V. C., and Djiewiatkowski, D. D. (1975): The effect of proteoglycans on the formation of fibrils from collagen solutions. *Arch. Biochem. Biophys.*, 170:698–709.
25. Ogston, A. G. (1966): On water binding. *Fed. Proc.*, 25:986–989.
26. Ogston, A. G. (1970): The biological functions of the glycosaminoglycans. In: *Chemistry and Molecular Biology of the Intercellular Matrix, Vol. 3*, edited by E. A. Balazs, pp. 1231–1240. Academic Press, New York.
27. Okayama, M., and Holtzer, H. (1976): Differences among sulfated proteoglycans synthesized in non-chondrogenic cells, presumptive chondroblasts and definitive chondroblast. *J. Cell Biol.*, 70:382a.
28. Okayama, M., Yoshimura, M., Muto, M., Chi, J., Roth, S., and Kaji, A. (1977): Transformation of chicken chondrocytes by Rous Sarcoma virus. *Cancer Res.*, 37:712–717.
29. Okayama, M., Yoshimura, M., Muto, M., Huang, J., and Kaji, A. (1976): *Unpublished results*.
30. Pratt, R. M., Larsen, M. A., and Johnston, M. C. (1975): Migration of cranial neural crest cells in a cell-free hyaluronate-rich matrix. *Dev. Biol.*, 44:298–305.
31. Praus, R., and Brettschneider, I. (1971): Glycosaminoglycans in the developing chicken cornea. *Ophthalmol. Res.*, 2:367–373.
32. Saito, H., Yamagata, T., and Suzuki, S. (1968): Enzymatic methods for the determination of small quantities of isomeric chondroitin sulfates. *J. Biol. Chem.*, 243:1536–1542.
33. Satoh, C., Duff, R., Raff, F., and Davidson, E. A. (1973): Production of mucopolysaccharides by normal and transformed cells. *Proc. Natl. Acad. Sci. U.S.A.*, 70:54–56.
34. Schubert, M., and Hamerman, D. (1968): *A Primer on Connective Tissue Biochemistry*. Lea & Febiger, Philadelphia.
35. Scott, J. E. (1968): Ion binding in solutions containing acid mucopolysaccharides. In: *The Chemical Physiology of Mucopolysaccharides*, edited by G. Quintarelli, pp. 171–187. Little, Brown, Boston.
36. Serafini-Fracassini, A., Wells, P. J., and Smith, J. W. (1970): Studies on the interactions between glycosaminoglycans and fibrillar collagen. In: *Chemistry and Molecular Biology of the Intercellular Matrix, Vol. 2*, edited by E. A. Balazs, pp. 1201–1215. Academic Press, New York.
37. Solursh, M. (1976): Glycosaminoglycan synthesis in the chick gastrula. *Dev. Biol.*, 50:525–530.
38. Solursh, M., Vaerewych, S. A., and Reiter, R. S. (1974): Depression by hyaluronic acid of glycosaminoglycan synthesis by cultured chick embryo chondrocytes. *Dev. Biol.* 41:233–244.
39. Stoolmiller, A. C., and Dorfman, A. (1969): The biosynthesis of hyaluronic acid by streptococcus. *J. Biol. Chem.*, 244:236–246.
40. Struck, D. K., and Lennarz, W. (1976): Utilization of exogenous GDP-mannose for the synthesis of mannose-containing lipids and glycoproteins by oviduct cells. *J. Biol. Chem.*, 251:2511–2519.
41. Swann, D. A. (1968): Studies on hyaluronic acid. II. The protein component(s) of rooster comb hyaluronic acid. *Biochim. Biophys. Acta*, 160:96–105.
42. Szirmai, J. A. (1966): Effect of steroid hormones on the glycosaminoglycans of target connective tissue. In: *The Amino Sugars, Vol. 2B*, edited by R. W. Jeanloz, and E. A. Balazs, pp. 129–154. Academic Press, New York.
43. Terry, A. H., and Culp, L. A. (1974): Substrate-attached glycoproteins from normal and virus-transformed cells. *Biochemistry*, 13:414–425.
44. Toole, B. P. (1969): Solubility of collagen fibrils formed *in vitro* in the presence of sulphated acid mucopolysaccharide-protein. *Nature*, 222:872–873.
45. Toole, B. P. (1972): Hyaluronate turnover during chondrogenesis in the developing chick limb and axial skeleton. *Dev. Biol.*, 29:321–329.
46. Toole, B. P. (1973): Hyaluronate and hyaluronidase in morphogenesis and differentiation. *Am. Zool.*, 13:1061–1065.
47. Toole, B. P. (1975): Binding and precipitation of soluble collagens by chick embryo cartilage proteoglycan. *J. Biol. Chem.*, 251:895–897.

48. Toole, B. P. (1976): Morphogenetic role of glycosaminoglycans in brain and other tissues. In: *Neuronal Recognition,* edited by S. H. Barondes, pp. 275–329. Plenum Press, New York.

49. Toole, B. P., and Gross, J. (1971): The extracellular matrix of the regenerating newt limb: Synthesis and removal of hyaluronate prior to differentiation. *Dev. Biol.,* 25:57–77.

50. Toole, B. P., Jackson, G., and Gross, J. (1972): Hyaluronate in morphogenesis: Inhibition of chondrogenesis *in vitro. Proc. Natl. Acad. Sci. U.S.A.,* 69:1384–1386.

51. Toole, B. P., and Linsenmayer, T. F. (1975): Proteoglycan-collagen interaction: Possible developmental significance. In: *Extracellular Matrix Influences on Gene Expression,* edited by H. C. Slavkin, and R. C. Greulich, pp. 341–346. Academic Press, New York.

52. Toole, B. P., and Lowther, D. A. (1968): The effect of chondroitin sulfate protein on the formation of collagen fibrils *in vitro. Biochem. J.,* 109:857–866.

53. Toole, B. P., and Lowther, D. A. (1968): Dermatan sulfate-protein: Isolation from and interaction with collagen. *Arch. Biochem. Biophys.,* 128:567–578.

54. Toole, B. P., and Trelstad, R. L. (1971): Hyaluronate production and removal during corneal development in the chick. *Dev. Biol.,* 26:28–35.

55. Trelstad, R. L., Hayashi, K., and Toole, B. P. (1974): Epithelial collagens and glycosaminoglycans in the embryonic cornea. Macromolecular order and morphogenesis in the basement membrane. *J. Cell Biol.,* 62:815–830.

56. Varma, R., Varma, R. S., Allen, W. S., and Wardi, A. H. (1974): On the carbohydrate-protein linkage group in vitreous humor hyaluronate. *Biochim. Biophys. Acta,* 362:584–588.

57. Wiebkin, O. W., and Muir, H. (1973): The inhibition of sulphate incorporation in isolated adult chondrocytes by hyaluronic acid. *FEBS Lett.,* 37:42–46.

58. Zwilling, E. (1968): Morphogenetic phases in development. *Dev. Biol.* (Suppl.) 2:184–207.

Cell and Tissue Interactions, edited by
J. W. Lash and M. M. Burger. Raven
Press, New York, 1977.

Membrane Involvement in Cell-Cell Interactions: A Two-Component Model System for Cellular Recognition That Does Not Require Live Cells

Max M. Burger and J. Jumblatt

*Department of Biochemistry, Biocenter, University of Basel,
CH 4056 Basel, Switzerland*

This meeting has focused on embryonal aspects of cell-cell interactions. Our knowledge of these interactions at the molecular level, particularly with regard to mechanisms of cellular recognition, is still quite rudimentary. Any model system for cellular recognition that can be described in molecular terms should provide a valuable conceptual basis for the more complicated embryonal systems.

This chapter will first characterize and analyze the types of embryonal cell-cell interactions involved in morphogenesis. Subsequently, the molecular components of sponge cell recognition—a model recognition system—will be described. Special emphasis will be given to the question of to what extent cellular recognition requires the living state or can be demonstrated to occur between dead cells or in an acellular, purely molecular system.

Cell-cell recognition in the embryo is only one of several processes that occur during morphogenesis. Morphogenetic processes can be dissected into at least three parts: (a) movement, (b) recognition of the final cellular distribution, and (c) fixation of that final state. This applies to single cells as well as to groups of cells that can move around as a unit.

Movement

Single cells can migrate randomly during embryogenesis until they find and recognize their final target. Besides random movement, cells or groups of cells can also reach their target via chemotactic routes laid down in the form of gradients of small or macromolecular substances (24,25).

Another mode of reaching the target is by contact guidance, as has been suggested, for example, for the migration of neurons in the fetal neocortex (16). However, such a mechanism tends to beg the question of how the orientation of the guiding element is originally obtained. Contact guidance could be provided by the extracellular matrix or by neighboring cells. If the information for the guided track is present on the surfaces of neighboring cells encountered by the migrating cells on their way to the target, then this information may be coded in the form of macromolecules arranged in gradient patterns. This viewpoint would narrow the gap between those who are exclusively subscribing to chemotactic gradients as a guide for direction and those advocating that the information for the displacement of cells and cell groups during morphogenetic movement is surface bound.

A cell or a group of cells must "realize" in some manner that it has reached its final destination after a migratory period. This knowledge can come from within or from its environment. An internal signal to halt migration could be based, for instance, on an internal clock defining the total migratory time or distance. However, more examples can be found in which groups of cells reach their final destination based on information obtained from the environment.

Recognition

The morphogenetic migratory phase is terminated by a recognition step that determines the final position of a given cell in the embryo. One can question the need for a separate recognition step after migration, since chemotaxis can, on theoretical grounds at least, replace recognition. Thus, migrating cells will terminate their journey wherever the concentration gradient to which they respond is highest. If they are kept long enough at a given location, such cells are likely to be trapped physically by gap and other cellular junctions or by extracellular matrix.

Despite such theoretical objections, nature provides us with examples at the level of lower eukaryotes where a surface-mediated recognition process follows chemotaxis and guarantees a secure specific juxtaposition of cells. The best illustration is furnished by the slime molds — organisms that may also be considered as a model system *par excellence* for developmental chemotaxis. While migrating toward a center emitting the chemotactic material, slime mold cells begin to develop additional species-specific aggregating capabilities (2), which are thought to be due to the appearance of sugar-specific lectin-like macromolecules on the cell surface (1). Similarly, embryonal organ morphogenesis in higher animals may involve, in addition to chemotactic guidance systems, such recognition mechanisms that function at the end of the migratory phase.

While we will be concerned primarily with mechanisms of recognition mediated by cell surface molecules, we feel that any novice to the study of cell-cell interactions — particularly if he has a biochemical background — should be aware that the current emphasis in the field on surface recognition is partly a reflection of the need for simple, experimentally testable systems. It is thus worthwhile to consider some other possible mechanisms that deserve equal attention as soon as they can be tested better.

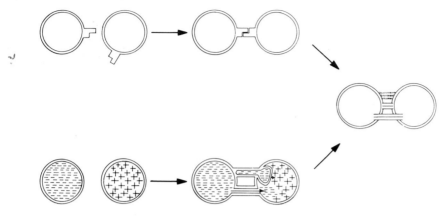

FIG. 1. Two possible mechanisms for recognition. The upper portion of the figure shows recognition occurring via the interaction of specific surface molecules. In the lower portion, an alternative mechanism is depicted in which informational signals are passed between cells via vesicular exchange **(top)** or via gap or other junctions **(bottom).** Such connections are likely to have a short life span and can be considered as mutual probing, frequently seen at the ruffling edges of moving cells. Both types of *specific* interactions can lead either to secondary, stabilizing linkages (e.g., intercytoplasmic, intermembranous, or extracellular linkages as depicted by the cell doublet to the right), or to dissociation of the interacting cells (not shown). (From ref. 3.)

For instance, recognition between cells may be mediated by exchange of intracellular information, perhaps in the form of informational molecules (Fig. 1). Such information could conceivably pass via gap junctions to neighboring cells. Not only ions, but small metabolites could spread very rapidly through such a network of gap junctions (9). Larger molecules, particularly RNA or protein, would have to find another route for exchange, the most obvious but not sufficiently analyzed mechanism being exchange via pinocytotic or even phagocytotic vesicles (Fig. 1).

Termination of the migration process at the final destination is only one among several possible consequences of recognition that can occur *in vivo*. Thus, recognition between cells may lead to the triggering of differentiation and the production of tissue-specific proteins. In some cases, recognition leads to fusion of cells, as for instance in the formation of a multinucleated myotube from the mononucleated precursors.

Fixation of the Proper Cell Distribution

In many cases, recognition forces alone may not be sufficient to immobilize the cells at their final position within the embryo. For example, it is well known from reaggregation studies of embryonal neural cells that specific recognition is only a temporary phenomenon and is lost after a few days during embryonal life. After recognition has taken place, proper cell and tissue distributions can be fixed by three types of nonspecific intercellular linkages (3). Intercellular connections joining cytoplasmic elements can be formed, such as gap and other types of junctions (Fig. 1). In addition to continuous cytoplasmatic channels, filamentous cytoskeleton bridges may also provide strong associative forces between cells, although the molecular continuity from cell to cell has so far not been established sufficiently. Secondly, intermembranous connections that lack specificity can be formed, but these are capable of holding the membranes together. Lastly, extramembranous fibers, such as collagen and elastin, are major contributors to the forces that join cells together. They can also form a sheath of material around groups of cells or parts of organs, thereby encapsulating and fixing them.

For the time being, it is not possible to assess the relative importance of these three types of intercellular bridges, nor can a temporal sequence be established if some of them occur together. It is likely, however, that the formation of gap junctions is a relatively rapid phenomenon, as has been established from *in vitro* experiments on sponge cell aggregation (10).

SPECIES-SPECIFIC SPONGE CELL RECOGNITION: A SYSTEM INVOLVING AT LEAST TWO SURFACE COMPONENTS

Wilson (22,23) first reported that suspensions of sponge cells obtained by passing pieces of sponge tissue through bolting cloth (*mechanical dissociation*) would reaggregate in a species-specific manner. Humphreys (7) and Moscona (12) extended these studies by devising improved techniques for dissociating sponge cells in the absence of divalent cations (*chemically dissociated cells*). The supernatant of such chemically dissociated cells contained macromolecular aggregation factors, which promoted species-specific reaggregation of sponge cells.

Assay of Sponge Cell Aggregation

The assay we have adopted is a variation of the endpoint method described by Humphreys for titration of aggregation factor activity (7). Cells are incubated with serial dilutions of aggregation factor on a rotary shaker for 20 min at room temperature, allowing aggregation to reach a macroscopic equilibrium. During this assay period, aggregation is first observed at the highest concentration of factor, then at progressively lower concentrations until a stable endpoint (indicating the *degree* of cellular aggregation in response to factor) is attained. If one records the time-dependent change in the endpoint at short intervals during the assay, it is possible to determine and compare the approximate *rates* of cellular aggregation. As shown in Fig. 2A, mechanically dissociated cells show an apparently faster aggregation rate than chemically dissociated cells in response to the same con-

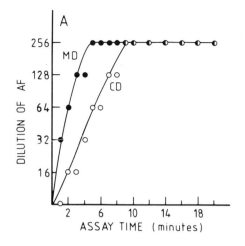

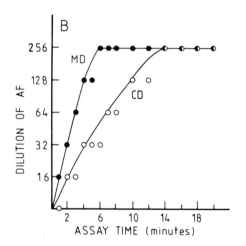

FIG. 2. Factor-promoted aggregation of mechanically and chemically dissociated *Microciona* cells. Suspensions of chemically and mechanically dissociated sponge cells were prepared as previously described (3,7). Cell reaggregation in response to serially diluted aggregation factor was assayed at room temperature by the endpoint method of Humphreys (7). To follow the time course of aggregation, the progressive change in the endpoint was recorded at 1- to 2-min intervals during the assay. Note that the curves connect the earliest points at which aggregation was observed at each dilution of aggregation factor. **A:** Live *Microciona* cells (2×10^7/ml). **MD,** mechanically dissociated, **CD,** chemically dissociated. **B:** Glutaraldehyde-fixed *Microciona* cells (2×10^7/ml). Cells from the same preparation as in **A,** fixed in 1% (v/v) glutaraldehyde, 12 hr, 4°C.

centrations of factor. Obviously, for such comparisons to be meaningful, each time-dependent change in the endpoint must be roughly equivalent for both cell types in terms of actual numbers of cells aggregated. In our case, this requirement is met due to the short assay period and moderately high shear forces that minimize background adhesion. Thus, a visible shift in the endpoint corresponds to incorporation of approximately 70% (± 15) of single cells into aggregates.

The rate difference we have noted between aggregation of mechanically and chemically dissociated cells (Fig. 2A) is in basic agreement with earlier studies of Moscona (12), who followed both increase in aggregate size and entry of single cells into aggregates. Despite such rate differences, mechanically and chemically dissociated cells show a similar *degree* of aggregation in response to a given amount of factor, when measured either by the endpoint technique (Fig. 2A) or as percent of single cells aggregated above background (12). The main advantage of the assay we currently employ is the ease and sensitivity with which differences in aggregation rates can be detected (above background) without regard to aggregate *size* — a parameter often influenced by events subsequent to primary adhesion. Whether the different aggregation rates of mechanically and chemically dissociated cells are due to qualitative or quantitative differences in the binding of exogenous aggregation factor, differences in surface topology, or possibly the existence of secondary, stabilizing forces, must now be clarified. A simple hypothesis is that the binding of factor to sponge cells is accelerated in the case of mechanically dissociated cells by factor molecules already present on the cell surface. To test this idea will require a careful analysis of binding of factor to cells, although such a study is further complicated by the tendency of factor molecules to associate into gel-like aggregates under the conditions of the assay (6). A more detailed discussion of parameters influencing the factor-induced aggregation of sponge cells will be presented in a later section.

Aggregation Factor

Humphreys isolated a high molecular weight fraction from cells dissociated in calcium-free seawater, which was necessary to reaggregate such "chemically dissociated" cells in the cold (7). Similar isolation and characterization of aggregation factors from different sponge species has been reported by others (5,11). Henkart et al. (6) and Cauldwell et al. (4) have recently described the factor from *Microciona parthena* to be 20×10^6 daltons and to consist of almost exactly 50% carbohydrate and 50% protein. Since biochemical homogeneity could not be established in rigorous terms due to the fact that sedimentation, exclusion chromatography, and precipitation with Ca^{++} were the only procedures used for purification, one cannot yet rule out the possibility of biologically relevant heterogeneity in both affinity and molecular structure of the aggregation factor. The most purified fraction of factor molecules examined by electron microscopy displayed a "sunburst" configuration with a central ring of 80 nm diameter and approximately 15 outwardly extending "arms" 110 nm in length. Both "ring" and "arm" structures contained equal proportions of protein and carbohydrate.

A Second Component: Baseplate

Based on the following experimental evidence, we postulated a second type of molecule on the surface of sponge cells that could act as a receptor for the aggregation factor. Chemically dissociated cells, washed free of factor, could be further treated with hypotonic salt solution and thereby lose their responsiveness to aggregation factor (21). The supernatant of such

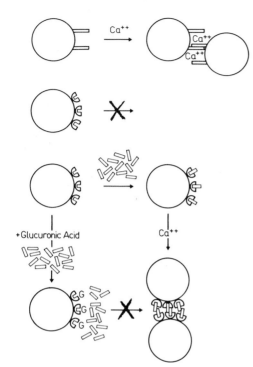

FIG. 3. Aggregation of agarose beads conjugated with baseplate or with aggregation factor. As depicted in the upper portion of the figure, beads conjugated with aggregation factor **(bars)** are found to aggregate rapidly in the presence of Ca^{++}. This aggregation can occur even in the presence of glucuronic acid—a hapten sugar that specifically inhibits the interaction of the factor with the cell surface. Beads conjugated with baseplate (depicted as grooved, kidney-shaped receptors, **second row**) do not aggregate spontaneously in the presence of Ca^{++} but are efficiently aggregated when aggregation factor is added **(rows 3 and 4, right)**. This factor-promoted aggregation of the beads can be prevented, however, by glucuronic acid **(lines 3 and 4, left)**. (From ref. 3.)

a hypotonically treated cell suspension contained a protein, termed baseplate, which had at least some of the expected properties of a receptor for the aggregation factor. The baseplate component was not only capable of inhibiting the factor-promoted aggregation of fresh, chemically dissociated cells but could also partially restore the responsiveness of hypotonically treated cells to factor.

The question arose as to whether the apparent interaction between the factor and baseplate was restricted to living cells, or whether it could also be demonstrated in an acellular assay system. We therefore covalently coupled the crude baseplate preparation to Sepharose beads and found the beads to be as efficiently aggregated by factor as were the intact cells. Further analysis revealed that this cell-free aggregation system shared inhibitors and several other qualitative aspects (sequence of addition of factor or inhibitor, etc.) with the aggregation of live cells (21). Figure 3 schematically summarizes the outcome of these experiments with insolubilized baseplate and aggregation factor.

Purification of Baseplate

The baseplate component released from hypotonically treated cells has been partially purified by gel filtration, as shown in Table 1. The degree of

TABLE 1. *Partial purification of baseplate from Microciona prolifera*

Fraction	Specific activity[a]	Reconstitution activity[b]

Cells (chemically dissociated)
    0.15 M NaCl, 15 min 22°C
    600 × g, 10 min		

Supernatant ↓ 12,000 × g, 15 min	−	N.D.
Crude baseplate ↓ 105,000 × g, 90 min	1.5	+
Supernatant ↓ Sephadex G200	2.2	+
Pooled inhibitory fractions (G200) ↓ Sephadex G100	6.4	+
Pooled inhibitory fractions	16.0	N.D.

[a] Relative inhibitory units per microgram protein.
[b] Ability to restore aggregation to hypotonically treated cells.
N.D., not determined.

purification was calculated on the basis of a semiquantitative assay of the capacity of baseplate to inhibit the aggregation of chemically dissociated cells by aggregation factor (8). On gel filtration, baseplate was eluted from Sephadex G200 and G100 columns as a single, symmetrical peak of activity in the molecular weight range of 45,000 to 60,000 daltons (8). An additional fivefold purification over that shown in Table 1 has been attained by including an initial $(NH_4)_2SO_4$ precipitation step (50 to 75% saturated), yielding a fraction that is substantially, although not completely, purified as judged by SDS-Gel electrophoresis (not shown).

Our results to date indicate the baseplate component to be a soluble, heat-labile, periodate-sensitive polypeptide (possibly a glycoprotein) with a molecular weight of approximately 50,000. Its solubility in seawater and release from cells under relatively mild conditions suggest that the baseplate be tentatively classified as a peripheral membrane protein. As such, its association with the cell surface may involve other (perhaps integral) membrane proteins. The two activities of baseplate that have been described, namely the capacity to (a) inhibit sponge cell aggregation by factor and (b) to reconstitute the responsiveness of hypotonically treated cells to factor may turn out to be independent functions residing in separate regions of the molecule.

Most of our attention so far has been focused on the nature of baseplate-factor rather than baseplate-cell interaction. Insights on baseplate-factor interaction have come from the cell-free bead experiments described above

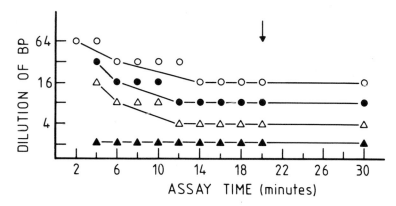

FIG. 4. Inhibition of factor-promoted *Microciona* cell aggregation by baseplate. A partially purified (approximately 40 times) fraction of baseplate was serially diluted into the wells of a Linbro plate containing 100 μl of buffered Mg^{++}-free seawater (2 mM Ca^{++}). Fifty microliters (13 μg) of purified aggregation factor (titer = 64 units/ml) was added to each well. After a 10-min incubation at room temperature, chemically dissociated cells were added, and aggregation was allowed to proceed on a rotary shaker for 30 min under standard assay conditions (70 rpm, room temperature). The plate was removed and inspected at 2-min intervals during the assay, and at each time point the dilution of baseplate showing visible *inhibition* of aggregation in comparison to controls (no baseplate) was recorded **(ordinate)**. The amounts of baseplate present in the first well (fourfold dilution) were as follows: 28 μg (——○——), 14 μg (——●——), 7 μg (——△——), and control lacking baseplate (——▲——). In our routine assay for baseplate inhibition of factor activity, the *inhibitory titer* is read at 20 min **(arrow)**.

and from studies of inhibitory effects on baseplate factor activity. Figure 4 describes an assay of baseplate inhibition, in which baseplate was serially diluted and preincubated with a fixed amount of factor, chemically dissociated cells added, and the endpoint of inhibition recorded at intervals during 20 min of shaking. For quantitation, a *unit of inhibition* is defined as that amount of baseplate required to visibly inhibit one unit of factor activity (7). At sufficient concentrations, baseplate has been found to inhibit both the rate *and* extent of cell aggregation by factor. When the effects of baseplate are followed with time (Fig. 4), it can be seen that inhibition is initially observed at a very low concentration (high dilution) of baseplate. As the assay progresses, the degree of inhibition (endpoint) appears to decline, i.e., inhibition is only visible at increasingly higher concentrations of baseplate. Finally, the endpoint of inhibition becomes stable after 12 to 15 min of assay and shows direct concentration dependence. The time course of inhibition indicates that the inhibitory effect of baseplate is not due to enzymatic or other irreversible inactivation of factor or cells, since inhibition appears to decrease rather than increase with time. Moreover, we have found that aggregation factor incubated with high concentrations of baseplate can be recovered in active form after centrifugation and washing (Weinbaum, Jumblatt, *unpublished results*). Since the bead ex-

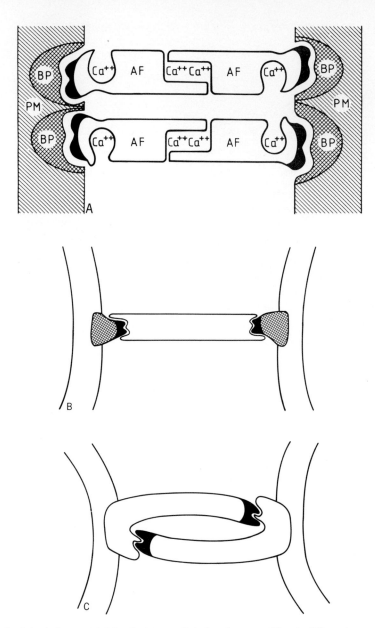

FIG. 5. A: A tentative model for factor-mediated cell recognition in *Microciona prolifera*. (See text for detailed explanation and discussion.) **PM,** plasma membrane; **AF,** aggregation factor; **BP,** baseplate component. **Shaded areas,** plasma membrane and cell surface. **Dark areas,** carbohydrate moieties of factor molecule that interact with surface sites, namely baseplate. **Hatched areas,** surface receptors including baseplate component. **Clear areas,** proteoglycan subunits of factor molecule. **B.** Inversion of polarity: An alternative model for ligand-mediated recognition in which the recognizing component lies between the cells, and the antigenic or recognized component on the cell surface. Black and hatched areas, carbohydrate. Clear areas, multivalent protein ligand. **C.** A third type of ligand-mediated recognition in which recognizing and recognized components are present on the surface of the same cell. The complementary components can be distinct molecules, or domains of the same molecules as shown here. Black and clear areas, complementary interacting signs (proteins and carbohydrates). (From ref. 3.)

periments indicate that baseplate binds directly to factor (Fig. 3), it is likely that baseplate inhibits aggregation by blocking one or more sites on the factor molecule needed to interact with the cell surface or with other factor molecules.

A tentative model for the recognition observed in the sponge *Microciona* is depicted in Fig. 5. In accordance with the available evidence, the factor molecule is considered to be multivalent and to consist of subunits that require Ca^{++} ions for stable association. The baseplate component serves as a carbohydrate-specific attachment site for a functional, specific interaction of factor with the cell surface. The fact that aggregation factor is dissociated from cells by removal of divalent cations (i.e., during chemical dissociation) raises the possibility that Ca^{++} ions are directly involved in baseplate-factor interaction (Fig. 5A), in addition, one should keep in mind that surface-mediated cell recognition in higher organisms and even in other sponge species may involve alternative mechanisms in which carbohydrate moieties on the cell surface are linked by protein ligands between cells (Fig. 5B), or both recognized and recognizing entities reside in the same molecule (Fig. 5C). We are currently investigating the possibility that both factor and baseplate molecules interact with other, yet undetected components of the cell surface.

VALENCY CONSIDERATIONS FOR CELL-CELL RECOGNITION

The concept of ligand receptor interaction, as applied, for instance, to hormone-cell surface interactions, has been widely adopted for cell-cell recognition studies. This conceptual approach has merit as long as one does not attach the expectation that quantitative aspects, such as affinities and rates, are likely to be similar in the two cases. There are good reasons to believe that affinities in particular, and rates as well, may differ by orders of magnitude between the expectations of the hormone biochemist and the requirements for cell-cell recognition.

This conclusion was reached partly through consideration of the interaction of the sponge aggregation factor with its cellular receptors. Glucuronic acid was found to inhibit the aggregation of *Microciona prolifera* cells, particularly in response to aggregation factor. Since galacturonic acid (which could bind Ca^{++} equally well) produced only poor inhibition, and since a crude glucuronidase preparation was found to inactivate aggregation factor, we suggested that glucuronic acid may be part of the recognition site on the aggregation factor (19). The amount of glucuronic acid necessary to inhibit factor-promoted aggregation was not only high (between 2×10^{-2} and 10^{-3} M) but varied from summer to summer, and sometimes from batch to batch of sponges.

Such poor or variable hapten inhibition does not automatically imply a very high affinity between factor and receptor site. In fact, the contrary situation may be true, i.e., the K_m for a single site interaction between ligand

and cell could be extremely high and still allow cells to associate tightly via a large number of relatively weak interactions. For the specific binding of a hormone molecule to its receptor, a K_m in the neighborhood of 10^{-7} to 10^{-9} might be appropriate. Since many assumptions have to be made about the size of the ligand, its shape, the surface charge of the cell, etc., it is not possible to accurately calculate the minimal K_m for single site interactions necessary for specific association between cells. If the number of coupling sites is high (for instance, 10^5 as a conservative estimate), it should be obvious that a K_m no greater than 10^{-2} to 10^{-3} would be sufficient for the single site.

This can be illustrated by another simple example. While it is possible to bind isolated glycoproteins to a column substituted with wheat germ lectin and elute them using N-acetyl-glucosamine, cells bound to the same wheat germ lectin beads could not be eluted with the hapten sugar even at very high concentrations. However, when the same experiment was performed with beads that were less substituted with lectin, the number of bridges that could be established between bead and cell was reduced, therefore allowing the cells to be eluted with low concentrations of hapten (26).

The requirement for multiple-site interaction between cells is consistent with the low overall reaction rate compared to insulin binding, for example. Recognition should, therefore, be considered as composite rates, which include many types of secondary interactions, such as cooperative effects due to site alignment. Model studies with receptors covalently attached to beads, as well as integrated into liposomes, are badly needed for any further kinetic evaluation of the cell-cell recognition reaction.

The amplification obtained due to the multiplicity of single-site interactions present between specifically adhering cells has a conservative function, since it permits recognition to be based on rather small differences between surface receptors. Thus, minor differences in the chemistry of the receptor sites, which would be considered insignificant by the enzyme chemist who considers single substrates only, are amplified to the extent that they become functionally useful for entire cells. One should not expect, therefore, that the information for surface-mediated recognition between cells is encoded exclusively in surface proteins. Small changes in the arrangement of the few carbohydrates making up the oligosaccharides of surface glycoproteins and glycolipids, could, in principle, give rise to sufficient diversity for the kinds of cell-cell recognition found either in sponges or in embryonal tissues.

SPONGE CELL RECOGNITION IN COMPARISON TO OTHER SYSTEMS

A Single Versus Multistep Mechanism

Most of the classic aggregation assays used to study cell sorting in higher organisms are performed over long periods of time during which cell-cell recognition may be only one of many steps leading to the final equilibrium distribution. For example, when cells from two organs are mixed together, they form clumps that consist of cells from both organ types in a random arrangement. Only hours later is one cell type found on the inside and the other on the outside of the clump (18; for general reference see also Moscona and Hausman, *this volume*). These older findings are now interpreted as being the result of damage inflicted to the surface of the cells during dissociation, leading to unspecific clumping. Later on, when the cells have repaired their surfaces, they can recognize each other and sort out within the clump formed originally by unspecific association.

Such a division of the process into two steps may not necessarily be due to damage alone. It is possible that unspecific forces play an important role in bringing cells together, both *in vivo* and *in vitro,* and only then can the secondary recognition processes (usually requiring live cells) begin, which give rise to the final, specific distribution of the cells. Even in the species-specific recognition of sponges, there is an example for such a two-stage process, although it may not be representative for all sponges. Müller et al. (13) found that cells from a Mediterranean sponge reaggregate in the absence of factor to a certain clump size that can be increased only by addition of the specific factor. The first stage of aggregation could be due to remnants of factor that were not removed when dissociating the cells, but this is unlikely in view of the fact that the primary aggregation phase is unspecific and proceeds in the presence of proteolytic enzymes.

In most assay systems used during the last few years, rates of aggregation are compared between a sample that obtained an aggregation promoting factor and another that did not. In the event of multistep recognition in such an assay system, it is difficult to assess the role of the factor in promoting the final pattern of selective cell-cell associations. Promotion of the rate of aggregation could occur in a number of indirect ways, such as by promotion of microvilli formation, promotion of an alignment of specific or nonspecific adhesive sites, or by induction or even stabilization of specific cell interactions produced by other mechanisms.

Sponge cell recognition represents a relatively simple system in which a number of possible steps in the recognition process could be ruled out, especially those requiring metabolic activities or other properties of cells in the living state. As will be discussed below, species-specific sponge cell recognition can occur not only at low temperatures but also after fixation with glutaraldehyde. This system should be contrasted, therefore, with most of the higher animal recognition systems where at least one of the two partners in the recognition process must be alive.

Size of the Promoting Factor

Most of the sponge aggregation factors that have been isolated so far are in the molecular weight range of millions of daltons. Similarly, the multiple mating factors from yeast are in the range of 10^6 daltons. Both resemble the proteoglycans of higher animals with respect to size, composition, and to some extent structure. A third aggregation factor from the alga *Chlamydomonas,* also a mating factor, again seems to be of very large size (17).

In contrast, organ- or tissue-specific aggregation factors isolated from higher animals seem to be of smaller size and probably correspondingly reduced valency. The best example is the neuroretinal cell aggregation promotion factor, which has a molecular weight of approximately 50,000 daltons (Moscona and Hausman, *this volume*). Nevertheless, an evolutionary trend toward reduction in the size of aggregation factors does not appear to be general, since the slime mold lectins isolated by Barondes' group (1) are between 100,000 and 250,000 daltons, and even among marine sponges active factors as small as 20,000 daltons have been reported (13). Moreover, higher organisms (vertebrates and mammals) contain specific (14) as well as unspecific aggregation factors (15), which have glycosaminoglycan character and molecular weights of several million daltons.

Temperature Dependence

Factor-promoted retinal cell aggregation requires an elevated temperature, as does factor-independent aggregation. Factor binding can occur at 4°C, but aggregation begins 2 hr after the cells are brought back to 37°C and reaches a maximum 12 hr later (for reference see Moscona and Hausman, *this volume*). This sequence clearly suggests that factor-promoted aggregation of retina cells is a multistep process, in which the factor could function as a specific ligand or as an inducer of a reaction subsequent to factor binding. In this regard, Moscona has recently indicated that, although the factor is specific in promoting aggregation of retina tissue only, its binding to the trypsinized cell might be unspecific. In this case, specificity must be contributed by secondary reactions between the cells that have already bound the factor.

In contrast to retinal cell aggregation, the specific reaggregation of both mechanically and chemically dissociated sponge cells can take place at 4°C, suggesting a simpler mechanism. At higher temperatures, secondary interactions occur between sponge cells, as in cell aggregates from higher organisms, but do not seem to contribute to the specificity. Corollary evidence for the existence of unspecific, secondary "forces" in sponge cell aggregation at elevated temperatures is provided by the finding that sponge cells deprived of factor will not form gap junctions (low resistance junctions) but can establish such connections within minutes once aggregation factor is added (10). A follow-up study of this interesting observation using morphometry as well as a detailed evaluation of temperature profiles might yield additional valuable information.

Hemagglutination as well as cell agglutination promoted by some lectins (e.g., wheat germ agglutinin) are somewhat analogous processes that can occur at low temperatures (4°C) without loss of specificity, although at a reduced rate. Since most agglutination reactions are carried out under relatively low shear forces, it has been speculated that the mechanisms involved for cell-cell adhesion are likely to be different from those operating during retina cell reaggregation where high shear forces are usually applied (20). Although this point of caution is well taken when comparing the results of assays using different shear forces, we do not believe that it is relevant to the comparisons we are making here. Sponge aggregation carried out at 4°C under shear forces equal to those of the retina cell assay results in clumps of similar size.

Requirement for the Living State for Cell Recognition

In contrast to the reaggregation of embryonic cells from higher organisms, the mechanism of sponge cell recognition does not appear to involve metabolic, biosynthetic, or other dynamic activities of living cells but can readily be demonstrated using dead or fixed cells. Moscona was first to show that formaldehyde-fixed sponge cells retained their capacity to aggregate in

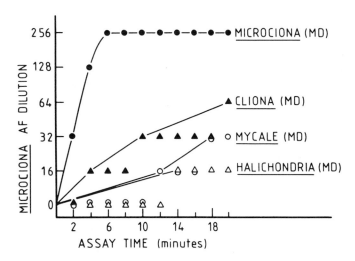

FIG. 6. Species specificity of factor-promoted aggregation of sponge cells after glutaraldehyde fixation. Suspensions of mechanically dissociated cells from four species of sponges were prepared and fixed with 1% glutaraldehyde, as previously described (8; see legend to Fig. 2B). Progressive cell aggregation in response to serially diluted aggregation factor from *Microciona prolifera* was measured, as explained in Fig. 2. The curves indicate the earliest time points at which aggregation was observed at each dilution of factor. It should be noted that at least some of the apparently nonspecific aggregation seen in this assay results from the physical trapping of cells within the gel-like polymers of factor molecules formed at high factor concentrations (see refs. 4 and 6).

response to aggregation factor and to bind homologous factor to their surfaces (12). We have employed glutaraldehyde (1%) to fix both mechanically and chemically dissociated cells from *Microciona* and other species. As shown in Fig. 2B, fixed *Microciona* cells closely resemble living cells in the *extent* of aggregation in response to factor, as measured by the endpoint assay. The most striking effects of fixation on factor-mediated cellular reaggregation are (a) a decrease in the aggregate size at equilibrium and (b) a small decrease in the apparent aggregation rate (Fig. 2B; ref. 12). We do not know, at present, whether these effects reflect the inability of fixed cells to form stabilizing secondary adhesions (e.g., gap junctions, ref. 10), or a loss of other dynamic properties (e.g., receptor mobility, formation of microvilli, surface deformability, etc.).

In agreement with earlier results of Moscona, glutaraldehyde-fixed cells show a high degree of species specificity in response to aggregation factor (Fig. 6). As we have reported previously, the species specificity of the *Microciona* aggregation factor activity appears to be quantitative rather than strictly qualitative, depending on the cell type (16). In addition, we have found that the aggregation of fixed *Microciona* cells could be inhibited by glucuronic acid or soluble "baseplate" receptor (data not shown), in accordance with our results using live cells (19). We are thus confident that the primary mechanism of species-specific adhesion is not destroyed by glutaraldehyde treatment and that the molecular interactions involved can be studied further using fixed cells as a convenient assay system.

CONCLUSIONS

Sponge cell recognition, although species-specific, may be a working model for some of the steps of cell- and tissue-specific recognition that occur during embryogenesis of higher organisms.

In the best studied species of sponge, a large proteoglycan-like aggregation factor appears to interact directly with a low molecular weight surface protein. It was pointed out that such large ligands possess a great number of potential recognition sites having, perhaps, relatively weak affinities for the partner site—a situation in which the chemical aspects of site-site interactions will be difficult to establish. However, such low affinities are sufficient for biological function, since they are amplified by the multivalent properties of the ligand.

Several important differences between the sponge and higher animal recognition systems were discussed. The requirements for elevated temperature and a living state for at least one of the two cells in most higher animal systems, together with other observations, indicate that this type of recognition is based on a more complicated, multistep process. The fact that sponges can aggregate in a species-specific manner at 4°C, and even after glutaraldehyde fixation, suggests a relatively simple mechanism that can be elucidated in molecular terms. Such studies will hopefully contribute to our understanding of the more complicated processes of recognition that occur during embryogenesis in higher animals.

ACKNOWLEDGMENTS

We would like to acknowledge the help of other colleagues formerly and still associated with this sponge project, which is carried out to a large part at the Marine Biological Laboratory in Woods Hole (Drs. G. Weinbaum, R. Turner, and W. Kuhns). This work is supported by Grant 3.720/76 from the Swiss National Science Foundation.

REFERENCES

1. Barondes, S. H., and Rosen, S. D. (1976): Cellular recognition in slime molds. Evidence for its mediation by cell surface species-specific lectins and complimentary oligosaccharides. In: *Surface Membrane Receptors,* edited by R. P. Bradshaw, W. A. Frazier, R. C. Merrell, D. I. Gottlieb, and R. A. Hogue-Angeletti, pp. 39–55. Plenum Press, New York.
2. Beug, H., Katz, F. E., Stein, A., and Gerisch, G. (1973): Quantitation of membrane sites in aggregating *Dictyostelium* cells using tritiated univalent antibody. *Proc. Natl. Acad. Sci. U.S.A.,* 70:3150–3154.
3. Burger, M. M., Turner, R. S., Kuhns, W. J., and Weinbaum, G. (1975): A possible model for cell-cell recognition via surface macromolecules. *Philos. Trans. R. Soc. Lond.* [Biol.], 271:379–393.
4. Cauldwell, C., Henkart, P., and Humphreys, T. (1973): Physical properties of sponge aggregation factor: A unique proteoglycan complex. *Biochemistry,* 12:3051–3055.
5. Gasic, G. J., and Galanti, N. L. (1966): Proteins and disulfide groups in the aggregation of dissociated cells of sea sponges. *Science,* 151:203–235.
6. Henkart, P. S., Humphreys, S., and Humphreys, T. (1973): Characterization of sponge aggregation factor: A unique proteoglycan complex. *Biochemistry,* 12:3045–3050.
7. Humphreys, T. (1963): Chemical and *in vitro* reconstruction of sponge cell adhesion. I. Isolation and functional demonstration of the components involved. *Dev. Biol.,* 8:27–47.
8. Jumblatt, J. E., Weinbaum, G., Turner, R. S., Ballmer, K., and Burger, M. M. (1976): Cell surface components mediating the reaggregation of sponge cells. In: *Surface Membrane Receptors,* edited by R. A. Bradshaw, W. A. Frazier, R. C. Merrell, D. I. Gottlieb, and R. A. Hogue-Angeletti, pp. 73–86. Plenum Press, New York.
9. Loewenstein, W. R. (1966): Permeability of membrane junction. *Ann. N.Y. Acad. Sci.,* 137:441–472.
10. Loewenstein, W. R. (1967): On the genesis of cellular communication. *Dev. Biol.,* 15:503–520.
11. Margoliash, E., Schenck, J. R., Hargie, N. P., Burokas, S., Richter, W. R., Barlow, G. H., and Moscona, A. A. (1965): Characterization of specific cell aggregating materials from sponge cells. *Biochem. Biophys. Res. Commun.,* 20:383–388.
12. Moscona, A. A. (1968): Cell aggregation: Properties of specific cell ligands and their role in the formation of multicellular systems. *Dev. Biol.,* 18:250–277.
13. Müller, W. E. G., Müller, I., and Zahn, R. K. (1974): Two different aggregation principles in reaggregation process of dissociated sponge cells (*Geodia cydonium*). *Experientia,* 30:899–902.
14. Oppenheimer, S. B. (1975): Functional involvement of specific carbohydrates in teratoma cell adhesion factor. *Exp. Cell Res.,* 92:122–126.
15. Pessac, B., and Defendi, V. (1972): Cell aggregation: Role of acid mucopolysaccharides. *Science,* 175:898–900.

16. Rakic, P. (1974): Mode of cell migration to the superficial layers of fetal monkey neocortex. *J. Comp. Neurol.*, 145:61–83.
17. Snell, W. J. (1976): Mating in *Chlamydomonas*. A system for the study of specific cell adhesion. I. Ultrastructural and electrophoretic analysis of flagellar surface component involved in adhesion. *J. Cell Biol.*, 48:48–49.
18. Steinberg, M. S. (1962): Mechanisms of tissue reconstruction by dissociated cells. II. Time course of events. *Science*, 137:762–763.
19. Turner, R. S., and Burger, M. M. (1973): Involvement of a carbohydrate group in the active site for surface guided reassociation. *Nature*, 244:509–510.
20. Walther, B. T. (1976): Mechanisms of cell agglutination by Concanavalin A. In: *Concanavalin A as a Tool*, edited by H. Bittiger and H. P. Schnebli, pp. 231–248. John Wiley & Sons, New York.
21. Weinbaum, G., and Burger, M. M. (1973): A two component system for surface guided reassociation of animal cells. *Nature*, 244:510–512.
22. Wilson, H. V. (1907): On some phenomena of coalescence and regeneration in sponges. *J. Exp. Zool.*, 5:245–258.
23. Wilson, H. V. (1911): On the behavior of the dissociated cells in hydroids, alcyonania and asterials. *J. Exp. Zool.*, 11:281–338.
24. Wolpert, L. (1969): Positional information and the spatial patterns of cellular differentiation. *J. Theor. Biol.*, 25:1–47.
25. Wolpert, L., Hornbruch, A., and Clarke, M. R. B. (1974): Positional information and positional signalling in *Hydra*. *Am. Zool.*, 14:647–663.
26. Zabriskie, D., Ollis, D. F., and Burger, M. M. (1975): Activity and specificity of covalently immobilized wheat germ agglutinin toward cell surfaces. *Biotechnol. Bioeng. Symp.*, 15:981–992.

Cell and Tissue Interactions, edited by
J. W. Lash and M. M. Burger. Raven
Press, New York, 1977.

Biological and Biochemical Studies on Embryonic Cell-Cell Recognition

A. A. Moscona and R. E. Hausman

*Departments of Biology, Pathology, and the Committee on Developmental Biology,
The University of Chicago, Chicago, Illinois 60637*

The mechanisms of cell-cell recognition and selective cell affinities enable cells in the embryo to identify one another, to adhere selectively, and to become organized into the multicellular patterns giving rise to tissues and organs (15,24,31,37,39,40,48,50,56,62). Identification of these mechanisms is important not only for the understanding of normal embryonic development but also of developmental abnormalities: Malfunctions in these processes can lead to defective morphogenesis and result in congenital malformations; also, the disruption of tissue architecture in neoplasia may reflect changes in the expression of cellular affinities.

The early vertebrate embryo consists largely of a mass of multiplying cells; in order to construct tissues and organs, the different cells must first become spatially sorted out, segregated, and assembled into specific groupings. Many tissues and organs arise in the embryo from cells that originate away from their final locations; these "precursor" cells leave their initial places and migrate to their target sites where they assemble, aggregate, and become organized into tissue primordia (Fig. 1). Embryonic cells are capable of these positional rearrangements because they possess on their surfaces communicational mechanisms that enable cell recognition and selective cell adhesion into tissue-forming groups. These mechanisms evolve in the embryo progressively and coordinately with cell differentiation. It has been suggested (37,39,40) that, as the diverse kinds and types of cells arise and evolve during embryogenesis, cell surfaces become differentially specified or encoded with molecular "labels;" these labels project outwardly the changing identities and characteristics of the cells, and they determine the recognition properties of the cells and their preferential affinities with respect to one another and to constituents in the extracellular environment.

CELL AGGREGATION *IN VITRO:* EXPERIMENTAL CONSTRUCTION OF TISSUES FROM CELLS

For obvious technical reasons, the mechanisms of embryonic cell associations are difficult to explore in the intact embryo; therefore, in order to

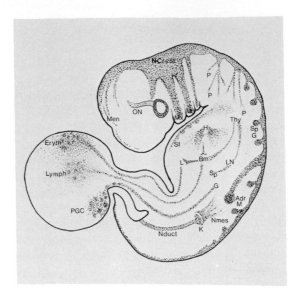

FIG. 1. Diagrammatic representation of some of the cellular traffic involved in the formation of tissues and organs during mammalian embryogenesis. The diagram combines events taking place at different stages of development. The "pathways" connect the original sites of cells with their destinations; they do not accurately depict the actual routes of the cells. The progenitors of erythroid cells **(Eryth)** originate in the yolk-sac endoderm, enter blood vessels, and their progeny are transported by the circulation to the subsequent sites of erythropoiesis (**L**, liver; **Bm,** bone marrow). The progenitors of lymphoid cells are transported to the bone marrow **(Bm)**, thymus **(Thy)**, lymph nodes **(LN)**, and spleen **(Sp)**. The primordial germ cells **(PGC)** migrate actively from the yolk sac to the germinal ridges and colonize them to form the gonads **(G)**. Cells from the neural crest **(NCrest)**—a temporary structure along the dorsal axis of the early embryo—migrate in swarms or individually, and their progeny give rise to a variety of structures, among them: meninges **(Men)**, cartilages of the embryonic skull **(Mx, Md)**, pigment cells **(P)**, spinal ganglia **(SpG)**, and adrenal medulla **(AdrM)**. Axons that grow out of retinal ganglion cells form the optic nerves **(ON)**, which advance toward the visual centers in the brain. The nephric ducts **(Nduct)** elongate toward aggregations of nephric mesenchyme cells **(Nmes)** to jointly form the kidney rudiments **(K)**. The sternum **(St)** is formed by mesenchymal cells that migrate from the dorsal region of the embryo. The heart **(H)** arises from aggregations of cells originating in the "heart-forming territory." The intricate translocations of cells involved in histogenesis of the nervous system are not included in this diagram, nor are other examples of morphogenetic cell migrations and aggregations (e.g., those resulting in the formation of hair follicles, teeth, limb rudiments, etc.).

analyze these problems at the cellular and biochemical levels, *in vitro* experimental systems have been developed and widely employed. These systems are based on reaggregation *in vitro* of cell suspensions from chick and mouse embryos. Such experiments start with the preparation of cell suspensions from embryonic tissues or organs by dissociation with trypsin. The dispersed cells are then reaggregated into multicellular aggregates where they reform their characteristic histological patterns. Thus, it is possible to construct *in vitro* multicellular complexes from cellular building blocks, in effect to "resynthesize" tissues from cells and thus to study, under controlled experimental conditions, mechanisms that mediate the association of cells into morphogenetic patterns (37,42).

As an example of this approach, we will review briefly experiments on reaggregation of neural retina cells from 7- to 10-day chick embryos. Treatment of freshly isolated tissue with trypsin causes the cells to detach from one another so that they can be dispersed into a suspension (31,34). When the cell suspension is rotated on a gyratory shaker, the cells form aggregates of progressively increasing size (34). The rate of cell reaggregation and the size of cell aggregates can be variously measured, and such data provide information on the mutual adhesiveness of the cells in the suspension with reference to a given set of experimental parameters (34). This procedure and its variations (49,57) lend themselves well to analyses of endogenous and exogenous parameters affecting the formation of morphogenetic cell-cell associations.

Early in this work it was established that morphogenetic reaggregation of dissociated cells is reversibly prevented by suboptimal temperatures and by inhibitors of macromolecular synthesis, and, hence, it requires energy-dependent and biosynthetic processes. In all likelihood, these processes are needed, in part, for regeneration of cell surface components removed in separating the cells, which are essential for morphogenetic cell associations (43–45). Another highly striking aspect of cell aggregation is the role of elongated filopodial processes by means of which separated cells make initial contacts. These processes bridge cells across considerable distances and bring them together, and thus appear to serve an important role in cell assembly into multicellular complexes (5).

Within the emergent aggregates, the various types of cells are initially interspersed; however, they begin to reshuffle and progressively sort themselves out: The different types of retinocytes gradually become positioned and aligned in correct locations relative to one another and reconstruct the histotypic architecture of the retina. The reconstructed tissue continues to differentiate morphogenetically and biochemically; eventually synapses are formed and various products characteristic of neurodifferentiation are detectable (37,53,60).

Such tissue reconstruction *in vitro* from cell suspensions has been described for cells from practically every embryonic tissue (32,35,37). It is particularly noteworthy that even cells dissociated from embryonic cerebrum and cerebellum (avian and mammalian) are capable of reconstructing in aggregates histological patterns characteristic of brain tissues (11,54), and that such aggregates can attain advanced morphological and biochemical differentiation (7,8,51,52).

The above comments referred to reaggregation of cells derived from a single tissue. The ability of the diverse types of cells that belong to the same tissue to become histologically organized is reflective of *cell-type-specific cell recognition*. However, in the course of embryogenesis, it is essential that cells giving rise to different tissue become spatially segregated into separate, distinct groupings. This suggests the existence of mechanisms for *tissue-specific cell recognition* [or "tissue affinities," originally envisaged by Holtfreter (23)]. And, in fact, if one combines in a single suspension cells from two embryonic tissues, say, retina cells and liver cells, they tend to segregate preferentially into discrete groupings within which they reconstruct their characteristic histological patterns. Initially, the cells form "composite" aggregates, but they soon sort out into distinct regions consisting of retina cells or liver cells. Examination of a large variety of such binary embryonic cell combinations (31,32,34,35,37,42,57) has confirmed, as a general principle, the propensity of heterologous cells to segregate into distinct groupings, in contrast to the preference of homologous cells (i.e., of cells belonging to the same tissue or tissue system) to associate with one another.

Therefore, it is operationally useful to distinguish two major categories of embryonic cell recognition (37):

(1) *Tissue-specific cell recognition,* which reflects mechanisms that enable cells belonging to different tissues to segregate from one another and to assemble preferentially into tissue-forming groupings.

(2) *Cell-type-specific cell recognition,* which reflects mechanisms that mediate the histological organization of the diverse cells that make up a single tissue. This category can be subdivided into: (a) *homotypic cell recognition,* i.e., affinities between cells with identical phenotypes, and (b) *allotypic cell recognition,* i.e., affinities between different phenotypes that are developmentally destined to establish functional associations (for example, contacts between different kinds of neurons; nerve-muscle connections, etc.).

This is obviously a simplified and abbreviated classification of the modes of cell recognition during embryonic development, but it aids in formulating experimentally testable questions concerning mechanisms of cell associations.

THE CELL-LIGAND HYPOTHESIS

In examining these problems, we have been guided by the working hypothesis that the mechanisms of embryonic cell recognition and of cell affinities involve interactions of specific cell surface constituents which function as intercellular ligands or cell-cell receptors (33,35–40). This hypothesis suggests that the specificity of cell recognition is determined by the molecular characteristics of the ligands, and also by their topographical arrangements on the cell surface, i.e., by a "code" that reflects both the qualitative and the organizational properties of the units of which it consists.

The hypothesis postulated that cells belonging to different tissues might be specified by different cell ligands in a manner conducive to tissue-specific recognition of cells; on the other hand, type-specific cell recognition (conducive to histological organization of cells within a single tissue) might depend largely on the topographical-temporal characteristics of the ligands, i.e., on the nature of the patterns displayed by ligands on cell surfaces at different stages of cell differentiation. In a broad sense, the hypothesis implies that embryonic cells with complementary ligand displays on their surfaces (complementarity being dependent on both qualitative and structural matching) would show positive recognition and associate morphogenetically; while noncomplementarity (including absence of the appropriate ligands) would result in negative recognition, i.e., in transient or nonselective adhesion, or nonadhesion of the cells. Finally, it should be stressed that the term "cell ligand" does not necessarily imply an entity consisting of a single molecule, but includes the possibility of a multimeric complex (homo- or heteromeric). Also, it should be mentioned that other molecular components exist between cells ("extracellular materials"), some of which may play a role in nonspecific cell adhesions (including sticking of cells to tissue-culture dishes), in stabilizing cell contacts (intercellular "cement substances"), in providing the extracellular matrix ("ECM; 33), etc.

CELL-AGGREGATING FACTORS

The cell-ligand hypothesis raised certain testable questions. The possibility that cells from different embryonic tissues may possess qualitatively different surface-constituents has been examined by immunological techniques (14). These studies demonstrated the existence of tissue-specific antigenic disparities between surfaces of cells from different tissues of the chick embryo. Conceivably, some of these determinants could be involved in cell recognition and thus might represent the postulated tissue-specific cell ligands.

A more direct approach to this problem is represented by attempts to isolate from embryonic cells materials with the activity of the postulated tissue-specific cell ligands. The assumption was that addition of such materials to cell suspensions would enhance cell reaggregation, i.e., cause cells to aggregate faster and to form larger aggregates than the controls. Most importantly, the cell-aggregating effect of these materials was expected to be tissue-specific.

Evidence for a retina-specific, cell-aggregating factor derived from embryonic chick neural retina cells was first reported in 1962 (27,35); subsequently, a cerebrum-specific, cell-aggregating factor was derived also from cerebrum cells of mouse and chick embryos (10,12). These factors were detected in the supernatant medium collected from primary cultures of retina cells and of cerebrum cells. The rationale for this procedure was that, under these culture conditions, the cells release into the medium various macromolecules, some of which represent components that normally are associated with the cell surface.

Addition of the retina-derived supernatant to freshly prepared suspensions of embryonic chick retina cells caused a striking enhancement in cell aggregation: the cells aggregated at a faster rate than in the controls, and the resulting aggregates were significantly larger. Most importantly, this enhancement was obtainable only with retina cells, and not with cells from other tissues. Furthermore, the reaggregated cells became histotypically organized, i.e., the factor did not simply clump or agglutinate the cells, but enhanced their morphogenetic associations. Finally, this factor could be obtained only from cells isolated from embryos younger than 13 days, and only pre-13-day retina cells were responsive to the aggregation-enhancing effect of the factor; therefore, both the availability of this factor and cell responsiveness to it appear to be limited to developmental stages during which cells in the neural retina are undergoing histological organization (17,18,35).

BIOCHEMICAL CHARACTERIZATION OF THE
RETINA CELL-AGGREGATING FACTOR

As a step in this study, we sought to establish the molecular identity of the cell-aggregation-enhancing factor in the supernatant from retina cell cultures. When analyzed by SDS-polyacrylamide gel electrophoresis, the supernatant yielded 30 to 40 protein-containing bands (17). The retina-specific cell-aggregating activity was associated with proteins that localized on the gel in a region corresponding to a 50,000 to 60,000 molecular weight (29).

By isoelectric focusing, these proteins were separated into seven to eight discrete peaks; the retina cell-aggregating activity was present only in the peak that focused at pH 4. Subsequent analysis demonstrated that this peak contained a glycoprotein consisting of approximately 400 amino acids and less than 20% (by weight) of sugar residues; the latter were identified as N-acetyl-glucosamine, galactose, mannose, and sialic acid. Determination of Stoke's radius, sedimentation coefficient, and a partial specific volume indicated that the cell-aggregating activity resided in a molecule having a molecular weight of 50,000 ± 6,000 in solution; its frictional ratio was 1.3 (18).

RETINA CELL-AGGREGATING GLYCOPROTEIN: A CELL
SURFACE CONSTITUENT?

Metabolic labeling with both amino acids and glucosamine and other tests (17–19) suggested that the retina cell-aggregating glycoprotein was synthesized by the cells. Furthermore, it was released into the culture medium in primary cell cultures, probably as a result of turnover and exteriorization of cell surface components—processes that are known to occur in cultured cells (25,59,61). Finally, the fact that this glycoprotein was produced and/or released, under these conditions, only by retina cells from pre-13-day embryos, suggested its being a developmentally regulated product. However, the possibility arose that this glycoprotein might be peculiar to cells in culture, rather than a constituent of cells *in vivo*. Therefore, in order to determine whether such molecules might play a role in morphogenetic cell associations in the embryo, it was essential to determine if they existed in embryonic tissues *in vivo*, and if they were associated there with cell membranes. Could a retina-specific, cell-aggregating factor be isolated directly from cell membranes prepared from noncultured embryonic retina tissue? To answer this question (20), we prepared cell membranes from retina tissue using a technique devised by Hemminki (22) for isolation of plasma membranes from rat brain. This procedure yields in the rat membrane preparations contaminated less than 10% with microsomal and mitochondrial markers.

Briefly, batches of 400 to 500 retinas were rapidly dounced on ice in 0.6 M sucrose. The membranes were subsequently purified by a two-step partitioning in sucrose gradients; this procedure yielded a "final pellet" of membrane vesicles consisting of about 100 μg of protein. Electron microscopy of this pellet (20) revealed it to contain predominantly smooth membrane vesicles.

This membrane preparation was extracted at 4°C with *n*-butanol (28); the aqueous phase of the extract was dialyzed, concentrated, and assayed against suspensions of freshly dissociated embryonic cells for its ability to enhance cell aggregation. At protein concentrations of 1 to 5 μg/ml, this material enhanced the reaggregation of embryonic neural retina cells but not of cells from other embryonic tissues (Fig. 2).

Using this procedure, a cerebrum-cell-aggregating factor was also obtained from a cell-membrane preparation of embryonic chick cerebrum. It enhanced the reaggregation of embryonic cerebrum cells but not of cells from other brain regions or from other tissues tested. By similar procedures, a cell-aggregating activity specific for spinal cord cells was also obtained from embryonic chick spinal cord (16) (Fig. 2).

The retina-cell-aggregating factor derived by butanol extraction from retina cell membranes was purified by column chromatography and isoelectric focusing. The activity localized in a protein, probably a glycoprotein, with a pI of 4 and a molecular weight of about 50,000. This protein contains less than 20% sugar residues by weight. In all properties so far examined, this material appears to be similar, and possibly identical, to the retina-cell-aggregating glycoprotein previously purified from the supernatant medium of cell cultures (20).

To further examine whether this protein was associated predominantly

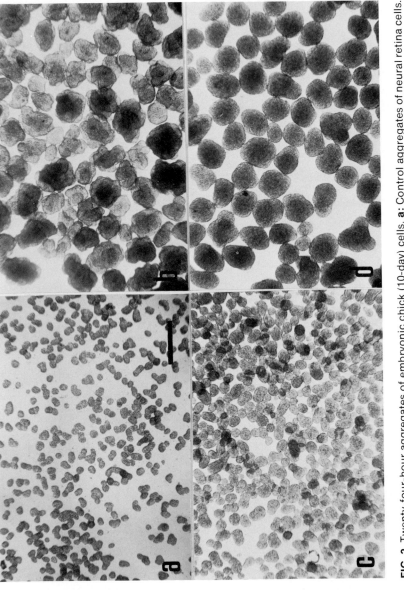

FIG. 2. Twenty-four-hour aggregates of embryonic chick (10-day) cells. **a:** Control aggregates of neural retina cells. **b:** Aggregates of retina cells formed in the presence of 2 μg/ml of the retina membrane-derived cell-aggregating factor. **c:** Control aggregates of cerebrum cells. **d:** Aggregates of cerebrum cells formed in the presence of 3 μg/ml of the cerebrum cell-aggregating factor. **Bar:** 500 μm. (Modified after ref. 16.)

with the cell surface, other fractions of the retina homogenate collected from the sucrose gradient were subjected to the butanol extraction procedure and assayed for the presence of cell-aggregating activity; no activity was detected, except in the "cytosol" fraction where activity was present at less than 5% of the level found in the membrane preparation. Therefore, the bulk of the activity obtainable by this procedure resides in the fraction containing the highest amount of purified cell membranes (20).

The above results suggest that the retina-cell-aggregating protein is associated with the cell membrane, a finding consistent with previous information (17), and with the concept of cell ligands. When released into the culture medium, or after its extraction with butanol, this protein appears to be a monomer with a 50,000 ± 6,000 molecular weight (18); however, it is possible that its activity at the cell surface (i.e., its cell-aggregating activity) may involve a multimeric arrangement and interactions with other cell surface macromolecules.

From a developmental point of view, it is highly significant that the retina cell-aggregating protein isolated from retina tissue by butanol extraction of cell membranes [like that purified from cell cultures (18)] cannot be obtained in detectable amounts from retinas older than 12 days of embryonic life, and also that the aggregation of cells older than 12 days is not enhanced by this factor (20). These results are consistent with the age-dependent decline in the capacity of dissociated retina cells to reaggregate (35); they suggest that the functional presence of the cell-aggregating protein is limited to the stages of development when the cells undergo morphogenetic reshuffling and organization (37).

MODE OF ACTION ON CELL AGGREGATION

Cell aggregation *in vitro* involves a gradual increase in the size of the aggregates by continuous accretion of cells and small aggregates. The effect of the cell-aggregating glycoprotein is to increase the rate of cell aggregation and the size of the aggregates; the basic kinetic pattern of the process is not altered. The effect of the protein is progressive and becomes gradually discernible within 2 hr of aggregation (19,20). The glycoprotein binds to cells at both 37°C and 4°C (19,20,26). However, in contrast to the effects of cell-agglutinating lectins or antisera, there is no rapid clumping of the cells. Thus, it is unlikely that the effect of the factor is due to simple "cross-linking" of the cell surfaces, but may require changes in cell surface properties, or in the organization of the protein molecules after their binding to the cell surface. As mentioned earlier, ongoing protein synthesis is necessary for normal cell aggregation; it is also required for the effect of the cell-aggregating glycoprotein (17). Since carbohydrate residues may be implicated in various cell interactions (3,6,47), we sought to determine whether the integrity of the carbohydrate portion of the retina cell-aggregating glycoprotein was essential for its effect. Cell-aggregating activity is rapidly destroyed by treatment with trypsin (18,26); therefore, the polypeptide chain is essential for activity. However, treatment of the glycoprotein with neuraminidase, β-galactosidase, galactose-oxidase, and also with periodate (under conditions that preferentially oxidize sugar residues) did not destroy its aggregation-enhancing activity (18,19). Thus, those sugar residues in the glycoprotein that are accessible to the above treatments seem not to be essential for enhancing *in vitro* cell reaggregation. These results do not rule out a role of the less exposed sugars; also, it is possible that after the treated glycoprotein binds to the cell surface, the carbohydrate portion is "repaired." However, it is noteworthy that the retina-cell-aggregating glycoprotein appears not to function by way of a galactosyltransferase mechanism (13,19).

The possibility that cell associations are multistep and multicomponent processes (37) has been considered also by other investigators (1,2,30,50). Additionally, it is becoming increasingly apparent that extracellular matrix macromolecules (glycoaminoglycan and proteoglycans) may function in cell organization and morphogenesis (21,33,58). However, their specific role in relation to cell recognition and histotypic cell affinities remains to be elucidated.

SUMMARY COMMENTS

The studies reviewed briefly in this chapter suggest the following working hypothesis: The cell-aggregating proteins isolated from embryonic cells may represent constituents of the tissue-specific cell-ligand mechanisms postulated to function in membrane-mediated embryonic cell affinities and selective cell adhesion. Accordingly, there appear to exist on embryonic cell surfaces tissue-specific proteins that function in morphogenetic cell recognition; for convenience of reference, the term *cognins* has been proposed for proteins with this particular function (39–41).

Assuming that *tissue-specific* cell affinities are determined by the molecular characteristics of tissue-specific cognins or their functional complexes (cell ligands), what is the mechanism of *cell-type-specific* recognition which enables different cells within a single tissue to become histologically arranged? Two theoretical possibilities might be considered (40). One is that each cell type is specified by a particular chemical "variant" of the tissue-specific cognin. Such a possibility was considered by Sperry (55) in his original concept of unique "chemospecifications" of different neurons. An analogous situation might be represented by the diversity and heterogeneity of antibodies generated from the products of a relatively small number of genes (9). Another possibility is that type-specific selective cell affinities are determined by differences in the topographic arrangement of cognin arrays (or ligand patterns) on cell surfaces. Unlike the first possibility, this one would not require a very large spectrum of unique chemical differences between cell surfaces. One can readily imagine that differences in the display patterns of ligands on cell surfaces, and topographical-temporal variations in the complementarities of these patterns, could furnish cells within a tissue with surface "information" conducive to histological positioning, alignment, and organization of the cells. These are presently largely speculative notions; however, they are amenable to testing and their exploration might help to advance our understanding of morphogenetic cell interactions.

ACKNOWLEDGMENTS

This work was supported by Research Grant HD01253 from the National Institute of Child Health and Human Development and, in part, by funds to the University of Chicago Cancer Center (1-PO1-Ca14599).

REFERENCES

1. Balsamo, J., and Lilien, J. (1974): Functional identification of three components which mediate tissue-type specific embryonic cell adhesion. *Nature*, 251:522–524.
2. Balsamo, J., and Lilien, J. (1975): The binding of tissue-specific adhesive molecules to the cell surface. A molecular basis for specificity. *Biochemistry*, 14:167–171.
3. Barondes, S. H., and Rosen, S. D. (1976): Cell surface carbohydrate-binding proteins: Role in cell recognition. In: *Neuronal Specificity*, edited by S. H. Barondes, pp. 331–351. Plenum Press, New York.
4. Bennett, D., Boyse, E. A., and Old, L. J. (1972): Cell surface immunogenetics in the study of morphogenesis. In: *Cell Interactions: Proceedings of the Third LePetit Colloquim, 1971*, edited by L. G. Silvestri, pp. 247–263. North-Holland Publishing Co., Amsterdam.
5. Ben-Shaul, Y., and Moscona, A. A. (1975): Scanning electron microscopy of aggregating embryonic neural retina cells. *Exp. Cell Res.*, 95:191–204.
6. Bosmann, H. B. (1971): Platelet adhesiveness and aggregation: The collagen:glycosyl, polypeptide:N-acetylgalactosamyl and glycoprotein:galactosyl transferases of human platelets. *Biochem. Biophys. Res. Commun.*, 43:1118–1124.
7. Crain, S. M., and Bornstein, M. B. (1972): Organotypic bioelectric activity in cultured reaggregates of dissociated rodent brain cells. *Science*, 176:182–184.
8. Crain, S. M., Raine, C. S., and Bornstein, M. B. (1975): Early formation of synaptic networks in culture of fetal mouse cerebral neocortex and hippocampus. *J. Neurobiol.*, 6:329–336.
9. Edelman, G. M. (1967): Antibody structure and diversity: Implications for theories of antibody synthesis. In: *The Neurosciences*, edited by G. C. Quarton, T. Melnechuk, and F. O. Schmitt, pp. 188–200. The Rockefeller University Press, New York.
10. Garber, B. B., and Moscona, A. A. (1969): Enhancement of aggregation of embryonic brain cells by extracellular materials from cultures of brain cells. *J. Cell Biol.*, 43:41a.
11. Garber, B. B., and Moscona, A. A. (1972): Reconstruction of brain tissue from cell suspensions. I. Aggregation patterns of cells dissociated from different regions of the developing brain. *Dev. Biol.*, 27:217–234.
12. Garber, B. B., and Moscona, A. A. (1972): Reconstruction of brain tissue from cell suspensions. II. Specific enhancement of aggregation of embryonic cerebral cells by supernatant from homologous cell cultures. *Dev. Biol.*, 27:235–271.
13. Garfield, S., Hausman, R. E., and Moscona, A. A. (1974): Embryonic cell aggregation: Absence of galactosyltransferase activity in retina-specific cell-aggregating factor. *Cell Differ.*, 3:215–219.
14. Goldschneider, I., and Moscona, A. A. (1972): Tissue-specific cell-surface antigens in embryonic cells. *J. Cell Biol.*, 53:435–449.
15. Grobstein, C. (1954): Tissue interactions in the morphogenesis of mouse embryonic rudiments *in vitro*. In: *Aspects of Synthesis and Order in Growth*, edited by D. Rudnick, pp. 233–267. Princeton University Press, Princeton, N.J.

16. Hausman, R. E., Knapp, L. W., and Moscona, A. A. (1976): Preparation of tissue-specific cell-aggregating factors from embryonic neural tissues. *J. Exp. Zool.,* 198:417–422.
17. Hausman, R. E., and Moscona, A. A. (1973): Cell-surface interactions: Differential inhibition by proflavine of embryonic cell aggregation and production of specific cell-aggregating factor. *Proc. Natl. Acad. Sci. U.S.A.,* 70:3111–3114.
18. Hausman, R. E., and Moscona, A. A. (1975): Purification and characterization of the retina-specific cell-aggregating factor. *Proc. Natl. Acad. Sci. U.S.A.,* 72:916–920.
19. Hausman, R. E., and Moscona, A. A. (1976): *In vitro* studies on embryonic cell associations. In: *Tests of Teratogenicity in Vitro,* edited by J. Ebert, and M. Marois, pp. 171–185. North-Holland Publishing Co., Amsterdam.
20. Hausman, R. E., and Moscona, A. A. (1976): Isolation of retina-specific cell-aggregating factor from membranes of embryonic neural retina tissue. *Proc. Natl. Acad. Sci. U.S.A.,* 73:3594–3598.
21. Hay, E. (1977): Interaction between the cell surface and extracellular matrix in corneal development. In: *Cell and Tissue Interactions,* edited by J. Lash and M. M. Burger. *(This volume.)*
22. Hemminki, K. (1973): Purification of plasma membranes from immature brain. *FEBS Lett.,* 38:79–82.
23. Holtfreter, J. (1939): Gewebeaffinität, ein Mittel der embryonalen Formbildung. *Arch. Exp. Zellforsch.,* 23:169–209.
24. Holtfreter, J. (1943): Properties and functions of the surface coat in amphibian embryos. *J. Exp. Zool.,* 93:251–323.
25. Hughes, R. C., Sanford, B., and Jeanloz, R. W. (1972): Regeneration of the surface glycoproteins of a transplantable mouse tumor cell after treatment with neuraminidase. *Proc. Natl. Acad. Sci. U.S.A.,* 69:942–945.
26. Lilien, J. E. (1968): Specific enhancement of cell aggregation *in vitro. Dev. Biol.,* 17:657–678.
27. Lilien, J. E., and Moscona, A. A. (1967): Cell aggregation: Its enhancement by a supernatant from cultures of homologous cells. *Science,* 157:70–72.
28. Maddy, A. H. (1966): The properties of the protein of the plasma membrane of ox erythrocytes. *Biochim. Biophys. Acta,* 117:193–200.
29. McClay, D. R., and Moscona, A. A. (1974): Purification of the specific cell-aggregating factor from embryonic neural retina cells. *Exp. Cell Res.,* 87:438–442.
30. Merrell, R., Gottlieb, D. I., and Glaser, L. (1975): Embryonic cell surface recognition: Extraction of an active plasma membrane component. *J. Biol. Chem.,* 250:5655–5659.
31. Moscona, A. A. (1952): Cell suspensions from organ rudiments of chick embryos. *Exp. Cell Res.,* 3:535–539.
32. Moscona, A. A. (1957): The development *in vitro* of chimeric aggregates of dissociated embryonic chick and mouse cells. *Proc. Natl. Acad. Sci. U.S.A.,* 43:184–194.
33. Moscona, A. A. (1960): Patterns and mechanisms of tissue reconstruction from dissociated cells. In: *Developing Cell Systems and Their Control,* edited by D. Rudnick, pp. 45–70. Ronald Press Co., New York.
34. Moscona, A. A. (1961): Rotation-mediated histogenetic aggregation of dissociated cells: A quantifiable approach to cell interactions *in vitro. Exp. Cell Res.,* 22:455–475.
35. Moscona, A. A. (1962): Analysis of cell recombinations in experimental synthesis of tissues *in vitro. J. Cell Comp. Physiol.,* 60 (Suppl. 1): 65–80.
36. Moscona, A. A. (1968): Cell aggregation: Properties of specific cell ligands and their role in the formation of multicellular systems. *Dev. Biol.,* 18:250–277.
37. Moscona, A. A. (1974): Surface specification of embryonic cells: Lectin receptors, cell recognition and specific cell ligands. In: *The Cell Surface in Development,* edited by A. A. Moscona, pp. 67–99. John Wiley & Sons, New York.
38. Moscona, A. A. (1975): Embryonic cell surfaces: Mechanisms of cell recognition and morphogenetic cell adhesion. In: *Developmental Biology—Pattern Formation, Gene Regulation, Vol. II,* edited by Daniel McMahon, and C. Fred Fox, pp. 19–39. W. A. Benjamin, New York.
39. Moscona, A. A. (1976): The cell surface and cell recognition in embryonic morphogenesis. Studies on experimental synthesis of tissues from cells. In: *From Theoretical Physics to Biology,* edited by M. Marois, pp. 151–168. North-Holland Publishing Co., Amsterdam.

40. Moscona, A. A. (1976): Cell recognition in embryonic morphogenesis and the problem of neuronal specificities. In: *Neuronal Recognition*, edited by S. H. Barondes, pp. 205–226. Plenum Press, New York.
41. Moscona, A. A., Hausman, R. E., and Moscona, M. (1975): Experiments on embryonic cell recognition: In search for molecular mechanisms. In: *Proceedings 10th FEBS Meeting, Paris, Vol. 38*, edited by Y. Raoul, pp. 245–256. North-Holland Publishing Co., Amsterdam.
42. Moscona, A. A., and Moscona, M. H. (1952): The dissociation and aggregation of cells from organ rudiments of the early chick embryo. *J. Anat.*, 86:287–301.
43. Moscona, A. A., and Moscona, M. H. (1966): Aggregation of embryonic cells in a serum-free medium and its inhibition at suboptimal temperatures. *Exp. Cell Res.*, 41:697–702.
44. Moscona, M. H., and Moscona, A. A. (1963): Inhibition of adhesiveness and aggregation of dissociated cells by inhibitors of protein and RNA synthesis. *Science*, 142:1070–1071.
45. Moscona, M. H., and Moscona, A. A. (1966): Inhibition of cell aggregation *in vitro* by puromycin. *Exp. Cell Res.*, 41:703–706.
46. Moyer, W. A., and Steinberg, M. S. (1976): Do rates of intercellular adhesion measure the cell affinities reflected in cell-sorting and tissue-spreading configurations? *Dev. Biol.*, 52:246–262.
47. Roseman, S. (1970): The synthesis of complex carbohydrates by multiglycosyltransferase systems and their potential function in intercellular adhesion. *Chem. Phys. Lipids*, 5:270–297.
48. Roseman, S. (1974): The biosynthesis of complex carbohydrates and their potential role in intercellular adhesion. In: *The Cell Surface in Development*, edited by A. A. Moscona, pp. 255–272. John Wiley & Sons, New York.
49. Roth, S. A., and Weston, J. A. (1967): The measurement of intercellular adhesion. *Proc. Natl. Acad. Sci. U.S.A.*, 58:974–980.
50. Rutishauser, U., Thiery, J. P., Brachenbury, R., Sela, B. A., and Edelman, G. M. (1976): Mechanisms of adhesion among cells from neural tissues of the chick embryo. *Proc. Natl. Acad. Sci. U.S.A.*, 73:577–581.
51. Seeds, N. W. (1971): Biochemical differentiation in reaggregating brain cell culture. *Proc. Natl. Acad. Sci. U.S.A.*, 68:1858–1861.
52. Seeds, N. W., and Vatter, A. E. (1971): Synaptogenesis in reaggregating brain cell culture. *Proc. Natl. Acad. Sci. U.S.A.*, 68:3219–3222.
53. Sheffield, J. B., and Moscona, A. A. (1970): Electron microscopic analysis of aggregation of embryonic cells: The structure and differentiation of aggregates of neural retina cells. *Dev. Biol.*, 23:36–61.
54. Sidman, R. L. (1974): Cell-cell recognition in the central nervous system. In: *The Neurosciences: Third Study Program*, edited by F. A. Schmitt, and F. G. Worden, pp. 743–758. MIT Press, Cambridge, Mass.
55. Sperry, R. W. (1943): Visuomotor coordination in the Newt (*Triturus viridscens*) after regeneration of the optic nerve. *J. Comp. Neurol.*, 79:33–55.
56. Spiegel, M. (1954): The role of specific surface antigens in cell adhesion. I. The reaggregation of sponge cells. *Biol. Bull.*, 107:130–148.
57. Steinberg, M. S. (1964): The problem of adhesive selectivity in cellular interactions. In: *Cellular Membranes in Development*, edited by M. Locke, pp. 321–366. Academic Press, New York.
58. Toole, B. (1977): Developmental roles of hyaluronate and chondroitin sulfate proteoglycans. In: *Cell and Tissue Interactions*, edited by J. Lash, and M. M. Burger. (*This volume.*)
59. Truding, R., Shelanski, M. C., and Morell, P. (1975): Glycoproteins released into the culture medium of differentiating murine neuroblastoma cells. *J. Biol. Chem.*, 250:9348–9354.
60. Vogel, Z., Daniels, M. P., and Nirenberg, M. (1976): Synapse and acetylcholine receptor synthesis by neurons dissociated from retina. *Proc. Natl. Acad. Sci. U.S.A.*, 73:2370–2374.
61. Warren, L., and Glick, M. C. (1968): Membranes of animal cells. II. The metabolism and turnover of the surface membrane. *J. Cell. Biol.*, 37:729–746.
62. Weiss, P. (1947): The problem of specificity in growth and development. *Yale J. Biol. Med.*, 19:235–278.

Cell and Tissue Interactions, edited by
J. W. Lash and M. M. Burger. Raven
Press, New York, 1977.

A Multicomponent Model for Specific Cell Adhesion

Jack Lilien and Richard Rutz

Department of Zoology, University of Wisconsin, Madison, Wisconsin 53706

The dissection and analysis of the mechanisms directing morphogenetic rearrangements has been a well-recognized challenge to developmental biologists. For many years, beginning with Townes' and Holtfreter's classic studies with dissociated amphibian embryos (14), attention has focussed on the *in vitro* capabilities of dissociated embryonic cells to reaggregate and sort out histotypically (10,13). However, with the recent explosion of interest in a variety of cell surface phenomena, technologies more appropriate for defining the molecular basis of histospecific cell readhesions are being introduced. While assays for readhesion, rather than reaggregation or sorting out, greatly simplify the experimental variables, the challenge remains the same — to understand the mechanisms of *in vivo* morphogenetic movements.

At the moment, attention is being focused by several laboratories on a single tissue type (6,9–12) — the chick embryo neural retina. Other tissue types must, of course, be used, most importantly to exploit and understand the cardinal fact of specificity. On the basis of previously published evidence (1–3), our own laboratory has constructed a working model of neural retina cell adhesion consisting of three components, each of which exhibits histotypic specificity. The components are: (a) a cell surface receptor, (b) a glycoprotein ligand, and (c) a protein-containing agglutinin. As discussed below and elsewhere (3), the evidence suggests that the ligand interacts with the receptor through its oligosaccharide moiety and interacts with the agglutinin through its peptide moiety.

In this chapter we shall discuss the recent development of a new assay for components presumably involved in adhesive recognition. This assay entails the agglutination of glutaraldehyde-fixed cells. We have chosen this particular approach for several reasons: (a) by fixing the cells, complications in the interpretation of effects directly relevant to the adhesive mechanism as opposed to effects on the endogenous synthesis of adhesive components are eliminated; (b) this technique enables us to assess adhesive capacity subsequent to various treatments in prior culture and thus to study the acquisition of specific adhesive capacity following preparation of single cells; (c) since cells tested immediately following single cell preparation with trypsin are not agglutinable under our assay conditions, we may attempt a step-by-step reconstruction of the adhesive mechanism; and (d) agglutinability can be easily quantitated by using the Coulter Counter to monitor the number of single cells remaining in suspension over time.

THE SPECIFICITY OF AGGLUTINATION

Our first indication that agglutination of fixed cells might be a valuable tool in dissecting the specific adhesive mechanism came with the finding that freshly trypsinized cells, which had been exposed to culture medium conditioned by neural retina tissue prior to fixation, agglutinated when maintained in suspension over a monolayer of homologous cells. Cells fixed immediately after dispersal were not agglutinated under these same conditions. Most importantly agglutination was tissue-specific: Neural retina and cerebral lobe cells fixed after exposure to homologous tissue culture medium were not agglutinated over monolayers of heterotypic cells (3). Thus, agglutination appeared to mimic the adhesion of live cells in the critical aspect of specificity.

The next tasks in the development of the assay have been to replace the live cell monolayer with a cell-free supernatant, characterize the parameters governing agglutination, and then to use this assay to determine the relative roles of the component present in tissue culture medium (TCM) versus the component(s) contributed by the monolayer culture medium (MCM).

THE REQUIREMENT FOR A LIVE CELL MONOLAYER

Our initial experiments were performed using monolayers of 10-day neural retina cells in Eagle's basal medium. Once the cells were adherent, the

TABLE 1. *The specificity of MCM-mediated agglutination*

Cell type[a]	MCM source	Percent agglutination[b]
Retina	Retina	29.1
Retina	Cerebrum	5.1
Cerebrum	Cerebrum	25.6
Cerebrum	Retina	7.4

[a] Single cells dispersed from the tissue following a tryptic digestion (see ref. 3) were incubated in TCM from homotypic monolayers for 30 min followed by fixation in 1.4 or 2% glutaraldehyde, washed twice in 0.2 M glycine and four times in 0.01 M phosphate-buffered saline (0.15 M), pH 7.2, and used immediately.

[b] Agglutination of 1×10^6 fixed cells was carried out in 30 mm Falcon plastic dishes in 3 ml of MCM at 37°C on gyratory shaker at 70 rpm. After 18 hr of incubation, the number of free single cells was determined utilizing a Coulter Counter. The results are recorded as the percentage of cells incorporated into agglutinates.

cultures were washed, fresh BME including the various fixed-cell preparations was added, and the cultures rotated at 70 rpm for 24 hr. Medium from 24-hr monolayer cultures alone was ineffective in mediating agglutination, suggesting the possibility that the component(s) contributed by the monolayer were extremely labile. This appears to be the case since activity can be preserved by the addition to the monolayer medium immediately on collection of inhibitors of protease activity such as PMSF or TLCK. The specificity of the agglutinating activity present in this cell-free monolayer-conditioned medium exactly parallels that of live cell monolayers (Table 1).

CHARACTERISTICS OF AGGLUTINATION

In all of the experiments that follow, cells fixed following exposure to TCM were compared to cells treated similarly but without exposure to the TCM. Maximal agglutination occurs over a period of 17 hr (Fig. 1) and is temperature-dependent with an optimum at 37°C (Fig. 2). These characteristics distinguish this process from plant-lectin-mediated agglutination of retina cells, which is maximal in 30 min at 22°C. In addition, both Con A and WGA agglutinate fixed, freshly dispersed cells (7). Observations on agglutinability versus the concentration of MCM indicate that no single component is limiting, since maximal agglutination is achieved at a concentration of 75% MCM (v/v) (Fig. 3).

One further characteristic of the agglutinating activity is particularly revealing. MCM dialyzed against EDTA followed by dialysis against physiological salt solution is completely inactive. Addition of either Ca^{++} or Mg^{++} restores activity with optimal agglutination occurring at concentrations between 1 and 4 mM. These results are interesting for two obvious

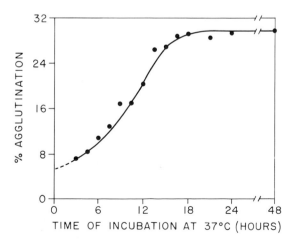

FIG. 1. Time course of agglutination. Conditions were the same as described in the legend to Table 1.

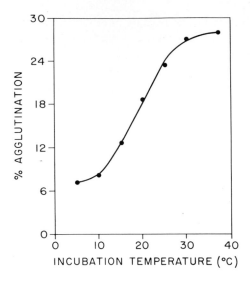

FIG. 2. The effect of incubation temperature on agglutination. All other conditions
were the same as described in Table 1.

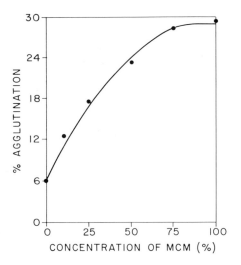

FIG. 3. The effect of MCM concentration
on agglutination. Conditions as per Table 1.

reasons: (a) since a Ca^{++} or Mg^{++} requirement for the readhesion of live
cells has been repeatedly and widely documented, these results give us
additional confidence that what we are measuring is indeed relevant to live
cell adhesion, and (b) more importantly, since MCM activity does exhibit this
requirement but interaction of TCM with cells does not, a strong suggestion
is provided as to where in the adhesive assembly the divalent cations function.

While these particular characteristics of the assay cannot as yet be interpreted in molecular terms, optimization of the assay conditions has allowed
us to pursue several other lines of experimentation.

DISTINCTION OF TCM VERSUS MCM

As is already obvious, TCM and MCM appear to contain separate and distinct activities: TCM contains a component that interacts with freshly dispersed cells rendering them agglutinable by MCM. This relationship between prior exposure to TCM and agglutinability by MCM can be quantitatively evaluated. As shown in Fig. 4, following exposure to dilutions of ^{3}H-glucosamine-labeled TCM, agglutinability by MCM increases with increasing amounts of radioactivity bound to the cells and with increasing protein content of the TCM. Agglutinability reaches a plateau with respect to either measured variable, which suggests saturation of cell surface sites.

The activities present in these two preparations can be further distinguished in several other ways. We have previously reported that neural retina TCM contains a tissue-specific component which binds at the cell surface (1,2). We have also reported that TCM contains a tissue-specific component which inhibits plant lectin-induced capping of cell surface receptors (8). Both of these activities depend on the integrity of an oligosaccharide moiety (2,8) and reside in a single glycoprotein of approximately 95,000 daltons (Hermolin, Cook, and Lilien, *in preparation*). Monolayer-conditioned medium contains less than 20% of the activity of TCM, as measured by its ability to inhibit the capping of Con A receptors. Conversely TCM is completely inactive when assayed for its ability to agglutinate either freshly dispersed fixed cells or cells fixed after a prior exposure to TCM. This is true even when protease inhibitors are added to the TCM.

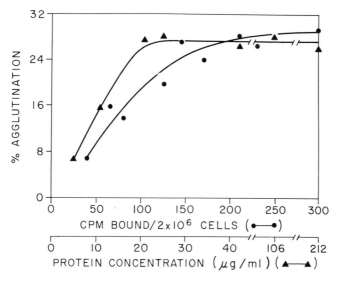

FIG. 4. Agglutination as a function of TCM protein concentration during preincubation prior to fixation and as a function of counts bound during preincubation with a TCM preparation labeled with ^{3}H-glucosamine (1). Agglutination conditions as per Table 1.

RELATIVE ROLES OF TCM AND MCM

What is the nature of the interactions between the components in TCM and MCM? Do they interact directly with one another, or indirectly through some other cell surface-mediated molecular mechanism? We have used several experimental approaches to attack this question, the results of which strongly suggest that the interaction is direct and that the TCM glycoprotein interacts in a polar fashion with the oligosaccharide portion interacting with a cell surface receptor and the peptide portion interacting with the agglutinating component(s) present in MCM.

TCM was digested with either pronase or β-N-acetyl hexosaminidase. These preparations were used in three different ways: (a) as inhibitors of concanavalin A-induced capping of surface receptors, (b) as effectors of MCM-mediated agglutinability by incubation with freshly dispersed cells prior to glutaraldehyde fixation, and (c) as inhibitors (i.e., competitors) of MCM-mediated agglutinability of fixed cells exposed to undigested TCM prior to fixation. The results are as follows: (a) glycosidase digestion, but not pronase digestion, destroys the ability of TCM to inhibit capping (Table 2). Pronase digestion also does not destroy the specific binding of ^{3}H-glucosamine-labeled TCM (2). We interpret these results to indicate that the functional binding of the TCM glycoprotein to the cell surface is through the oligosaccharide portion of the molecule and not the peptide portion. (b) Both glycosidase and pronase digestion destroy the ability of TCM to render cells agglutinable by MCM (Table 2). (c) Pronase-digested TCM is ineffective in inhibiting MCM-mediated agglutination of intact TCM-treated fixed cells. However, both intact and glycosidase-digested TCM are effective inhibitors (Fig. 5). Thus, soluble TCM behaves as a competitor of MCM agglutinating activity only if it retains its peptide portion with or without an

TABLE 2. *The effect of digestion of TCM with pronase[a] or β-N-acetyl hexosaminidase[b] on capping of Con A receptors and MCM-mediated agglutinability*

	Capping inhibition[c]	MCM agglutinability[d]
TCM	+	+
Pronase TCM	+	−
β-N-Acetyl hexosaminidase TCM	−	−

[a] Immobilized protease (enzite agarose, Miles) was used at 0.1 units/ml TCM and incubated at 37°C, pH 7.0, for 3 hr. Enzyme was removed by centrifugation prior to assay.

[b] Purified β-N-acetyl hexosaminidase (*Turbo cornutus;* Miles) was used at 0.01 units/ml of TCM and incubated at 30°C, pH 4.2, for 30 min. Prior to use, the pH was adjusted to 7.2.

[c] Capping inhibition was assayed as previously described (8).

[d] Agglutination was scored visually.

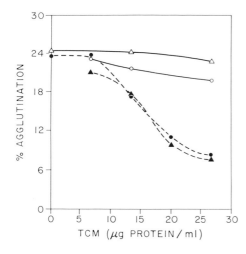

FIG. 5. The effect of various soluble TCM preparations on MCM-mediated agglutination. ●——●, neural regina TCM; △——△, cerebral lobe TCM; ○——○, protease-digested TCM (see legend to Table 1 for digestion conditions); ▲——▲, β–N–acetyl hexosaminidase-digested TCM (see legend to Table 2 for digestion conditions). Protein determinations were made prior to digestion. Agglutination conditions as per Table 1.

intact oligosaccharide portion. We interpret this to mean that TCM interacts directly with MCM through the peptide portion.

We have also used TCM from cerebral lobes as a competitor in MCM-mediated agglutinability of retina cells and find that it is 50 times less effective at equivalent protein concentrations than neural retina TCM substantiating the specificity of this interaction (see Fig. 5).

MCM was also tested for its ability to interfere with the effect of TCM on the capping of Con A surface receptors. The fact that MCM did not interfere independently confirms that MCM does not interact with the same portion of TCM responsible for cell surface binding. Significantly, this fact also excludes the possibility that MCM functions in our assay simply as a soluble form of the cell surface receptor for TCM.

DO THE ASSAYS FOR SOLUBLE TCM REFLECT NORMAL CELL-BOUND PROPERTIES OF THE GLYCOPROTEIN?

The last experiments to be described suggest that the glycoprotein component present in TCM is normally present and functions at the cell surface but is degraded or removed during dispersal of the tissue to single cells with trypsin. Recall that TCM contains activity both to inhibit capping and to participate in MCM-mediated agglutinability. The experiment was to allow fresh trypsin-dispersed cells to repair in cell culture for various periods of time. At each time point, cells were fixed and assessed for agglutinability by MCM (repaired cells are not agglutinable in the absence of MCM). Superimposed on the graph of the results (Fig. 6) is the loss with time in culture of the ability to cap Con A receptors. Through the first 2 hr in culture, agglutinability increases, and the ability to cap decreases. The two curves exhibit similar time courses, the implication being that cell repair

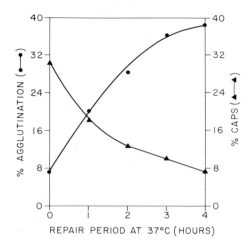

FIG. 6. Agglutination and ability to form Con A-induced caps as a function of prior culture period. Following trypsinization, single cells were cultured in Eagle's basal medium with 2× the usual concentration of nonessential amino acids, an extra 2 mg/ml glucose, and 0.5% gentamycin at 37°C in Falcon petri dishes at 5×10^6 cells/ml, 3.5 ml/dish. At the indicated times, cells were either scored for their ability to form Con A-induced caps (7) or fixed as per Table 1 and assayed for MCM-mediated agglutinability.

following trypsinization includes functional replacement of cell-bound TCM activity and that a single molecular species may be responsible for both phenomena.

EVALUATION AND PERSPECTIVES

Many uncertainties remain with regard to the three-component working model. While the work indicating that the TCM activity is a single glycoprotein is nearing completion, we do not yet know whether the agglutinating activity in MCM resides in a single peptide-containing species. Until this is known, the evidence remains circumstantial as to its mode of action. Moreover, we cannot as yet distinguish whether the requirement for Ca^{++} or Mg^{++} is for maintenance of MCM activity or to participate in the molecular interaction between the glycoprotein ligand and the agglutinin.

While we cannot at present determine what role, if any, these molecular species play in morphogenesis, we (5) and others (4) have noted the close parallels between cell movement and capping. Based on this similarity, we (8) have suggested that the glycoprotein in TCM may play a role in stabilizing cellular architecture during and subsequent to morphogenetic rearrangements. The present collection of *in vitro* assays, when combined with access to purified components, should eventually allow a description of the spatial and temporal patterns of appearance and function of adhesive components, yielding considerable insight into morphogenetic processes.

ACKNOWLEDGMENT

Studies conducted in the laboratory of Dr. Lilien were supported by a grant from NSF.

REFERENCES

1. Balsamo, J., and Lilien, J. (1974): Embryonic cell aggregation: Kinetics and specificity of binding of enhancing factors. *Proc. Natl. Acad. Sci. U.S.A.,* 71:727–731.
2. Balsamo, J., and Lilien, J. (1975): The binding of tissue-specific molecules to the cell surface. A molecular basis for specificity. *Biochemistry,* 14:167–171.
3. Balsamo, J., and Lilien, J. (1974): Functional identification of three components which mediate tissue-type specific adhesion. *Nature,* 251:522–524.
4. DePetris, S., and Raff, M. C. (1973): Fluidity of the plasma membrane and its implications for cell movement. In: *Locomotion of Tissue Cells. Ciba Foundation Symposium 14.* Associated Scientific Publishers, New York.
5. Lilien, J., Balsamo, J., McDonough, J., Hermolin, J., Cook, J., and Rutz, R. (1977): Adhesive specificity among embryonic cells. In: *Surfaces of Normal and Cancer Cells,* edited by R. O. Hynes. Wiley International, London (*in press*).
6. McClay, D. R., and Baker, S. R. (1975): A kinetic study of embryonic cell adhesion. *Dev. Biol.,* 43:109–122.
7. McDonough, J., and Lilien, J. (1975): Spontaneous and lectin induced redistribution of cell surface receptors on embryonic chick neural retina cells. *J. Cell Sci.,* 19:357–368.
8. McDonough, J., and Lilien, J. (1975): Inhibition of cell surface receptor mobility by factors which mediate specific cell-cell interactions. *Nature,* 256:216–217.
9. Merrel, R., Gottlieb, D. I., and Glaser, L. (1975): Embryonal cell surface recognition: Extraction of an active plasma membrane component. *J. Biol. Chem.,* 250:5655–5659.
10. Moscona, A. A., and Hausman, R. E. (1977): Biological and biochemical studies on embryonic cell recognition. In: *Cell and Tissue Interactions,* edited by J. Lash, and M. M. Burger. (*This volume.*)
11. Roth, S., McGuire, E. J., and Roseman, S. (1971): An assay for intercellular adhesive specificity. *J. Cell Biol.,* 51:525–535.
12. Rutishauser, U., Thiery, J. P., Brackenbury, R., Sela, B. A., and Edelman, G. M. (1976): Mechanisms of adhesion among cells from neural tissues of the chick embryo. *Proc. Natl. Acad. Sci. U.S.A.,* 73:577–581.
13. Steinberg, M. S. (1970): Does differential adhesion govern selfassembly processes in histogenesis? Equilibrium configuration and the emergence of a hierarch among populations of embryonic cells. *J. Exp. Zool.,* 173:395–434.
14. Townes, P. L., and Holtfreter, J. (1955): Directed movements and selected adhesion of embryonic amphibian cells. *J. Exp. Zool.,* 128:53–120.

Cell and Tissue Interactions, edited by
J. W. Lash and M. M. Burger. Raven
Press, New York, 1977.

Cell-Cell Recognition in the Embryonal Nervous System

Luis Glaser, Roger Santala, David I. Gottlieb, and Ronald Merrell

Department of Biological Chemistry, Division of Biology and Biomedical Sciences, Washington University, St. Louis, Missouri 63110

It is now clearly established that many hormones act by binding to the external surface of a cell and do not need to penetrate into the cell in order to have a metabolic effect. Target cells have on their surface a specific receptor that binds to the hormone, and this binding triggers the metabolic events that are characteristic of the particular hormone and the particular target cell. Cell-cell recognition can be considered in very similar terms.

When two cells bind to each other, the binding represents the interaction of a receptor on the surface of one type of cell with a ligand on the surface of the second cell. The cells can either be of the same type, or heterologous cells. An example of the first type is the binding of dissociated embryonal cells to each other, while an example of the second type would be synapse formation between a neuronal axon and a muscle cell.

The initial cell-cell binding is then followed by both morphological and biochemical changes. The induction of these changes as a consequence of cell to cell binding are the analogues of the hormone-induced metabolic changes. The reason for emphasizing the analogy between cell to cell binding and hormone action is that cell to cell binding, in addition to simply holding cells together, appears to have metabolic consequences that have not been extensively explored. Yet, a study of these metabolic and morphological effects could allow us at least in principal to distinguish between biologically meaningful and artifactual binding between cells.

The chemistry of cell adhesion is poorly understood. In the two cases of sponges (1,2,8,9,17) and slime molds (3,19,20,27), there is evidence that cell recognition involves the binding of a carbohydrate to a receptor (30) on an adjacent cell. The precise mechanism of cell adhesion has not been completely elucidated in either of these systems.

The chemical basis of cell-cell recognition in other systems is even less clear. One of the most popular systems for the study of cell-cell adhesion has been the study of the reaggregation of single cells obtained by dissociation of embryonal tissues (7,15,24). In this system it was first shown by Roth and Weston (21,23) and Roth et al. (22) that initial adhesion events occurred with a specificity such that cells from any given organ prefer to adhere to each other as compared to heterologous cells. Although it had previously been shown that, in mixed aggregates, cells would sort out according to the tissue of origin (for review, see refs. 15,28), it is unclear in such experiments whether one is measuring only the affinity of cells for each other or whether other variables such as chemotaxis also play a role. Not until the Roth, Weston, and Roseman experiments was initial adhesive specificity clearly demonstrated.

Several laboratories are currently investigating cell to cell adhesion with dissociated embryonal cells. Very frequently these laboratories use quite different methods to study cell adhesion. Because the process is complex, it is quite possible that different methods will give somewhat different results. Examples of these methods are a study of aggregate size after 24 hr aggregation (7,15,16), use of antibodies to detect aggregation components (24), measurement of the ability of single cells to bind to large aggregates (22) or to monolayers (4,29), and the use of plasma membranes as probes of cell surface adhesive specificity (5,11,12).

Our own approach has been to assume a simple model for cell adhesion, namely that cells contain complementary ligands on their surface and that binding of one or more of these ligands between cells results in adhesion.

It seems necessary to assume that one of these ligands is a protein, since no other known biological molecule with the exception of nucleic acids would have the required binding specificity. The complementary ligand to this protein can be another protein, the carbohydrate portion of a glycoprotein, a glycolipid, or some combination of these ligands.

On the basis of this simple assumption, we have asked the following questions: (a) Can the binding specificity of cells be reproduced at the level of plasma membranes? (b) Does the cell surface specificity of embryonal cells change only from organ to organ, or does it also change as a function of time of development? (c) Can the molecules responsible for specific cell recognition be isolated from plasma membranes?

Before we attempt to summarize the current status of our work to try to answer some of these questions, a few comments regarding the use of single cells obtained from the embryonal nervous system are necessary.

A single cell suspension obtained from any one region of the embryonal nervous system may appear superficially as a uniform population of cells, of a certain diameter, but it clearly is derived from a mixture of cells of different types and in different stages of development. This cell heterogeneity greatly complicates the interpretation of cell to cell adhesion experiments.

One of the most important questions regarding the nervous system is the specificity of synapse formation, and it is tempting to consider cell to cell adhesion as if it were relevant to the known specificity of synapse formation. It should be clear that the specificity with which cell bodies adhere to each other may be very different from the specificity with which two cells connect via their axons. The relevance of cell to cell adhesive specificity to synaptogenesis is not known; it may either be very important, or it may be irrelevant.

Using chick retinal cells, we could show that plasma membranes prepared from such cells bind specifically to homologous neural retina cells, and also inhibit the aggregation of homologous cells (11).

The mechanism by which membranes prevent cell aggregation is not understood. The two most likely alternatives are: (a) the membranes only contain one of the two complementary ligands involved in cell adhesion and therefore cannot serve as a bridge between two cells coated with membrane vesicle; and (b) the membranes contain both complementary ligands, but the adhesion between cells and membranes or membranes and membranes is not strong enough to hold two or more cells together.

Using an inhibition of aggregation assay, we could show that there were changes in adhesive specificity both in the retina and tectum with time of development such that membranes from 8-day retina only inhibited the aggregation of homologous cells, but showed very little inhibition of aggregation of 7-day cells or 9-day cells (Fig. 1). Retinal plasma membranes also inhibit the aggregation of tectal cells, but tectal plasma membranes do not inhibit the aggregation of retinal cells.

Although these data suggest an absolute difference in cell surface specificity between cells at different days of development, the limitations of the assay suggest that this difference may be quantitative rather than qualitative. The assay selects for those cell surface adhesive components that can bind rapidly and extensively to cells and thereby prevent cell to cell adhesion. Membrane vesicles that bind slowly to cells, or in small numbers, will appear to have very little effect on cell to cell adhesion. The inhibition

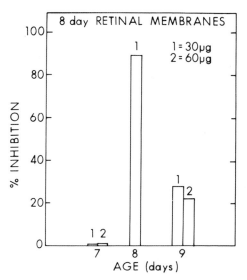

FIG. 1. Inhibition of retinal cell aggregation by plasma membranes. Neural retinal cells of the ages indicated were incubated with plasma membranes (30 or 60 μg of protein) for 30 min under standard aggregation conditions and the number of remaining single cells were counted. In control incubations 80 to 90% of the retinal cells had aggregated at 30 min (for details see ref. 5).

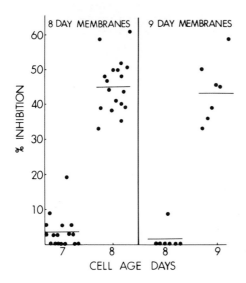

FIG. 2. Inhibition of tectal cell aggregation by plasma membranes. The figure shows the inhibition of tectal cell aggregation by either 8- or 9-day plasma membranes. Each point represents a separate experiment with different cell and membrane preparations. Submaximal quantities of membranes were used to facilitate cell counting. The experimental protocol was that used in ref. 13.

of cell aggregation assay will therefore only measure the presence of the most abundant ligand present on the cell surface and not measure less abundant ligands. In addition, the membrane inhibition assay can only be carried out with a limited amount of membranes, since very high membrane concentrations prevent counting of the cells. Thus, changes in affinity for certain ligands may be scored as a disappearance of such ligands from the cell surface because it is not possible to increase the membrane concentration high enough to bind to these sites.

In spite of the reservations regarding the interpretation of this assay, it has provided reproducible evidence for changes in cell surface specificity with developmental age. For example, the data in Fig. 2 show the results of replicate experiments with different cell and membrane preparations that show differences in cell surface specificity among 7-, 8-, and 9-day cells obtained from the optic tectum.

We have been interested in studying changes in cell surface specificity in tissue culture and have concentrated on the temporal transition between day 7 and day 8 optic tectum. In experiments carried out in collaboration with M. W. Pulliam and R. Bradshaw, we could show that aggregated cell cultures prepared from 7-day optic tectum cells showed a transition to 8-day specificity only in the presence of mouse submaximal nerve growth factor (NGF) (Fig. 3). The high concentration of NGF required in this system (10^{-7} M), as well as a variety of other considerations, suggested that NGF was acting in this system as an analogue of a natural trophic factor (13). This supposition was strengthened by the observation that some lots of fetal calf serum contained high activity of this trophic factor, which could not be accounted for by its content of NGF.

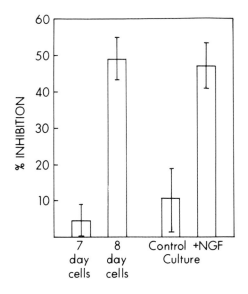

FIG. 3. Appearance of 8-day tectal specificity in tissue culture. Seven-day tectal cells (5×10^6) were kept in a rotating culture for 24 hr either in the presence or absence of NGF (10^{-7} M) **(bars on right).** The tectal aggregates were dissociated, and the ability of plasma membranes prepared from 8-day chick tectum to inhibit the aggregation was determined. As control, the same membranes were assayed with 7- and 8-day tectal cells obtained from chick embryos **(bars on left).** The result shows the average of 40 separate experiments; the bars show standard deviations. For details see ref. 13.

The finding of a trophic effect on the development of cell surface specificity in the embryonal nervous system is of considerable interest. Trophic effects on the development of other neuronal characteristics in cell culture have been noted in a number of instances (for example, refs. 14,18) in addition to the well-known effect of NGF on the maintenance of sympathetic neurons. The present finding is complicated because we have not identified the natural trophic factor, and the assay is very clumsy, thus making the purification of the natural cofactor very difficult.

It was possible using lithium diiodosalicylate to obtain "proteins" from acetone powder of plasma membranes of retina and tectum that inhibited cell aggregation, but the quantity of these proteins was too small to allow their purification (12). Because of the great cellular heterogeneity of the embryonal systems, we have to consider the possibility that any given tissue contains several recognition systems of different specificity. There are no data at the moment that estimate the number of adhesive specificities present in one organ. However, it is clear that quantitatively not all the cells in one organ have identical adhesive behavior. Using a new monolayer adhesion assay (4), we could show the presence of a gradient of adhesive specificity in the neural retina such that dorsal cells prefer to adhere to monolayers prepared from ventral cells, while ventral cells prefer to adhere to monolayers prepared from dorsal cells (Fig. 4). This dorsal-ventral gradient may be related to retina-tectal connectivity as has been discussed in great detail by S. Roth (*this volume*) and by ourselves elsewhere (6). In the context of this paper, it illustrates one proven example of the heterogeneity in cell surface adhesive components of neuronal cells.

The isolation of adhesive components from cells is clearly a difficult

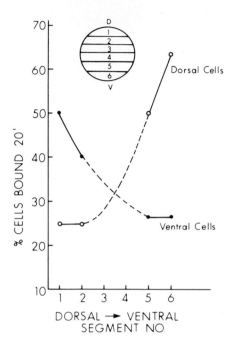

FIG. 4. Gradient of adhesive specificity in the neural retina. The neural retina from 12-day-old chick embryos was separated into six strips along the dorsoventral axis such that strip 1 came from the most dorsal area and strip 6 from the most ventral. Cell monolayers were prepared from cells in areas 1, 2, 5, and 6, and the ability of extreme dorsal (area 1) and extreme ventral (area 6) cells to adhere to such monolayers was determined. The results are the average of five to six experiments done in triplicate. The data summarize observations shown in Table 1 of ref. 6.

task, not only because of the limited availability of material, but also because the adhesive components are likely to be hydrophobic molecules. Hydrophobic molecules are usually fractionated in the presence of detergents, but detergents are generally incompatible with an assay, which depends on the interaction of these molecules with live cells.

We have recently turned to a model system using cultured neuronal cells that can be obtained in large quantity but show a relatively broad adhesive specificity within the nervous system in the hope that the chemistry of cell adhesion can be studied more readily in such a system.

A number of cloned rat neuronal cell lines were kindly given to us by Dr. David Schubert of the Salk Institute (25). The adhesive properties of one of these cell lines, B103, have been examined in some detail.

Using a new monolayer adhesion assay (Fig. 5), we can show that B103 cells will adhere to a monolayer of neuronal cells obtained from either rat or chicken embryos but will adhere either very poorly or not at all to a monolayer of Chinese hamster ovary (CHO) cells, fibroblasts, or liver cells prepared either from rat or chick embryos (Fig. 6). These cells cannot distinguish among various areas of the chick embryo nervous system (Fig. 7). Thus, the adhesive specificity of these cells is quite broad, but they appear able to distinguish neuronal from nonneuronal cells in this assay. B103 cells will also adhere to rat neuronal cells as well as chick neural cells. This ability is not as surprising as it might appear. Previous data from other

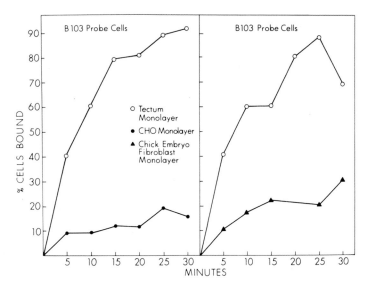

FIG. 5. Binding of B103 cells to tectum and fibroblast monolayers. The figure shows the time dependence of binding of radioactive B103 probe cells to monolayers of the cells indicated on the figure.

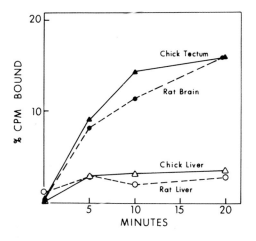

FIG. 6. Binding of ^{3}H-leucine B103 membranes to rat cerebral cortex and rat liver monolayers. Monolayers were prepared from 16-day-old rat embryos and 8-day-old chick embryos.

laboratories have shown that with embryonal cells, organ specificity overrides species specificity (15,21).

A plasma-membrane-enriched fraction prepared from B103 will bind preferentially to neuronal cells from the chick embryo (including B103 cells) but not to liver (Fig. 8) or fibroblasts, or to C6, an established rat glial cell line (data not shown).

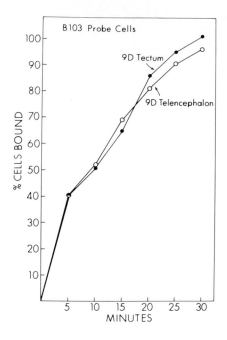

FIG. 7. Binding of B103 cells to tectum and telencephalon. Monolayers were prepared from 9-day chick embryos, and the binding of radioactive B103 cells was measured under standard conditions (4,6).

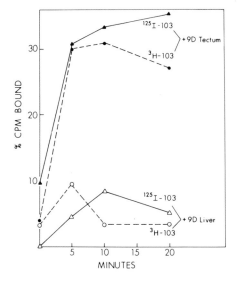

FIG. 8. Binding of B103 plasma membranes to neuronal cells. Plasma membranes were prepared from B103 cells either labeled metabolically with [³H]D-glucosamine or with ¹²⁵I with lactoperoxidase. The figure shows the time course of binding to these membranes to cells.

The binding of membranes to neural cells shows the following characteristics. (a) It is abolished by treatment of the plasma membranes with trypsin (Fig. 9); (b) it is abolished at low temperature; (c) it is abolished by mild fixation of the cells with either glutaraldehyde or formaldehyde; or (d) it is about 70% inhibited by treatment of the cells with trypsin.

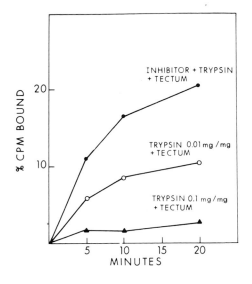

FIG. 9. Trypsin inactivation of B103 membranes. Plasma membranes were treated with the indicated ratios of crystalline trypsin to membrane protein for 10 min at 37°, and the ability of the membranes to bind to tectal cells was examined after addition of excess soybean trypsin inhibitor. The control received a mixture of trypsin and trypsin inhibitor.

The fact that treatment of either the cells or the membranes abolishes cell-membrane adhesion suggests that both the complementary adhesive components are proteins. It should be noted, however, that residual adhesive activity remains after exposure of cells to a high concentration of trypsin. Thus, it is possible that a trypsin-resistant component also contributes to adhesion in this system. However, this trypsin-resistant component becomes trypsin-sensitive in the plasma membranes. This would indicate that the three-dimensional arrangement of adhesive sites on the cell surface and on the isolated plasma membrane is different.

The fact that treatment of cells with low concentrations of formaldehyde or glutaraldehyde abolishes membrane-cell adhesion suggests one of two possible explanations. (a) Both recognition components are proteins, and formaldehyde or glutaraldehyde inactivate these proteins. (b) Treatment with these fixatives interferes with mobility of the adhesive components on the cell surface and may prevent adhesion because a clustering of adhesive sites is required to make a stable membrane-to-cell linkage. At the moment, we have no experimental evidence to distinguish these two possibilities, but the temperature sensitivity of the membranes to cell adhesion favors the second possibility.

Even in the adhesive system using neuronal cell lines, the chemistry is more complex than it would appear initially. We have recently obtained evidence that these neuronal cell lines contain more than one adhesive component of different specificity (Table 1). The presence of multiple components may explain why these cell lines can bind to embryonal neural cells of different regions without apparent specificity. The data that we have obtained are consistent with the assumption that only a part of the recognition system remains functional in the plasma membranes. Thus, if we

TABLE 1. *Binding specificity of different neuronal cell lines*

Cells	Plasma membranes			
	B103	B65	B50	C-6
B103	+	−	+	−
B65	+	+	+	+
B50	+	+	+	N.D.
C-6	−	−	N.D.	−
Tectum	+	+	+	+

N.D., not determined.

assume that Aa and Bb are pairs of complementary cell surface ligands, our data on cell lines B103 and B65 could be accounted for by the following scheme.

Cell line	Ligand expressed on cells	Ligand expressed on membranes
B103	Aa	a
	b	b
B65	Bb	b

Preliminary evidence suggests that other lines will in turn express different specificities than either B103 or B65.

It is of considerable interest that the membrane of the neuronal cell line expresses a much more restricted set of ligands than whole cells. Similarly, we have previously assumed that embryonal plasma membranes only express one of the complementary adhesive components. Although we do not understand why membranes only express a restricted number of ligands, this property appears to have potential use in dissecting the adhesive properties of cells.

In summary, we have been able to show that membranes specifically interact with cells by two methods: by membrane inhibition of cell aggregation and specific binding of membranes to cells. We believe that the neuronal cell lines will make a significant contribution to our understanding of cell adhesion because they are a uniform cell population that is available in relatively large quantities and show specific adhesion properties.

ACKNOWLEDGMENTS

Work in the authors' laboratory has been supported by NIH Grants GM-18405 and NSF BMS-75–22638. Dr. Santala was supported by Grant J-T2-GM-07157. Dr. Gottlieb was supported by Grant NS-10943. Dr. Merrell was supported by Grant 5 TO1-GM-60371. We are grateful to Dr.

M. Young of Harvard University for assay of nerve growth factor in fetal calf serum. We are particularly grateful to Dr. David Schubert of the Salk Institute for the kind gift of neuronal cell lines.

REFERENCES

1. Burger, M. M. (1977): Membrane involvement in cell-cell interactions: A two-component model system for cellular recognition that does not require live cells. In: *Cell and Tissue Interactions,* edited by J. Lash, and M. M. Burger. (*This volume.*)
2. Cauldwell, C. B., Henkart, P., and Humphreys, T. (1973): Physical properties of a sponge aggregation factor: A unique proteoglycan. *Biochemistry,* 12:3051–3055.
3. Frazier, W. A., Rosen, S. D., Reitherman, R. W., and Barondes, S. H. (1975): Purification and comparison of two developmentally regulated lectins from *Dictyostelium discoideum. J. Biol. Chem.,* 250:7714–7721.
4. Gottlieb, D. I., and Glaser, L. (1975): A novel assay of neuronal cell adhesion. *Biochem. Biophys. Res. Commun.,* 63:815–821.
5. Gottlieb, D. I., Merrell, R., and Glaser, L. (1974): Temporal changes in embryonal cell surface recognition. *Proc. Natl. Acad. Sci. U.S.A.,* 71:1800–1802.
6. Gottlieb, D. I., Rock, K., and Glaser, L. (1976): A gradient of adhesive specificity in developing avian retina. *Proc. Natl. Acad. Sci. U.S.A.,* 73:410–414.
7. Hausman, R. E., and Moscona, A. A. (1975): Purification and characterization of a retina specific aggregation factor. *Proc. Natl. Acad. Sci. U.S.A.,* 72:916–920.
8. Humphreys, T. (1963): Chemical dissolution and *in vitro* reconstruction of sponge cell adhesions. I. Isolation and functional demonstration of the compounds involved. *Dev. Biol.,* 8:27–47.
9. Margoliash, E., Schenk, J. R., Hargie, M. P., Burokas, S., Richter, W. R., Barlow, G. A., and Moscona, A. A. (1965): Characterization of specific cell aggregating materials from sponge cells. *Biochem. Biophys. Res. Commun.,* 20:383–388.
10. McGuire, E. J., and Burdick, C. L. (1976): Intercellular adhesive selectivity. I. An improved assay for the measurement of embryonic chick intercellular adhesion (liver and other tissues). *J. Cell. Biol.,* 68:80–89.
11. Merrell, R., and Glaser, L. (1973): Specific recognition of plasma membranes by embryonic cells. *Proc. Natl. Acad. Sci. U.S.A.,* 70:2794–2798.
12. Merrell, R., Gottlieb, D. I., and Glaser, L. (1975): Embryonal cell surface recognition extraction of an active plasma membrane component. *J. Biol. Chem.,* 250:5655–5659.
13. Merrell, R., Pulliam, M. W., Randono, L., Boyd, L. F., Bradshaw, R. A., and Glaser, L. (1975): Temporal changes in cell surface specificity induced by nerve growth factor. *Proc. Natl. Acad. Sci. U.S.A.,* 72:4270–4274.
14. Monard, D., Solomon, F., Rentsch, M., and Gysin, R. (1973): Glial induced morphological differentiation in neuroblastoma cells. *Proc. Natl. Acad. Sci. U.S.A.,* 70:1894–1897.
15. Moscona, A. A. (1965): Recombination of dissociated cells and the development of cell aggregates. In: *Cells and Tissues in Culture,* edited by E. N. Willmar, pp. 489–529. Academic Press, New York.
16. Moscona, A. A., and Hausman, R. E. (1976): Biological and biochemical studies on embryonic cell recognition. In: *Cell and Tissue Interactions,* edited by J. Lash, and M. M. Burger. (*This volume.*)
17. Muller, W. E. G., Muller, I., Zahn, R. K., and Kundie, B. (1976): Species specific aggregation factor in sponges. *J. Cell. Sci.,* 21:227–241.
18. Patterson, P. H., and Chun, L. L. Y. (1974): The influence of non-neuronal cells on catecholamine and acetylcholine synthesis and accumulation in cultures of dissociated sympathetic neurons. *Proc. Natl. Acad. Sci. U.S.A.,* 71:3607–3610.
19. Reitherman, R. W., Rosen, S. D., Frazier, W. A., and Barondes, S. H. (1975): Cell surface species-specific high affinity receptors for Discoidin: Developmental regulation in *Dictyostelium discoideum. Proc. Natl. Acad. Sci. U.S.A.,* 72:3541–3545.
20. Rosen, S. D., Kafka, J. A., Simpson, D. L., and Barondes, S. H. (1973): Developmentally regulated, carbohydrate-binding protein in *Dictyostelium discoideum. Proc. Natl. Acad. Sci. U.S.A.,* 70:2554–2557.

21. Roth, S. (1968): Studies on intercellular adhesive selectivity. *Dev. Biol.,* 18:602–631.
22. Roth, S., McGuire, E. J., and Roseman, S. (1971): An assay for intercellular adhesive specificity. *J. Cell. Biol.,* 51:525–535.
23. Roth, S., and Weston, J. A. (1967): The measurement of intercellular adhesion. *Proc. Natl. Acad. Sci. U.S.A.,* 58:974–980.
24. Rutishauser, U., Thiery, J. P., Brackenbury, R., Sela, B. N., and Edelman, G. M. (1976): Mechanism of adhesion among cells from neural tissues of the chick embryo, *Proc. Natl. Acad. Sci. U.S.A.,* 73:577–581.
25. Schubert, D., Heinemann, S., Carlisle, W., Tarikas, H., Kimis, B., Patrick, J., Steinbach, J. H., Culp, W., and Brandt, B. L. (1974): Clonal cell lines from the rat nervous system. *Nature,* 249:224–227.
26. Simpson, D. L., Rosen, S. D., and Barondes, S. H. (1974): Discoidin: A Developmentally Regulated Carbohydrate Binding Protein from *Dictyostelium discoideum. Biochemistry,* 13:3487–3493.
27. Simpson, D. L., Rosen, S. D., and Barondes, S. H. (1975): Pallidin, purification and characterization of a carbohydrate binding protein from *Polyspondylium pallidum. Biochem. Biophys. Res. Commun.,* 412:109–119.
28. Steinberg, M. S. (1963): Tissue reconstruction by dissociated cells. *Science,* 141:401–408.
29. Walther, B. T., Ohman, R., and Roseman, S. (1973): A quantitative assay for intercellular adhesion. *Proc. Natl. Acad. Sci. U.S.A.,* 70:1569–1577.
30. Weinbaum, C., and Burger, M. M. (1973): Two component systems for surface guided reassociation of animal cells. *Nature,* 244:510–512.

Cell and Tissue Interactions, edited by
J. W. Lash and M. M. Burger. Raven
Press, New York, 1977.

A Possible Enzymatic Basis for Some Cell Recognition and Migration Phenomena in Early Embryogenesis

Stephen Roth,* Barry D. Shur,** and Robert Durr*

*Department of Biology, The Johns Hopkins University, Baltimore, Maryland 21218; **Department of Developmental Genetics, Sloan-Kettering Institute for Cancer Research, New York, New York 10021*

In the related fields of embryology and developmental biology, the single, largest problem at present is: How do gene products control form? An impressive amount of elegant experimentation during the last several decades has established that morphogenesis occurs as a result of changes in cellular adhesion and motion and that these changes can take place in isolated cells as well, almost, as in undisturbed embryos. This finding allows the primary question to be restated: How do gene products control changes in cellular adhesion and motility? One hypothesis, originating from experiments on isolated cells (7,9), suggests that intercellular adhesive recognition could result from enzyme-substrate interactions between cell surface glycosyltransferases and their appropriate glycoprotein and glycolipid acceptors. This chapter will summarize data that show a clear correlation between migrating cell types in early chick embryos and cell surface transferase-acceptor complexes (8,10,11). Additionally, the chapter will include a brief account of experiments that implicate this same, general mechanism in early, sperm-egg interactions, i.e., the first adhesive recognition phenomenon to occur in embryogenesis (2).

The glycosyltransferases catalyze the transfer of monosaccharides from monosaccharide-nucleotide derivatives to specific acceptor molecules according to the general scheme illustrated in Fig. 1. Virtually all of these enzymes are bound to membranes when they are in cells, although a number of soluble enzymes have been described in various biological fluids (see ref. 12 for a review). For the most part, the cellular transferases are localized in the Golgi apparatus or in the Golgi-associated membrane structures. However, in a broad variety of cell types, glycosyltransferases are also located on the external surfaces of plasma membranes (12). In this position, they have been implicated in cell adhesion and recognition, growth control, differentiation, lectin-binding, and repair of surface glycosides. In theory, there are a num-

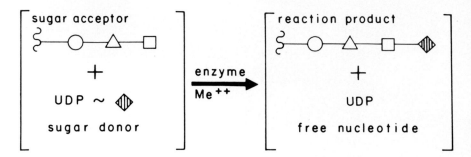

FIG. 1. Schematic representation of a glycosyltransferase reaction in which the monosaccharides are depicted as geometric shapes and the aglycone as a wavy line. The enzyme catalyzes the transfer of the monosaccharide **(diamond)** from its nucleotide derivative to the nonreducing terminus of the acceptor molecule, here shown as a trisaccharide with a nonreducing square. Many but not all transferases require divalent metal ions for optimum activity.

ber of excellent reasons for expecting surface transferases to be important in cellular recognition and migration phenomena. First, many cellular interactions appear to depend on protein-carbohydrate matching (4). Evolutionarily, this would dictate the coordinate changing of at least two different genes as the organisms evolve. Further, this coordinate evolution would have to occur in two compounds that are of crucial importance to the development of the organisms involved. If, on the other hand, the carbohydrate moiety is the enzymatic product of the protein moiety, then a molecular complementarity is assured. A second, powerful advantage of enzyme-substrate mechanisms over protein-carbohydrate mechanisms results from the fact that cells undergoing active recognition and migration during morphogenesis must be able to form correct attachments, and they must be able to break these attachments as they move on. If the protein component of the recognition mechanism is an enzyme whose catalysis can be rigidly controlled by the cell, then letting go becomes a function of allowing an enzyme reaction to go to completion. A corollary of breaking attachments by catalysis is that the cellular substratum will be chemically altered. The alteration could make subsequent cellular attachments more or less likely.

In addition to purely theoretical arguments like the two listed above, there is a rapidly increasing body of experimental evidence that places transferases on the surfaces of many cell types. In the specific cases of the embryonic neural retina (9), cultured fibroblasts (5), human blood platelets (1,3) and the alga, *Chlamydomonas* (6), there are good data for the implication of these enzymes and their substrates in the adhesive recognition phenomena that are displayed by the respective cell types.

In order to test for the existence of transferase-acceptor complexes in the early chick embryo, living embryos were incubated with isotopically labeled sugar nucleotides. In the presence of these sugar donors, enzyme-substrate complexes, when present, should be forced to undergo catalysis with the subsequent addition of the labeled monosaccharide to the acceptor. Embryos incubated in such a fashion were assayed either in a liquid scintillation system for a quantitative evaluation of the degree of enzyme-acceptor activity present, or autoradiographically in order to determine the localization of these complexes. These experiments have been reported in their entirety previously (8,10,11).

Similar experiments have been carried out with mouse sperm and ovum combinations (2) and will be summarized here as well.

METHODS AND RESULTS

Gastrulating Chick Embryos

Early chick embryos (11 somites) were incubated both *in ovo* and *in vitro* with seven different sugar nucleotides: uridine diphosphate galactose (UDP gal), UDP-N-acetylglucosamine (UDPglcNAc), GDPfucose (GDPfuc), CMP-N-acetylneuraminic acid (CMPNAN), UDPglucose (UDPglu), UDP-N-acetylgalactosamine (UDPgalNAc), and UDPglucuronic acid (UDPGA). In all incubations, the final concentration of each sugar nucleotide was 20 μM. Only GDPfuc and UDPGA were labeled with ^{14}C and were used at a final specific activity of 50 mCi/mmole. All of the other sugar nucleotides were labeled with ^{3}H and were used at a final specific activity of 1 Ci/mmole. Unlabeled UDPgalNAc, UDP-^{3}H-galNAc, and UDP-^{3}H-gal were prepared in our laboratory. All other labeled and unlabeled sugar nucleotides were purchased from commercial sources and used only after being examined for purity on at least two chromatographic systems.

For liquid scintillation assays, embryos, and areae pellucidae were dissected free of the areae opacae and were incubated in 75 μl of Dulbecco's modification of Eagle's medium (DME) without serum. After appropriate time periods, the embryos were removed and deposited on borate-impregnated paper, which was then subjected to high-voltage electrophoresis. When hexosaminyltransferase activity was being assayed, the origins of the electrophoretograms were further subjected to ascending chromatography in 70% ethanol in order to remove electrophoretically immobile, free hexosamines that may have accumulated during the assays.

Most autoradiographic experiments were conducted with embryos that were incubated *in ovo* with 75 μl of the desired sugar nucleotide in DME placed above and below the embryo. After the incubations, the embryos were dissected from the eggs, washed gently, fixed, embedded, sectioned, and prepared for autoradiography according to standard methods.

Figure 2 shows sample sections of an embryo labeled *in ovo* with UDP-^{3}H-gal, along with a graphic reconstruction of this and many other, similarly labeled embryos. Heavy labeling is seen on the cells that are invaginating

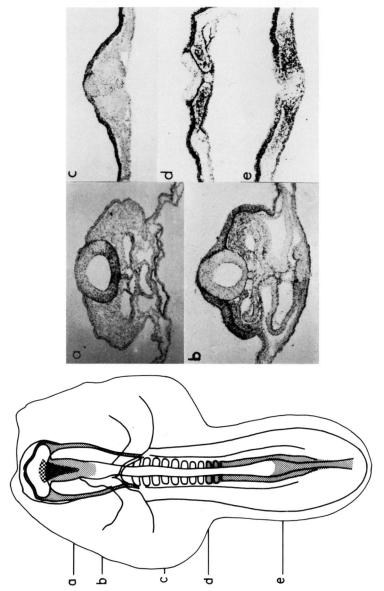

FIG. 2. Autoradiographic preparations of transverse sections of a 10-somite chick embryo incubated while alive in 20 μM UDP-^{3}H-galactose. The schematic diagram on the left shows the level of the sections (**a to e**) as well as the overall distribution of grains. Migrating primitive streak cells proximal to the midline label extensively under these conditions as does the ventral neural tube in the anterior portion of the embryo. Newly formed somites are much more heavily labeled than older somites.

into the primitive streak, although this label falls to undetectable levels as one follows the newly formed mesodermal layer distally. The youngest, most posterior somites are heavily labeled but the older, more anterior somites are relatively unlabeled. Hensen's node becomes labeled.

By contrast, embryos of identical age incubated with UDP-^{3}H-glcNAc show entirely different patterns of sugar incorporation. Here, label is incorporated only in the distal areas of the newly formed mesoderm, not in the medial areas. Anterior somites become labeled but not posterior somites. The heart-forming structures become heavily labeled, although they show little or no labeling with UDP-^{3}H-gal. Hensen's node is not labeled with UDP-^{3}H-glcNAc, although it does incorporate galactose from UDP-^{3}H-gal. Figure 3 shows sections of an embryo incubated with UDP-^{3}H-glcNAc.

Labeling patterns observed in embryos exposed to GDP-^{14}C-fuc are much simpler than the two patterns previously described. Fucose is incorporated on invaginating mesoderm cells, cells of the medullary plate, the neural tube, the cranial crest and the skin ectoderm.

In CMP-^{3}H-NAN incubations, grains are found in association with the neural tube only. Posteriorly, the entire neural tube is surrounded by grains, but as one follows the tube anteriorly, the incorporation disappears from the dorsal aspect of the neural tube until, most anteriorly, radioactivity may only be detected in association with the neural tube-notochord interface.

Embryos incubated identically with UDPgalNAc, UDPglu, or UDPGA show no patterns whatsoever, although the first two of these sugar donors produce a diffuse, general incorporation that is barely greater than background activity. UDPGA, on the other hand, yields no detectable activity.

Of great interest is the observation that migrating cell types show unusually high levels of transferase-acceptor activities. It has already been mentioned that migrating, presumptive mesenchyme cells show appreciable activity with UDPgal medially and with UDPglcNAc distally. Also, cranial crest cells show distinct incorporation from GDPfuc. Along similar lines, trunk neural crest cells, as they are leaving the dorsal area of the neural tube, show high levels of incorporation from both UDPgal and UDPglcNAc, as shown in Fig. 4. These same two sugar donors also label the migrating, primordial germ cells as they move from the blood-forming regions into the embryo proper. In fact, these cells label more heavily than any other cell type observed in these embryos. They can be unequivocally identified as primordial germ cells on the basis of their position, number, and size, and a sample section through the germ cell area is shown in Fig. 5.

Since, during the course of these incubations, the sugar nucleotides are broken down to various degrees, it is necessary to control for the possibility that the observed grain patterns result from cellular uptake of either the sugar-1-phosphates, which are the predominant breakdown products found in these incubations, or the free sugars, which never are found in amounts

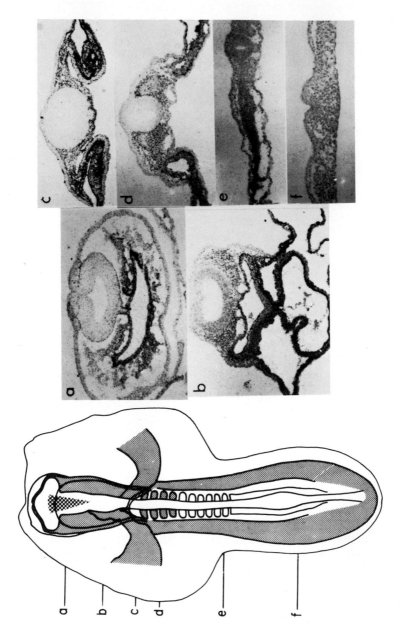

FIG. 3. Same preparations as in Fig. 2 except that this embryo was incubated in UDP-³H-**N**-acetylglucosamine. Here, the invaginated cells of the primitive streak become labeled only after they have migrated more distally from the midline. The heart is labeled heavily and, in contrast to the picture with UDPgalactose, only the mature somites show a clear grain pattern.

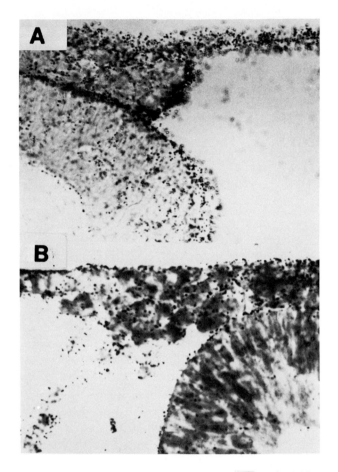

FIG. 4. Transverse section autoradiographs of 11 somite chick embryos incubated with 20 μM (a) UDP-^{3}H-galactose or (b) UDP-^{3}H-N-acetylglucosamine. This is the point of contact between the dorsal neural tube and the skin ectoderm showing the neural crest cells as they begin to migrate from the tube. They label well with both sugar donors.

greater than 11% of the total sugar nucleotide concentrations at t_0. These controls have been carried out in a variety of ways.

First, when embryos are incubated with labeled, free sugars or sugar phosphates at concentrations equal to those found at the end of a normal incubation with sugar nucleotide, no comparable grain patterns are seen with any of the compounds tested. In addition to low levels of incorporation detectable autoradiographically when embryos are incubated with labeled free sugars, scintillation assays show that the amount of radioactivity, in picomoles per embryo, is also severely reduced. Table 1 shows some of these data.

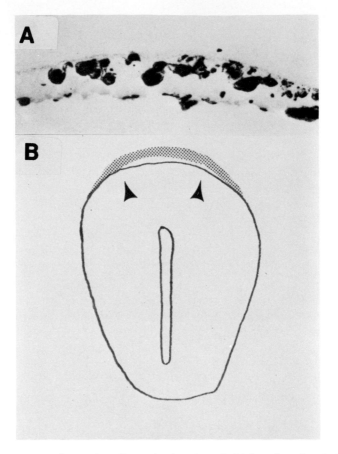

FIG. 5. Transverse section autoradiograph of a stage 4 chick embryo incubated with 20 μM UDP-³H-galactose. The primordial germ cells are extraordinarily well labeled as they migrate from the extraembryonic membranes into the embryo. **B:** Diagrammatic representation of the position of these labeled cells in a whole mount view.

Second, the monosaccharide transport inhibitors phloridzin, phloretin, and cytochalasin B affect uptake from free sugar incubations much more markedly than they do from incubations with sugar nucleotides.

Third, and most compelling, is the observation that grain patterns and picomoles incorporated per embryo are not affected when inhibitors of sugar nucleotide hydrolysis are added to the incubations. Using unlabeled, UDP-glcNAc, 5′AMP, and GDPmannose to completely prevent any breakdown of labeled UDPgal, UDPglcNAc, and GDPfuc, respectively, autoradiographic and scintillation assays yield the same results as when the phosphatase inhibitors are omitted.

Fourth, all of the described assays are conducted in DME, which contains 25 mM glucose. This represents a 1,000-fold molar excess of free glu-

TABLE 1. *Sugar incorporation contributed by free monosaccharides in sugar nucleotide incubations of intact, living chick embryos*

Labeled substrate	Unlabeled additions	Glycosylated product (pmoles/embryo)	Percent of activity with 20 μM sugar nucleotides	Percent of original isotope remaining after 4 hr
20 μM UDPgal	None	38.0[a]	100	
1 μM galactose	None	3.65, 4.30	10.4	97
1 μM galactose	17.8 μM UDPgal	4.22, 4.71	11.7	88
20 μM UDPglcNAc	None	38.2	100	
1 μM glcNAc	None	6.51, 4.54	14.4	100
1 μM glcNAc	19 μM UDPglcNAc	4.01, 5.23	12.0	85
20 μM GDPfuc	None	4.83	100	
1.8 μM fucose	None	0.351, 0.488	8.7	100
1.8 μM fucose	18.2 μM GDPfuc	0.390, 0.585	10.0	100

[a] Embryos were incubated and assayed electrophoretically, as described in ref. 3.

cose over the labeled sugar nucleotide, and any intracellular sugar derived from a labeled sugar nucleotide should be expected to enter high, internal pools and, thereby, have its specific activity drastically lowered. The clear inactivity of UDPglu as a labeled sugar donor with intact embryos makes it unlikely that sugar nucleotides are being taken up, since sugar nucleotides are utilized internally in detectable amounts. This is made even more unlikely, since acid hydrolysis of the incubated embryos show that the labeled products of sugar nucleotide incubations with UDPgal, UDPglcNAc, and GDPfuc are always galactose, glucosamine, and fucose, respectively.

In sum, incubations of more than 2,000 gastrulating chick embryos in seven different, labeled sugar nucleotides show that there are a rich variety of distributions of grain patterns that are specific for the sugar nucleotide being used, the age of the embryo, and the embryonic locale. Many of these patterns correlate well with areas undergoing active cell migration.

Sperm-Egg Interactions

Mouse spermatozoa and ova were assayed, singly and in combination, in an effort to determine the presence and function of glycosyltransferases on these cell types (2). Sufficient numbers of sperm could be obtained from mice to detect high levels of at least two transferase activities — UDPgalactose:N-acetylglucosamine galactosyltransferase and CMPNAN:asialofetuin sialyltransferase. Both of these activities are appreciable in the seminal fluid, although repeated washing of the sperm removes this supernatant activity. For this reason, the possibility that the activities to be described in this communication result from adsorbed enzyme and not intrinsic, surface enzymes cannot be ruled out. The sperm sialyltransferase was examined in detail for two reasons. First, although the sperm exhibits high activities toward the exogenous acceptor, asialofetuin, they show no detectable endogenous activity either autoradiographically or with the scintillation counter assay. Second, the sialyltransferase shows a definite temperature optimum at 34°C. Sperm alkaline phosphatase, beta-galactosidase, acid phosphatase, and the galactosyltransferase activity toward N-acetylglucosamine all show more orthodox temperature requirements. That is, they demonstrate increasing catalytic rates as the temperatures are raised and do not begin to show decreased rates until about 42°C.

Relatively few ova could be obtained for any given experiment, and this necessitated that transferase activities on eggs be assayed only autoradiographically. Such assays show that isolated mouse eggs (dissected from ampullae after being shed) possess detectable levels of endogenous, sialyltransferase activity. The cumulus cells often surrounding the eggs have somewhat higher levels of incorporated radioactivity. Figure 6 shows a brightfield micrograph of an egg incubated in ³H-CMPNAN on a slide in a humidified chamber at 37°C. After the incubation, the egg was fixed and prepared for autoradiography according to standard methods. The diffuse, but definite, grain pattern can be seen in the brightfield micrograph. All 10 eggs that were recovered from such incubations without sperm show this type of labeling.

By contrast, when eggs and sperm were incubated together under otherwise identical conditions, 9 of 12 eggs showed an additional sort of labeling pattern. In these, clear grain accumulations are present in densities that far exceed the general level seen in eggs incubated without sperm. Figure 7

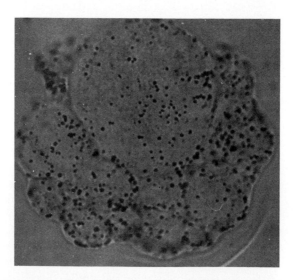

FIG. 6. Brightfield micrograph of a mouse ovum incubated in CMP-³H-N-acetylneuraminic (sialic) acid and then processed for autoradiography. Grains are evenly distributed over the egg itself as well as over the zona cells.

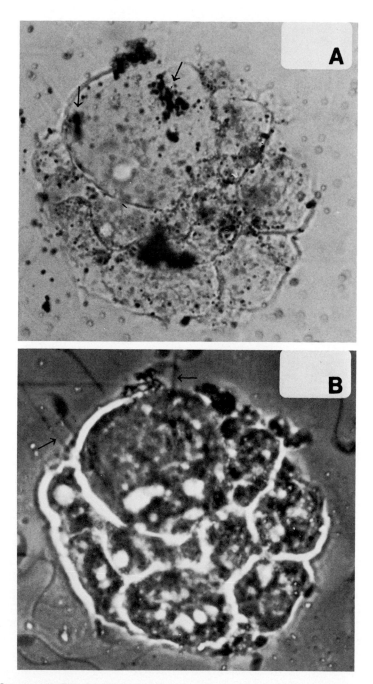

FIG. 7. Same as in Fig. 6 except that the egg was incubated in the presence of mouse sperm. The sperm are visible in the phase micrograph **(b),** and two sperm tails in particular are clearly associated with two of the three dense grain deposits visible in the brightfield micrographs **(arrows,** grain deposits and the sperm tails). No dense grain accumulations were ever observed on ova incubated in the sialyl donor in the absence of sperm.

shows phase and brightfield micrographs of such an egg. In this particular case, three dense grain accumulations can be seen. In two of these areas, visible in the brightfield micrograph (arrows), sperm heads have already penetrated the zona pellucida. The position of these sperm can be inferred by the sperm tail trajectories visible in the phase micrograph (arrows). No spermatozoan could be unambiguously correlated with the third, heavily labeled area on the pictured egg.

These, and other (2), data are consistent with a role for sperm sialyl-transferases and egg sialyl acceptors in the early interaction between these two cell types in fertilization.

DISCUSSION

Early chicken embryos possess extracellular, transferase-acceptor complexes of at least four different types: galactose, N-acetylglucosamine, fucose, and sialic acid. The positions of these enzyme-substrate complexes are highly specific and change consistently with the developmental age of the embryo. Virtually all known migrating cell populations in these embryos have at least one type of transferase-substrate complex associated with them. These migrating cells include the cranial and trunk neural crests, the invaginating primitive streak cells, and the primordial germ cells.

It is highly unlikely that the incorporation of monosaccharides that occurs when the embryos are incubated with sugar nucleotides is a result of uptake and internal utilization of free sugar cleaved from the sugar nucleotide. This possibility may be eliminated, since the degree of incorporation (in picomoles per embryo) and the grain patterns are not affected when sugar nucleotide breakdown is totally inhibited. Further, when this breakdown is not specifically inhibited and free sugar is liberated, subsequent incubations using these levels of labeled, free sugar cannot duplicate either the amount or the pattern of sugar incorporation using sugar nucleotides. Finally, when embryos that have been incubated with sugar nucleotides are hydrolyzed, the resulting free sugars are always the same as the sugars originally used as nucleotide sugars. This implies that the nucleotide sugars did not become exposed appreciably to the intracellular environment, since, if they did, at least some degree of epimerization would be expected.

It remains possible that particular cells of the early chick embryo can take up free sugars from sugar nucleotides by first splitting off the monosaccharide and then transporting it across the membrane without allowing it to enter the soluble pool on either side of the membrane. These hydrolytic and transporting enzyme systems would have to be specific for different sugar nucleotides and be distributed largely among the migrating cell populations. This alternative explanation of the data presented is not very likely, although, if it were true, it would not be uninteresting.

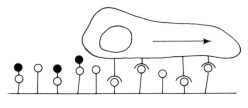

Carbohydrate Matrix

FIG. 8. Model suggesting that cells may migrate over cellular or noncellular substrates containing oligosaccharide chains using their surface transferases to interact with the substrate. If glycosylation were to occur predominantly at the trailing edge of the cell, the cell could have a mechanism to not only release itself from its substrate but, also, to modify it.

Returning to the simplest explanation of the data—the existence of extracellular transferase-acceptor complexes—the crucial question is, of course, do they have a function(s)? This question cannot be answered with conviction at the present time, although preliminary data indicate that, in fact, the enzymes may function to recognize oligosaccharide chains and guide migrating cells along their correct routes. This sort of a mechanism has been suggested to occur as schematically outlined in Fig. 8 (10). The data in support of such a concept come from experiments showing that the sugar nucleotides that are active in donating monosaccharides to the invaginating primitive streak cells (UDPgal and UDPglcNAc) are also teratogenic if millimolar levels are applied to embryos at the time of gastrulation. Sugar nucleotides that are not active donors to embryonic cell populations are not teratogenic under these conditions nor are the active sugar donors at later times, when gastrulation has ceased. Millimolar levels of the free sugars are also not similarly teratogenic. Figure 9 shows a control embryo

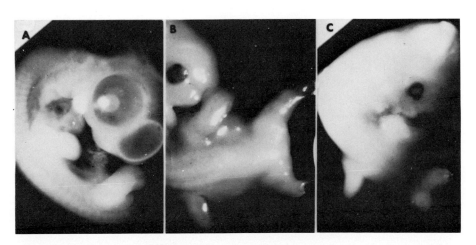

FIG. 9. Stage 25 embryos treated at the 10-somite stage with **(a)** Dulbecco's modified Eagle's medium (DME) or **(b and c)** with 5.0 mM UDPgalactose and 5.0 mM UDP-N-acetyl-glucosamine in DME. The trunks are severely stunted and are often twisted, the hind limbs are underdeveloped, and the embryos are microphthalmic. The same treatment causes no such defects if the embryos are treated at the 20-somite stage.

and two embryos incubated *in ovo* with UDPgal and UDPglcNAc. The two experimental embryos possess abnormal spinal flexures, highly stunted posterior limb buds, and are micropthalmic. Of similarly treated embryos, 11 of 15 showed this same pattern of developmental malformation, while only 1 of 18 control embryos were detectably malformed.

These, and other (10) data, imply that perturbation of naturally occurring, transferase-acceptor complexes in early chick embryos may interfere with the normal morphogenesis that takes place at this time. This, in turn, suggests a link between extracellular glycosyltransferases and embryonic morphogenesis. Such a link would be consistent with data obtained in many laboratories concerning a role for cell surface glycosyltransferases in cell recognition and specific, intercellular adhesion (12). The autoradiographic analysis of murine sperm-egg interactions (2) is also consistent with the possibility that sperm make use of exposed sialyltransferases to bind specifically to oligosaccharide sialyl acceptors on the zona pellucida of the egg.

The data presented and summarized in this chapter should not be construed as suggesting that all, surface-mediated cell interactions occur by mechanisms utilizing glycosyltransferases. On the other hand, there are many interacting systems where surface enzymes seem to be present and where, potentially, they may have some functional importance. For example, in the chick embryo these enzymes are interestingly distributed in time and space. An unequivocal demonstration of functional significance, however, has not been presented.

In short, a biochemical mechanism like the one posited here is consistent with a vast body of data implicating protein-carbohydrate interactions in a myriad of morphogenetic systems, it makes specific predictions which, in a number of cases, have been verified, and it is evolutionarily satisfying. At present, these facts are not true for any other, unifying hypothesis for a molecular mechanism for cell-cell interactions.

REFERENCES

1. Bosmann, H. B. (1971): Platelet adhesiveness and aggregation: The collagen glycosyl, polypeptide:N-acetylgalactosaminyl and glycoprotein:galactosyl transferases of human platelets. *Biochem. Biophys. Res. Commun.*, 43:1118–1124.
2. Durr, R., Shur, B., and Roth, S. (1977): Surface sialyltransferases on mouse eggs and sperm. *Nature*, 265:547–548.
3. Jamieson, G. A., Urban, C. L., and Barber, A. J. (1971): Enzymatic basis for platelet: collagen adhesion as the primary step in hemostasis. *Nature [New Biol.]*, 234:5–7.
4. Kent, P. W. (ed.) (1973): *Membrane Mediated Information*. American Elsevier, New York.
5. Lloyd, C. W., and Cook, G. M. W. (1974): On the mechanism of the increased aggregation by neuraminidase of 16c malignant rat dermal fibroblasts *in vitro*. *J. Cell Sci.*, 15:575–590.
6. McClean, R. J., and Bosmann, H. B. (1975): Enhancement of glycosyltransferase ecto-enzyme systems during *Chlamydomonas* gametic contact. *Proc. Natl. Acad. Sci., U.S.A.*, 72:310–313.
7. Roseman, S. (1970): The synthesis of complex carbohydrates by multiglycosyltransferase

systems and their potential function in intercellular adhesion. *Chem. Phys. Lipids,* 5:270–297.

8. Roth, S. (1973): A molecular model for cell interactions. *Q. Rev. Biol.,* 48:541–563.
9. Roth, S., McGuire, E. J., and Roseman, S. (1971): Evidence for cell surface glycosyltransferases. *J. Cell Biol.,* 51:536–547.
10. Shur, B. (1976): Temporally and spatially specific cell surface glycosyltransferase activities in gastrulating chick embryos. *Dev. Biol.,* 58 (*in press*).
11. Shur, B., and Roth, S. (1973): The localization and potential function of glycosyltransferases in chick embryos. *Am. Zool.,* 13:1129–1135.
12. Shur, B., and Roth, S. (1975): Cell surface glycosyltransferases. *Biochim. Biophys. Acta,* 415:473–512.

Cell and Tissue Interactions, edited by
J. W. Lash and M. M. Burger. Raven
Press, New York, 1977.

Cell Interactions in the Metastatic Process: Some Cell Surface Properties Associated with Successful Blood-Borne Tumor Spread

G. L. Nicolson,*† C. R. Birdwell,† K. W. Brunson,*
J. C. Robbins,† G. Beattie,† and I. J. Fidler‡

**Department of Developmental and Cell Biology, University of California, Irvine,
California, 92717; †Department of Cancer Biology, The Salk Institute for Bio-
logical Studies, San Diego, California 92112; and ‡Basic Research Program,
NCI-Frederick Cancer Research Center, Frederick, Maryland 21701*

Tumor metastasis in animals and man frequently occurs via the circulatory system, which allows malignant cells to spread to near and distant sites (6,17,40,56,60,69). The metastatic process begins when a primary tumor extends or invades into surrounding host tissues (local or primary invasion) (Fig. 1). Certain types of tumors, notably tumors of the skin and thyroid, rarely progress beyond this primary stage before detection. It is thought that loss of proper cell positioning and cell-tissue interactions may account for this aberrant behavior (42) which is one of the most important characteristics of the malignant phenotype.

Primary Invasion

Although the actual mechanism for primary invasion remains obscure, three quite different means of tumor invasion have been proposed. Eaves (11) has stated that mechanical factors are paramount in primary invasion. As the malignant primary tumor mass grows in size, it exerts mechanical pressure on surrounding normal tissues. Eventually, the normal tissues are thought to be displaced as tumor processes are literally injected into regions of tissue weakness. The second and more popular alternative is based on the observations that highly invasive tumors possess cell-released or cell surface-bound enzymes that degrade the normal tissue matrix (7,31,57,59). As the tumor expands into the regions of disrupted normal cell matrix, the host cells may be damaged mechanically or enzymatically. The third proposal depends on greater ameboid motility of the malignant cells, and hence, they simply infiltrate the more static normal tissues (12). Of course, these theories need not be exclusive—tumor cells may use combinations of the above and other mechanisms not yet elucidated to invade surrounding tissues.

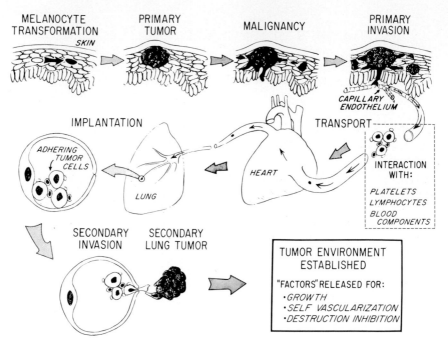

FIG. 1. Pathogenic sequence of melanoma metastasis. (From ref. 40.)

Detachment

Once primary malignant cell invasion has occurred, the next step appears to be cell detachment where individual tumor cells or tumor emboli separate from the primary tumor. Detachment has been proposed to occur because malignant cells have reduced adhesive forces holding them together. For example, Coman (5) found that malignant cell pairs could be more easily separated by mechanical means than normal cell pairs, suggesting that homotypic (adhesive) forces are reduced in malignant lesions. In another series of experiments to determine the differences in cell detachment, McCutcheon et al. (37) found that agitation released more cells from malignant human adenocarcinoma than an equivalent agitation of normal surrounding tissue. These experiments should be interpreted with caution, however, because of the problems in obtaining suitable normal cell controls to directly compare with tumor tissue. In addition, Weiss (65) has stressed that the release or detachment of malignant cells from a primary tumor could occur by a variety of mechanisms including adhesive weakness, local cell rupture, and enzymatic destruction, among others. Therefore, tumor cell detachment in malignant lesions may not reflect an actual decrease in homotypic adhesive properties of the tumor cells.

Transport

When malignant cells invade into surrounding tissues and cell detachment occurs, the tumor cells can enter the lymphatic system and travel to regional lymph nodes. This appears to be the most common route in initial tumor cell dissemination (17). Tumor cells need not be trapped in the first lymph node encountered during their transport. In an interesting series of experiments, Fisher and Fisher (21,24,25) found that tumor cells are not always trapped in the first lymph node but can quickly pass into the circulatory system.

Once tumor cells enter the circulation, whether by translymphatic passage or by direct invasion and entry, the disease can spread to distant sites and critical organs. Tumor cell presence in the blood does not, however, indicate that distant metastases will form. In an extensive review of the literature, Salsbury (56) has concluded that there is no decisive evidence that the mere presence of tumor cells in the circulation indicates a worse prognosis than their absence. In fact, the overwhelming majority of circulating cancer cells (from solid tumors) die quickly in the blood (4,56), and only a very small percentage survive to eventually form secondary tumors (13). The hostile circulatory environment and trauma of transcapillary passage probably account for considerable tumor cell death.

Arrest

During circulatory transport, tumor cells can undergo a variety of cellular interactions including aggregation with other tumor cells (homotypic aggregation) (46), platelets (28,62), lymphocytes (18), as well as interactions with noncirculating host cells (44,46). Some tumor cells are thromboplastic and elicit fibrin formation either during their circulation or soon after their arrest in capillary beds (1,3,62,66). If blood-borne tumor cells are aggregated by homotypic or heterotypic cell interactions or by soluble blood components into large emboli, their success in forming tumors after their arrest in the microcirculation should be increased. In fact, it has been experimentally demonstrated that larger emboli are more effective per input tumor cell in implantation and survival to form gross tumor colonies after intravenous injection (14). Thus, purely mechanical factors such as emboli size and deformability (as well as capillary diameter and deformability) should be important in implantation. Zeidman (70) and Zeidman and Buss (71) have demonstrated that the rates at which tumor cells or their cell emboli pass through capillary beds are not related to cell or emboli size but appeared rather to be related to their deformability during transcapillary transport. There are several reports where tumor cells passed quickly through the first capillary encountered, so strictly mechanical factors — although important — do not solely determine the distribution of arrested tumor cells (14,20,22,58).

In many experimental systems, the ultimate location of gross tumor colonies is nonrandom and does not correlate with the first capillary system encountered. Organ distribution after intravenous injection of plasmacytoma (48), reticulum cell sarcoma (47), histiocytoma (9), and melanoma cells (20,34) revealed that the tumor cells play an important role in determining their ultimate fate *in vivo*. This could occur, in part, by heterotypic adhesive interactions between circulating tumor cells and host capillary endothelial cells, which could lead to implantation at specific sites (40,44,45).

Secondary Invasion

After tumor cell arrest, most malignant cells eventually invade the surrounding capillary by diapedesis, extravasation, or destruction of endothelial cells and/or the underlying basement membrane (3). In addition, certain tumor cells after their arrest are quickly trapped or surrounded by a fibrin matrix (3,66), which may help to stabilize the arrested cells, but this does not seem to be a requirement for successful tumor survival (3,23,36). Under these conditions fibrinolysis via tumor cell production of plasminogen activator (47,61) could be detrimental to tumor cell survival, if fibrin formation protects the arrested tumor cells from fluid shear forces and host immunocytes.

Secondary invasion of the capillary endothelium and underlying basement membrane probably depends on the release of tumor and perhaps also host degradative enzymes. Cell surface-associated and secreted proteases (7,31,35,57,59,67) are known to be higher in tumor compared to surrounding normal tissues. When malignant cells reach an extravascular environment, they usually continue proliferation until growth is slowed due to nutrient and hormonal depletion (27). But in malignant lesions, vascularization of the micrometastases is probably stimulated by tumor angiogenesis factor, a glycoprotein that stimulates endothelial cell movement and division (26).

Host Immune Interactions

Host immunity against neoplasms should, in principal, destroy malignant cells before or during the metastatic steps discussed above, although in many systems immunological response against tumor cells enhances tumor growth (8,16,50) and successful experimental metastasis (16). These seemingly contradictory results can be reconciled by Prehn's (49) hypothesis that a weak host immunological response should enhance tumor growth, while a strong response is inhibitory. In experimental metastatic systems, this hypothesis has gained support from studies in which a low ratio of immune lymphocytes to tumor cells was found to be enhancing for melanoma lung colonization after intravenous injection, whereas a high ratio of immune lymphocytes to tumor cells inhibited lung tumor formation (16). At least some of the "tumor enhancement" in this system may be due to the ability of immune lymphocytes to clump melanoma cells in the circulation aiding in tumor cell arrest (18).

METASTASIS AND MODEL SYSTEMS
TO STUDY TUMOR SPREAD

To study the various steps in the metastatic process, it would be desirable to have sets of metastatic and equivalent low or nonmetastatic tumor cells in order to compare properties that may be important in metastasis. Fortunately, such systems exist where tumor cell line variants have been selected *in vivo* in syngeneic hosts for enhanced abilities to implant and survive to form gross organ tumors after intravenous injection of suspended individual tumor cells. Using C57BL/6 mice, Fidler (15) selected for tumor cell variants of syngeneic B16 melanoma by intravenous injection of B16 cells, and 2 to 3 weeks later gross pulmonary (pigmented) colonies were removed from lungs and tumor cells adapted to tissue culture (Fig. 2). The B16 variants that grew in culture after the first *in vivo* selection were established as a continuous line and were called B16-F1. Cells from this line were reinjected back into new syngeneic animals, and 3 weeks later a new group of lung tumor colonies were removed and cultured to yield B16-F2. With each succeeding cycle *in vivo,* the ability of the selected B16 lines to implant, survive, and form lung tumors increased (15,18). After 10 such selections, B16-F10

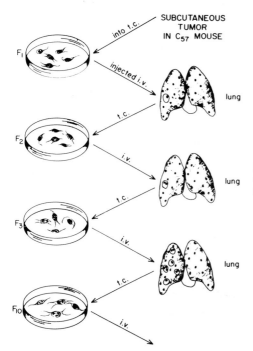

FIG. 2. *In vivo* selection of B16 melanoma cell lines for enhanced implantation, survival, and growth in C57BL/6 mice.

was obtained which forms significantly more gross lung tumors per input cell compared to B16-F1 after intravenous (18) or intracardiac (20) injection into syngeneic C57BL/6 mice. This difference has proven to be reasonably stable over 3 years of culture *in vitro* under nonselecting conditions. However, we have obtained some preliminary results that suggest that the differences between B16-F10 and B16-F1 in ability to form lung tumors, while still significant, may be decreasing with long-term culture. Therefore, it may be necessary at intervals to start fresh cultures from frozen stocks of early passage cells or repeat the *in vivo* selection to prevent loss of the selected properties by genetic drift in tissue culture.

Some of the advantages of the B16 variant system are: (a) the B16 melanoma arose spontaneously in skin (epithelial) of inbred C57BL/6 mice and has been maintained in syngeneic hosts; (b) the melanoma tumors can be observed easily in skin, lymph nodes, or other organs by eye due to the presence of melanin; (c) the B16 cells have a very characteristic morphology; (d) comparisons with dubious "normal" counterpart cells are unnecessary, since B16 variants of low and high metastatic potential are available; and (e) the B16-F cells can be grown easily at a variety of *in vivo* sites or grown *in vitro* in tissue culture.

The increase in the number of gross metastases per input tumor cell in the selected variants was not due to nonspecific trapping in the first capillary bed encountered. When [125]IUDR-labeled, viable B16-F1 and B16-F10 single cell suspensions were injected into the tail veins of C56BL/6 mice, 64% and 99% of these cells, respectively, were localized in the lungs (the first organ encountered) of recipients within 2 min (15,18,20). However, when the [125]IUDR-labeled cells were injected into the left ventricle of the heart, approximately 30% of the cells of either line were localized in the lungs 2 min later, and the remainder were present in blood and in all the major organs (20). By 1 day postinjection the numbers of viable melanoma cells in the lungs and other organs were the same, independent of the route of injection (20). In all cases the more metastatic cells from line B16-F10 implanted and survived at significantly ($p < 0.001$) higher rates compared to B16-F1 (15,18,20). Two weeks later, when gross experimental metastases were located and counted at autopsy, the same number of tumors was found in the lungs of mice receiving B16 melanoma cells by way of the tail vein (9 ± 3 and 79 ± 16 nodules for B16-F1 and B16-F10, respectively, per 25,-000 input cells) or left ventricle (10 ± 2 and 83 ± 10, for B16-F1 and B16-F10, respectively, per 25,000 input cells) (20). Although the initial tumor cell distributions, rates of vascular entry, and first organ capillary beds encountered were different, this did not change the ultimate outcome of experimental metastatic colonization. Melanoma cells entering into the circulation via the left ventricle had to pass through extrapulmonary capillary beds before entering the lungs, but this did not affect their ability to ultimately reach the lungs, survive, and form the same number of tumor colonies as melanoma cells entering the lung capillary beds first after tail vein injection (20). Therefore, the arrest and survival of melanoma cells destined to form lung tumors is not random, and these cells were able to recirculate and specifically arrest in the organ for which they were selected *in vivo*.

Organ Preference

The ability of B16-F10, but not B16-F1, to form exclusively lung colonies after injection suggested that selection for organ preference of B16 melanoma variants was possible. This was experimentally confirmed by intracardiac (left ventricle) injection of B16-F1 or -F10 into mice, and 3 weeks later scoring pulmonary and extrapulmonary tumor foci. Over several experiments B16-F1 formed extrapulmonary (most in liver, adrenal, and mesenteric lymph nodes) tumors after intracardiac injection in approximately 44% of mice injected (20,41), but no extrapulmonary tumor colonies were found after intracardiac injection of B16-F10 (20,41).

In another series of experiments a brain-preferring B16 line was selected from B16-F1 (41 and Brunson and Nicolson, *in preparation*). The line selected once for lung implantation and survival (B16-F1) was injected intracardially (left ventricle) into a large number of C57BL/6 mice, and the rare brain tumors removed and placed into tissue culture (B16-F1B1). In practice we failed to find brain tumors after tail vein injection; only intracardiac (left ventricle) injection was successful in producing brain lesions. This result could have been due to trivial reasons such as the lower concentrations of melanoma cells in the blood after tail vein compared to intracardiac injection, or the B16 cells destined to form lung tumors may implant more efficiently, stopping in the lungs after intravenous injection, leaving the other less successful B16 cells to circulate harmlessly.

B16 brain tumors were grown in culture and reintroduced intracardially into new syngeneic mice. After three *in vivo* selections using the intracardial route of injection, selections were attempted by tail vein injection. Occasional brain melanomas were placed into tissue culture, and subsequent selections were performed via tail vein injection for a total of eight *in vivo* selections to obtain line B16-F1B8. Injection of B16-F1B8 into animals yielded brain tumors in almost every animal (Table 1). Interestingly, B16-F1B8 yielded one-half the normal number of lung tumors compared to B16-F1, suggesting that brain preference selection might eventually result in loss or near loss of lung tumor formation (Table 1) (41). B16-F1B8 tumors seem to be located in the forebrain, and histology has revealed the presence of extravascular melanin-containing cells in the meninges. Subsequent studies after further brain selections should reveal whether the brain-preferring variants can home to particular brain regions.

TABLE 1. *Comparison of pulmonary and brain tumors found in C57BL/6 mice 2 weeks after intracardiac injection of B16 melanoma lines[a]*

Melanoma line	Average number of brain tumors per mouse (range)	Average number of lung tumors per mouse (range)
B16-F1	1 (0–1)	29 (11–46)
B16-F1B8[b]	12 (5–23)	13 (5–19)

[a] Mice were injected with 100,000 viable B16 melanoma cells into the left side of the heart and sacrificed 2 weeks later. Pigmented tumor colonies were scored with the aid of a dissecting microscope.
[b] Selected eight times *in vivo* for enhanced brain implantation, survival, and growth.
From ref. 41.

SOME PROPERTIES OF B16 MELANOMA METASTATIC VARIANT LINES

Several modifications in cell properties have been described for transformed compared to untransformed cells. In particular, specific changes at the cell surface generally correlated with neoplastic transformation (reviews: 39,52,54,55); however, few of these changes are probably relevant to metastasis (39). Using B16 variant lines of differing metastatic potential such as the B16-F series, it should be possible to determine which cell (and host) properties are important in tumor metastasis.

Surface Properties of B16 Variants

Although metastatic variant systems have been available for only a limited time, several observations have been made concerning their lectin agglutinability, surface labeling by lactoperoxidase-catalyzed iodination, surface secreted enzymes, some antigens, and glycopeptides released by trypsin/pronase digestion. Lectin agglutination was assessed as previously described (38) using affinity purified concanavalin A (Con A) *Ricinus communis I* agglutinin (RCA$_I$), wheat germ agglutinin (WGA), and *Limulus polyphemus* agglutinin (LPA). Agglutination with Con A was similar for B16 variants of low and high metastatic potential and was generally increased for the other lectins, although differences were never more than a few serial dilution endpoints at 22°C. Increases in WGA and LPA agglutinability can be explained by the greater quantities of sialic acid on the more metastatic B16 lines (2), while quantitative labeling with ^{125}I-RCA$_I$ failed to show a dramatic increase in RCA$_I$ receptors on the highly metastatic B16 variants (G. L. Nicolson, *unpublished data*).

Labeling of cell surface proteins of B16 lines grown *in vitro* or as intraperitoneal tumors by lactoperoxidase-catalyzed iodination revealed few differences in cell surface-located iodinatable proteins. These experiments

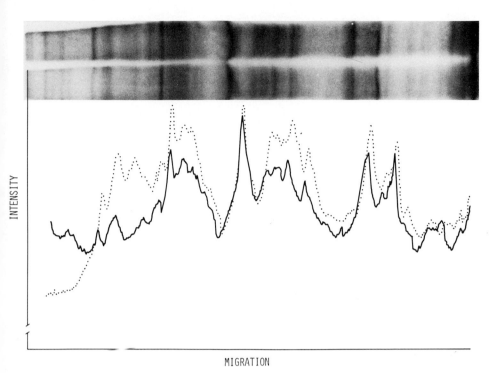

FIG. 3. Lactoperoxidase-catalyzed iodination (surface labeling) of B16 melanoma variants. The labeled B16 cells were dissolved in sodium dodecylsulfate and electrophoresed on polyacrylamide slab gels in sodium dodecylsulfate. **a:** Autoradiogram of surface labeled B16-F10 **(upper lane)** and B16-F1 **(lower lane).** Electrophoresis was from left to right in a 7.5% acrylamide gradient gel. **b:** Densitometric tracings of gel from **(a)**, above. **Dotted line,** F10; **solid line,** F1.

were performed with B16-F1 and -F10 grown and labeled in monolayer culture or labeled in suspension with ^{125}I (as in refs. 32,33,53). The labeled surface proteins were solubilized and separated by sodium dodecyl sulfate-polyacrylamide gel electrophoresis (SDS-PAGE) and identified by auto-radiography (Fig. 3a). Comparison of the densitometer scans of a surface-labeled B16 melanoma variant line after SDS-PAGE are shown in Fig. 3b, and it is readily apparent that there are no *major* differences in the surface-labeling patterns between the B16 variant lines. The minor differences that are seen in Fig. 3 were not consistent from experiment to experiment and were possibly due to minor differences in growth conditions, cell densities, etc. It should be stressed, however, that differences could also arise due to differential protease digestion of the cell peripheries of the B16 variants *in vivo*.

Protease treatment of transformed cells removed glycopeptide fragments which, after further pronase digestion, show altered properties when compared to similarly obtained glycopeptides from counterpart untransformed cells. Specifically, mixtures of glycopeptides (~4,000-dalton class) are of apparent higher molecular weight when chromatographed on Sephadex gel filtration columns (review: ref. 63). To assess the possible role of these peptides in malignant processes, B16 lines of low and high metastatic potential were metabolically labeled with ^{14}C- and ^{3}H-sugars. Mild trypsinization removed cell surface glycopeptides, which were further degraded by pronase and chromatographed on Sephadex G50 as described by Glick et al. (29). Examination of elution profiles obtained by cochromatography of peptide fragments from B16-F1 and B16-F10 again revealed few, if any, differences consistent with the findings of Warren et al. (64). Thus, the alterations in biological properties cannot be explained by *major* changes in the glycoproteins of the highly metastatic variants.

The glycolipids of low (B16-F1) and high (B16-F10) metastatic melanoma variants have been examined by extraction and thin-layer chromatography (68). Only minor changes in the glycolipid patterns were found with the most distinct change being the loss of ganglioside GD_{1a} in B16-F10. However, it is unlikely that these minor alterations in glycolipids could account for the altered implantation and adhesion properties (see below) of the highly metastatic B16 lines.

Adhesive Properties of B16 Variants

The importance of adhesive properties in metastatic tumor spread has been stressed by Coman (5) and Weiss (65). Obviously, cell detachment from the primary site and aggregation of tumor cells during transport are important in metastasis. When we examined the homotypic rates of adhesion of the B16 variant lines in qualitative or quantitative (45) assays, the highly metastatic B16 lines always adhered at faster rates than the less metastatic lines (Fig. 4). Similarly, heterotypic rates of adhesion of B16 variant cells to platelets (28), lymphocytes (18), and an endothelial cell line (Fig. 4) indicate that the highly metastatic B16 lines adhere at greater rates to circulating and noncirculating mouse cells. In experiments designed to test the hypothesis that organ cell recognition and adhesion may determine, at least in part, the organ specificity of blood-borne lung-selected B16 melanoma cells, purified, suspended organ cells were mixed with B16-F1, -F5, -F10, and -F13 cells and cell aggregation scored shortly thereafter. Within minutes the highly metastatic B16-F13 line aggregated lung cells into a single clump, whereas the B16-F1 line caused only slight lung cell aggregation (Table 2). Other suspended organ cells from nontarget tissues such as spleen and kidney were not aggregated above control levels by any of the B16 metastatic variants. These results suggest that target organ recognition may occur through cell surface adhesive interactions. The fact that virtually all lung cells were aggregated into one

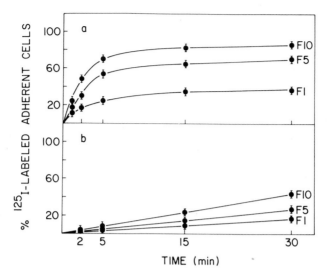

FIG. 4. Kinetics of B16 homotypic and heterotypic adhesions. **a:** Heterotypic rates of adhesion of ^{125}IUdR-labeled B16-F1, -F5, and -F10 variant lines to confluent monolayers of a mouse endothelial cell line. **b:** Homotypic rates of adhesion of B16 lines to confluent monolayers of the same B16 line. (From ref. 43.)

large, heterotypic clump by B16-F13 suggested that many, if not all, cells in the lung can be recognized by the highly metastatic B16 variant and that tissue-specific determinants exist on cells such as those of the vascular endothelium. Evidence for organ-specific determinants on vascular endothelial cells has been obtained by Pressman and Yagi (51) using antiorgan antibodies.

Enzymatic Properties of B16 Variants

Degradative enzymes released or surface-bound to B16 melanoma cells have been described by Bosmann et al. (2). They found that the more metastatic B16 variants had higher trypsin-like and cathepsin-like activities in culture, but only when cells were replated at low cell densities. Several glycosidases such as β-D-galactosidase, β-L-fucosidase, N-acetyl-β-D-galactosaminidase, and N-acetyl-β-D-glucosaminidase tended to be more active in the more metastatic B16 lines at low cell densities.

Many, but not all, transformed cell lines produce high levels of plasminogen activator, a serine protease released from cells that converts plasminogen to plasmin, which in turn hydrolyzes fibrin [review: Roblin et al. (55)]. We examined the release of plasminogen activator by the B16 metastatic variants and found no significant differences, irrespective of whether the assay was conducted in calf, dog, or B16 melanoma-bearing syngeneic C57BL/6 mouse serum (44). It should be borne in mind that

TABLE 2. *Aggregation of organ cell suspensions by B16 melanoma variants*[a]

B16 variant	Aggregation of organ cell suspension					
	No cells	RBC	Spleen	Brain	Liver	Lung
No cells	−	0	0	±	+	+
F1	0	±	±	±	+	+/++
F5	±	±	±	±	+	++
F10	+	+	+	+	+	+++
F13	+	+	+	+	++	++++

[a] Organs obtained from C57BL/6 mice were suspended in ice-cold Ca^{2+}–Mg^{2+}-free Hank's buffer containing 2 mM EDTA. Tissue was minced into ~ 1 mm^3 pieces and washed several times in the same ice-cold buffer to remove erythrocytes, macrophages, and other contaminating blood cells. The organ tissue blocks were carefully dissociated by repeated pipeting and filtering through gauze and then a coarse strainer at 4°C. The organ cell suspension contained some cell aggregates that were removed by differential centrifugation at 4°C. Eryhtrocytes remaining in the organ cell suspensions were lysed by resuspension in 0.155 M ammonium chloride–0.01 M potassium bicarbonate–0.002 M EDTA for 5 min at 4°C. Organ cells were then washed by centrifugation in Ca^{2+}–Mg^{2+}-free Hank's buffer and resuspended at a concentration of 3×10^6 cells per ml (2°C). Cell viabilities determined by dye exclusion were: spleen, >95%; brain, >75%; liver, >70%; lung, >80%. B16 melanoma lines F1, F5, F10, and F13 were grown to subconfluency in DMEM and harvested by use of Ca^{2+}–Mg^{2+}-free Hank's buffer containing 2 mM EDTA. B16 cells were washed in complete Hank's buffer and suspended at a concentration of 2×10^6 per ml in complete Hank's plus 1% fetal calf serum at 4°C. Cell viabilities determined by dye exclusion were 90–95%. The aggregation assay was performed by mixing 0.1 ml of organ cells plus 0.1 ml of B16 cell suspensions in Linbro FP-54 trays (18 mm wells) rotating at 1.5 Hz for 10 min at 22°C. Coded wells were independently scored by two investigators from 0 (no aggregation; all single cells) to ++++ (complete aggregation; no single cells) as described previously (38). Control wells contained 0.2 ml of the organ or B16 cell suspensions in complete Hank's plus 1% fetal calf serum.

enhanced fibrinolysis might be expected to inhibit—not enhance—metastasis by reducing the amount of fibrin in the tumor area. Hagmar (30) found that enhanced implantation was related to fibrin formation around the arrested tumor embolus, and it has been known for some time that prevention of clotting processes reduces experimental metastasis (23).

Invasive Properties of B16 Variants

The highly metastatic B16 melanoma lines are known to be more invasive *in vivo* than lines of low metastatic potential (18), so we examined this property *in vitro* in the absence of possible host immune responses by using the assay of Easty and Easty (10). This assay measures the ability of tumor cells to locally invade surrounding tissue and is monitored *in vitro* by assessing the abilities of tumor cells to infiltrate into the multicell layer of the chick chorioallantoic membrane (CAM) in tissue culture. Since the malignant B16 variants are seeded onto the exterior of the CAM instead of injecting them into tissue, purely mechanical forces should not

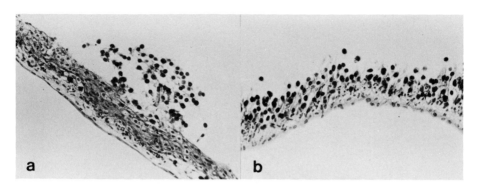

FIG. 5. *In vitro* invasion of B16 melanoma variants into isolated chick chorioallantoic membrane (CAM). **a:** A section of CAM 12 hr after attachment of B16-F1 cells. There is no evidence of invasion, but some of the B16-F1 cells have "peeled" off the CAM, probably as a result of fixation, dehydration, and embedding in paraffin. **b:** A section of CAM 12 hr after attachment of B16-F10 cells. The F10 cells have invaded deep into the CAM. ×200 before reduction.

contribute to invasion. B16 F1 and -F10 were allowed to adhere to the surfaces of CAM in culture, and at various times samples were removed and fixed for histological examination. The results showed that B16-F10 cells could invade into the CAM within 12 to 24 hr, while the low metastatic B16-F1 failed to enter the CAM during this period (Fig. 5). Thus, there is a difference in invasive potential of the B16 variants independent of host and mechanical factors.

Immunological Properties of B16 Variants

B16 melanoma cells possess on their surfaces a variety of antigens and host immune cell receptors. H-2 and Thy-1 alloantigen content have been determined by quantitative antisera absorption techniques (R. Hyman, *unpublished data*), while immunoelectron microscopy was necessary to detect the viral envelope antigens gp71 and p12 on the B16 variants (J. Ihle, *unpublished data*). Using these techniques, differences have not been seen between high and low metastatic B16 variants. These melanoma cells also possess lymphocyte receptors, because cytotoxic T-lymphocytes can kill B16 melanoma *in vitro* or *in vivo* (16). To study the effects of host cell-mediated immune reactivity on the pathogenesis of B16 melanoma, *in vitro* procedures were used to select sequentially B16 variants resistant to cell-mediated killing after continuous exposure to syngeneic immune lymphocytes (19). B16-F1, B16-F10, and their lymphocyte-resistant variants, which had been selected *in vitro* six times for lymphocyte resistance (B16-F1^{Lr-6} and B16-F10^{Lr-6}, respectively), were injected subcutaneously or intravenously into nonimmune syngeneic recipients. Although the growth patterns were similar for all four melanoma lines after subcutaneous injection, the incidence of experimental pulmonary metastases after intravenous injection was dramatically different. The lymphocyte-resistant lines B16-F1^{Lr-6} and B16-F10^{Lr-6} formed significantly fewer lung colonies than equal numbers of their parental cells, B16-F1 and -F10, such that the number of experimental pulmonary metastases formed by line B16-F10^{Lr-6} was similar to the number for B16-F1. Thus, in this system host lymphocyte interaction with the melanoma cells appears to enhance rather than inhibit the formation of metastases, and this might be due to the aggregation effects of immune lymphocytes on blood-borne melanoma cells leading to increased arrest and survival.

SUMMARY

The spread of tumors via blood-borne mechanisms is a complex multi-step process, and it appears to be dependent on characteristic tumor cell properties, but it is also dependent on host properties as well [see Fidler (17)]. In this brief chapter we have concentrated on metastatic tumor cell surface properties using as an experimental system a series of mouse melanoma variant cell lines selected *in vivo* for their enhanced abilities to implant, survive, and grow to form experimental metastases. Although few biochemical differences have been found between melanoma lines of low and high metastatic potential, some surface properties, notably homotypic and heterotypic adhesive characteristics as well as *in vitro* invasive properties are different. Since most of the biochemical assays used to analyze the melanoma variant lines were originally developed to study gross changes in cell surface moieties after neoplastic transformation, these methods may not be sensitive enough to detect subtle changes occurring with gradual shift to the malignant phenotype. Future studies will undoubtedly concentrate on mechanisms of tumor cell invasion, detachment, and implantation of blood-borne emboli and also on host properties such as immunity and angiogenesis. Using *in vivo* selected tumor cell variants of differing metastatic potential and secondary site preference, it should be possible to define some of the molecular properties important in malignancy.

ACKNOWLEDGMENTS

Supported by U.S. National Cancer Institute Contracts NO1-CO-25423 (to Litton Bionetics) and NO1-CB-33879 (to G.L.N.) and grants from the American Cancer Society (BC-211 to G.L.N.) and the U.S. Public Health Service (CA-15122 to G.L.N.). Dr. Birdwell is a Fellow of the Anna Fuller Fund for Cancer Research, and Dr. Robbins is a Fellow of the Leukemia Society of America, Inc.

REFERENCES

1. Baserga, R., and Saffiotti, U. (1955): Experimental studies on histogenesis of blood-borne metastases. *Arch. Pathol.,* 59:26–34.
2. Bosmann, H. B., Bieber, G. F., Brown, A. E., Case, K. R., Gersten, D. M., Kimmerer, T. W., and Lione, A. (1973): Biochemical parameters correlated with tumor cell implantation. *Nature,* 246:487–489.
3. Chew, E. C., Josephson, R. L., and Wallace, A. C. (1976): Morphologic aspects of the arrest of circulating cancer cells. In: *Fundamental Aspects of Metastasis,* edited by L. Weiss, pp. 121–150. North-Holland, Amsterdam.
4. Cliffton, E. E., and Agostino, D. (1962): Factors affecting the development of metastatic cancer. Effect of alterations in clotting mechanism. *Cancer,* 15:276–283.
5. Coman, D. R. (1944): Decreased mutual adhesiveness, a property of cells from squamous cell carcinomas. *Cancer Res.,* 4:625–629.
6. Coman, D. R. (1953): Mechanisms responsible for the origin and distribution of blood-borne tumor metastases: A review. *Cancer Res.,* 13:397–404.
7. Dresden, M. H., Heilman, S. A., and Schmidt, J. D. (1972): Collagenolytic enzymes in human neoplasms. *Cancer Res.,* 32:993–996.
8. Duff, R., Doller, E., and Rapp, F. (1973): Immunologic manipulation of metastasis due to Herpesvirus transformed cells. *Science,* 180:79–81.
9. Dunn, T. B. (1954): Normal and pathologic anatomy of the reticular tissue in laboratory mice, with a classification and discussion of neoplasms. *J. Natl. Cancer Inst.,* 14:1281–1433.
10. Easty, D. M., and Easty, G. C. (1974): Measurement of the ability of cells to infiltrate normal tissues *in vitro. Br. J. Cancer,* 29:36–49.
11. Eaves, G. (1973): The invasive growth of malignant tumors as a purely mechanical process. *J. Pathol.,* 109:233–237.
12. Enterline, H. T., and Coman, D. R. (1950): The ameboid motility of human and animal neoplastic cells. *Cancer,* 3:1033–1038.
13. Fidler, I. J. (1970): Metastasis: Quantitative analysis of distribution and fate of tumor emboli labeled with ^{125}I-5-iodo-2'-deoxyuridine. *J. Natl. Cancer Inst.,* 45:775–782.
14. Fidler, I. J. (1973): The relationship of embolic homogeneity, number, size and viability to the incidence of experimental metastasis. *Eur. J. Cancer,* 9:223–227.
15. Fidler, I. J. (1973): Selection of successive tumor lines for metastasis. *Nature [New Biol.],* 242:148–149.
16. Fidler, I. J. (1974): Immune stimulation-inhibition of experimental cancer metastasis. *Cancer Res.,* 34:491–498.
17. Fidler, I. J. (1975): Mechanisms of cancer invasion and metastasis. In: *Biology of Tumors: Surfaces, Immunology, and Comparative Pathology, Vol. 4: Cancer: A Comprehensive Treatise,* edited by F. F. Becker, pp. 101–131. Plenum Press, New York.
18. Fidler, I. J. (1975): Biological behavior of malignant melanoma cells correlated to their survival *in vivo. Cancer Res.,* 35:218–224.
19. Fidler, I. J., Gersten, D. M., and Budmen, M. B. (1977): Characterization *in vivo* and *in vitro* of tumor cells selected for resistance to syngeneic lymphocyte-mediated cytotoxicity. *Cancer Res. (in press).*
20. Fidler, I. J., and Nicolson, G. L. (1976): Organ selectivity for implantation survival and growth of B16 melanoma variant tumor lines. *J. Natl. Cancer Inst.,* 57:1199–1202.
21. Fisher, B., and Fisher, E. R. (1966): The interrelationship of hematogenous and lymphatic tumor cell dissemination. *Surg. Gynecol. Obstet.,* 122:791–798.
22. Fisher, B., and Fisher, E. R. (1967): The organ distribution of disseminated ^{51}Cr-labeled tumor cells. *Cancer Res.,* 27:412–420.
23. Fisher, B., and Fisher, E. R. (1967): Anticoagulants and tumor cell lodgment. *Cancer Res.,* 27:421–425.
24. Fisher, E. R., and Fisher, B. (1965): Experimental study of factors influencing the development of hepatic metastases from circulating tumor cells. *Acta Cytol.,* 9:146–158.
25. Fisher, E. R., and Fisher, B. (1967): Recent observations on concepts of metastasis. *Arch. Pathol.,* 83:321–324.

26. Folkman, J. (1974): Tumor angiogenesis. *Adv. Cancer Res.,* 19:331–358.
27. Folkman, J., and Hochberg, M. (1973): Self regulation of growth in three dimensions. *J. Exp. Med.,* 138:745–753.
28. Gasic, G. J., Gasic, T. B., Galanti, N., Johnson, T., and Murphy, S. (1973): Platelet-tumor cell interaction in mice. The role of platelets in the spread of malignant disease. *Int. J. Cancer,* 11:704–718.
29. Glick, M. C., Rabinowitz, Z., and Sachs, L. (1973): Surface membrane glycopeptides correlated with tumorigenesis. *Biochemistry,* 12:4864–4869.
30. Hagmar, B. (1972): Defibrination and metastasis formation: Effects of arvin on experimental metastases in mice. *Eur. J. Cancer,* 8:17–28.
31. Hashimoto, K., Yamanishi, Y., Maeyens, E., Dabbous, M. K., and Kanzaki, T. (1973): Collagenolytic activities of squamous cell carcinoma of the skin. *Cancer Res.,* 33:2790–2801.
32. Hogg, N. M. (1974): A comparison of membrane proteins of normal and transformed cells by the lactoperoxidase method. *Proc. Natl. Acad. Sci. U.S.A.,* 71:489–492.
33. Hynes, R. O. (1974): Role of surface alterations in cell transformation: The importance of proteases and surface proteins. *Cell,* 1:147–156.
34. Kinsey, D. L. (1960): An experimental study of preferential metastasis. *Cancer,* 13:674–676.
35. Koono, M., Ushijima, K., and Hayashi, H. (1974): Studies on the mechanisms of invasion in cancer. III. Purification of a neutral protease of rat ascites hepatoma cell associated with production of chemotactic factor for cancer cells. *Int. J. Cancer,* 13:105–115.
36. Ludatsher, R. M., Luse, S. A., and Suntzeff, V. (1967): An electron microscopic study of pulmonary tumor emboli from transplanted Morris hepatoma 5123. *Cancer Res.,* 27:1939–1952.
37. McCutcheon, M., Coman, D. R., and Moore, F. B. (1948): Studies on invasiveness of cancer. Adhesiveness of malignant cells in various human adenocarcinomas. *Cancer,* 1:460–467.
38. Nicolson, G. L. (1973): Neuraminidase "unmasking" and the failure of trypsin to "unmask" β-D-galactose-like sites on erythrocyte, lymphoma and normal and SV40-transformed 3T3 fibroblast cell membranes. *J. Natl. Cancer Inst.,* 50:1443–1451.
39. Nicolson, G. L. (1976): Transmembrane control of the receptors on normal and tumor cells. II. Surface changes associated with transformation and malignancy. *Biochim. Biophys. Acta,* 458:1–72.
40. Nicolson, G. L., Birdwell, C. R., Brunson, K. W., and Robbins, J. C. (1976): Cellular interactions in the metastatic process. In: *Membranes and Neoplasia: New Approaches and Strategies. J. Supramol. Struct.,* Suppl 1:237–244.
41. Nicolson, G. L., and Brunson, K. W. (1977): Organ specificity of malignant B16 melanomas: *In vivo* selection for organ preference of blood-borne metastasis. In: *GANN Monograph on Cancer Research,* 20:15–24.
42. Nicolson, G. L., and Poste, G. (1976): The cancer cell: Dynamic aspects and modifications in cell-surface organization. *N. Engl. J. Med.,* Part I, 295:197–203; Part II, 295:253–258.
43. Nicolson, G. L., Robbins, J. C., and Winkelhake, J. L. (1975): Tumor cell surfaces and metastasis: Dynamic changes in neoplastic membrane structure and their relationship to tumor spread. In: *Cellular Membranes and Tumor Cell Behavior,* edited by E. F. Walborg, pp. 81–127. Williams & Wilkins, Baltimore.
44. Nicolson, G. L., and Winkelhake, J. L. (1975): Organ specificity of blood-borne tumour metastasis determined by cell adhesion? *Nature,* 255:230–232.
45. Nicolson, G. L., Winkelhake, J. L., and Nussey, A. C. (1976): An approach to studying the cellular properties associated with metastasis: Some *in vitro* properties of tumor variants selected *in vivo* for enhanced metastasis. In: *Fundamental Aspects of Metastasis,* edited by L. Weiss, pp. 291–303. North-Holland, Amsterdam.
46. Ossowski, L., Unkeless, J. C., Tobia, A., Quigley, J. P., Rifkin, D. B., and Reich, E. (1973): An enzymatic function associated with transformation of fibroblasts by oncogenic viruses. II. Mammalian fibroblast cultures transformed by DNA and RNA tumor viruses. *J. Exp. Med.,* 137:113–126.
47. Parks, R. C. (1974): Organ-specific metastasis of a transplantable reticulum cell sarcoma. *J. Natl. Cancer Inst.,* 52:971–973.

48. Potter, M., Rahey, J. L., and Pilgrim, H. I. (1957): Abnormal serum protein and bone destruction in transmissible mouse plasma cell neoplasm (multiple myeloma). *Proc. Soc. Exp. Biol. Med.,* 94:327–333.
49. Prehn, R. T. (1971): Perspectives in oncogenesis: Does immunity stimulate or inhibit neoplasia? *J. Reticuloendothel. Soc.,* 10:1–12.
50. Prehn, R. T., and Lappé, M. A. (1971): An immunostimulation theory of tumor development. *Transplant. Rev.,* 7:26–54.
51. Pressman, D., and Yagi, Y. (1964): Chemical differences in vascular beds. In: *Small Blood Vessel Involvement in Diabetis Mellitus,* edited by M. D. Siperstein, A. R. Colwell, and K. Meyer, pp. 177–183. The American Institute of Biological Sciences, Washington, D.C.
52. Rapin, A. M. C., and Burger, M. M. (1974): Tumor cell surfaces: General alterations detected by agglutinins. *Adv. Cancer Res.,* 20:1–91.
53. Robbins, J. C., Hyman, R., Stallings, V., and Nicolson, G. L. (1977): Cell-surface changes in a *Ricinus communis* toxin-resistant variant of a murine lymphoma. *J. Natl. Cancer Inst.,* 58:1027–1033.
54. Robbins, J. C., and Nicolson, G. L. (1975): Surfaces of normal and transformed cells. In: *Biology of Tumors, Surfaces, Immunology, and Comparative Pathology, Vol. 4: Cancer: A Comprehensive Treatise,* edited by F. F. Becker, pp. 3–54. Plenum Press, New York.
55. Roblin, R., Chou, I-N., and Black, P. H. (1975): Proteolytic enzymes, cell surface changes and viral transformation. *Adv. Cancer Res.,* 22:203–259.
56. Salsbury, A. J. (1975): The significance of the circulating cancer cell. *Cancer Treatment Rev.,* 2:55–72.
57. Strauch, L. (1972): The role of collagenases in tumor invasion. In: *Tissue Interactions in Carcinogenesis,* edited by D. Tarin, pp. 399–434. Academic Press, New York.
58. Sugarbaker, E. D. (1952): The organ selectivity of experimentally induced metastasis in rats. *Cancer,* 5:606–612.
59. Sylvén, B. (1973): Biochemical and enzymatic factors involved in cellular detachment. In: *Chemotherapy of Cancer Dissemination and Metastasis,* edited by S. Garattini and G. Franchi, pp. 129–138. Raven Press, New York.
60. Tarin, D. (1976): Cellular interactions in neoplasia. In: *Fundamental Aspects of Metastasis,* edited by L. Weiss, pp. 151–187. North-Holland, Amsterdam.
61. Unkeless, J. C., Tobia, A., Ossowski, L., Quigley, J. P., Rifkin, D. B., and Reich, E. (1973): An enzymatic function associated with transformation of fibroblasts by oncogenic viruses. I. Chick embryo fibroblast cultures transformed by avian RNA tumor viruses. *J. Exp. Med.,* 137:85–111.
62. Warren, B. A. (1973): Environment of the blood-borne tumor embolus adherent to vessel wall. *J. Med.,* 4:150–177.
63. Warren, L., Fuhrer, J. P., and Buck, C. A. (1973): Surface glycoproteins of cells before and after transformation by oncogenic viruses. *Fed. Proc.,* 32:80–85.
64. Warren, L., Zeidman, I., and Buck, C. A. (1975): The surface glycoproteins of a mouse melanoma growing in culture and as a solid tumor *in vivo. Cancer Res.,* 35:2186–2190.
65. Weiss, L. (1967): *The Cell Periphery, Metastasis and Other Contact Phenomena, Vol. 7: Frontiers of Biology,* edited by A. Neuberger and E. L. Tatum, North-Holland, Amsterdam.
66. Wood, S., Jr. (1964): Experimental studies of the intravascular dissemination of ascitic V2 carcinoma cells in the rabbit, with special reference to fibrinogen and fibrinolytic agents. *Bull. Schweiz, Akad. Med. Wiss.,* 20:92–121.
67. Yamanishi, Y., Maeyens, E., Dabbous, M. K., Ohyama, H., and Hashimoto, K. (1973): Collagenolytic activity in malignant melanoma: Physiochemical studies. *Cancer Res.,* 33:2507–2512.
68. Yogeeswaran, G. (1975): The Salk Institute for Biological Studies Annual Report of Research Activities, pp. 27–29.
69. Zeidman, I. (1957): Metastasis: A review of recent advances. *Cancer Res.,* 17:157–162.
70. Zeidman, I. (1961): The fate of circulating tumor cells. I. Passage of cells through capillaries. *Cancer Res.,* 21:38–39.
71. Zeidman, I., and Buss, J. M. (1952): Transpulmonary passage of tumor cell emboli. *Cancer Res.,* 12:731–733.

Cell and Tissue Interactions, edited by
J. W. Lash and M. M. Burger. Raven
Press, New York, 1977.

Calcium and Cyclic Nucleotides as Universal Second Messengers

Howard Rasmussen

Departments of Internal Medicine and Cell Biology, Yale University School of Medicine, New Haven, Connecticut 06510

HISTORICAL PERSPECTIVE

In approaching a discussion of calcium and cyclic nucleotides, I should like to do so by reviewing the subject from the perspective of recent history, and particularly the history of hormone action.

Twenty years ago, a number of the more important mammalian hormones had been isolated and chemically characterized, and their effects upon specific cells or tissues identified. What was apparent from this work was that there was considerable diversity of chemical structure and of physiological or cellular effects within this group. It appeared that each specific hormone acted upon a restricted set of cell types to induce in them a particular response (49). Elucidation of the mode of hormone action thus appeared to represent a series of individual problems in biological research. The only theme that gave some philosophical unity to the field was the concept that those hormones that induced rapid responses in target cells, e.g., oxytocin on the myometrium, did so by interacting with receptors on the cell surface.

This concept of surface receptors for some hormones was an extension of a similar theory as to the mechanism by which neurotransmitters regulated postsynaptic events in the nervous system. In particular, in the case of events at the neuromuscular junction, it was proposed that interaction of acetylcholine with its receptors on the muscle cell membrane was the initial event in the contractile response. Hence, in one important aspect, the fields of general and endocrine physiology shared a common perspective.

There was also another. It was that calcium ions served as the coupling factor in excitation-contraction coupling in all forms of muscle whether contraction of the particular muscle was initiated by neural or hormonal means. Evidence in the past 20 years has only enlarged and reinforced this concept. However, it has also added an additional feature to the coupling role of calcium ion. This feature is that a change in calcium ion concentration not only couples contraction to excitation, but simultaneously couples metabolism to excitation so that appropriate metabolic events, e.g., glycogenolysis, are activated *pari passu* with the contractile events to supply metabolic energy in the form of ATP to meet the needs of this working tissue (40). Recognition of this additional coupling role of calcium is particularly relevant to the role of calcium in excitation-response coupling in the endocrine system (84).

A short time after the role of calcium as coupling factor in excitation-contraction response was established, evidence was forthcoming showing that calcium ions played a similar role in excitation-secretion coupling (33,34) whether that secretion was from synapses, exocrine glands, or endocrine glands. Development of this concept greatly extended the interface between general and endocrine physiology because a significant number of hormones have as a major physiological function the control of one or more secretory systems. Nonetheless, the concept did not have a marked impact upon studies into the mechanisms of hormone action because of a dramatic event, the discovery of 3',5'-cyclic adenosine monophosphate, cyclic AMP.

CYCLIC AMP, ADENYLATE CYCLASE, AND THE SECOND MESSENGER CONCEPT

In studying the mechanism by which epinephrine regulates hepatic glycogenolysis, Sutherland and Rall (99,100) uncovered a soluble intermediate that coupled events in a particulate cell fraction with those on isolated glycogen particles. Identification of this factor as cyclic AMP was the first step toward a revolution in endocrine research. An even more important step was the identification of adenylate cyclase as the enzyme that catalyzed the formation of cyclic AMP and inorganic pyrophosphate, PPi, from ATP:

$$\text{ATP} \xrightarrow[\text{(AC)}]{\text{Mg}^{2+}} 3'5' \text{ AMP} + \text{PPi}$$

and shortly thereafter the discovery of phosphodiesterase, an enzyme that catalyzed the hydrolysis of cAMP to 5'AMP

$$3'5'\text{AMP} \xrightarrow[\text{(PDE)}]{} \text{AMP}$$

Within a few years of these discoveries, it became apparent that this system of enzymes for regulating the formation and destruction of cAMP was widely distributed in nature, was organized in the geographic sense that the adenylate cyclase was localized within the plasma membrane of the cell, and phosphodiesterase primarily within the cell cytosol, and was involved in the action of a large number of peptide and amine hormones.

Accumulation of these data led Sutherland and associates (89,100) to propose a unifying concept of hormone action, the second messenger concept, and to develop a number of criteria to be used to establish whether or not a particular extracellular agent or messenger controlled cell function via the adenylate cyclase-cAMP system (Fig. 1).

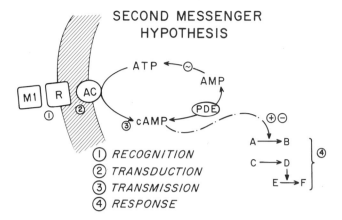

FIG. 1. A schematic representation of the second-messenger hypothesis and the four steps between recognition of a specific first messenger **(M1)** with a surface receptor **(R)** and intracellular response. **AC,** adenylate cyclase; **PDE,** phosphodiesterase; ⊕ ⊖ indicate positive and negative control, respectively, of an intracellular process.

The concept stated that particular peptide or amine hormones acted to initiate cellular response by a common mechanism: the activation of a membrane-bound adenylate cyclase and the generation of an intracellular or second messenger cAMP. Control of response involved both the activity of the membrane-bound cyclase and of the soluble phosphodiesterase. Specificity of response depended upon the specificity of hormone receptor interaction on the cell surface.

The criteria used to establish whether a particular extracellular messenger acted in this manner were: (a) addition of first messenger caused a rise in cAMP content of cell or tissue; (b) exogenous cAMP mimicked the effect of first messenger; (c) theophylline and other methyl xanthines (inhibitors of phosphodiesterase) enhanced the effectiveness of submaximal concentrations of first messenger; and (d) a particulate first-messenger-sensitive adenylate cyclase could be prepared from the particular tissue.

The importance of this concept to endocrine research and eventually to all of cell biology cannot be overemphasized. It immediately took what had apparently been a large number of individual problems in biological research, how specific peptide hormones stimulate the activity of particular tissues, and placed them in the context of a single problem. Simultaneously, it raised the possibility that other nonhormonal first messengers might regulate cell function in a similar fashion. It also put the problem of peptide hormone and/or first-messenger action into the context of an information transfer system: recognition → transduction → transmission → response (Fig. 1).

The mere fact that one could isolate a plasma membrane fraction that retained its recognition and transducing functions, i.e., the stimulation of cAMP production by specific first messengers, led to the first substantial biochemical evidence that hormone and other first-messenger receptors do exist on the cell surface (14,62,90), but of even greater importance to cell biology, that regulation of the binding properties, of the number, and of their coupling to the adenylate cyclase (or other membrane transducers) are all means by which the response of particular cell types to specific first messengers are modulated (61,92).

In the initial enthusiasm for this simple and elegant concept, its limitations were lost sight of, much work was done to prove that it was correct, but little to show that it might be wrong or insufficient. As is so often the case, the pendulum has recently swung in the opposite direction. It is ironic that substantial evidence against this concept has been obtained in the first two systems from which evidence for its existence was obtained: the action of epinephrine on the liver and of ACTH on the adrenal cortex (21,38,71,101). Upon hindsight two major limitations are apparent: (a) as first put forth, it proposed a highly stereotyped response system that excluded any plasticity of response in spite of the fact that both in the nervous system and in the organization of the endocrine system at the organismal level, plasticity is a key organizational element; and (b) it ignored that large body of evidence showing that calcium ion was a common coupling factor between excitation and response (81). It was in fact this latter feature that first attracted our attention. As the list of systems, in which the adenylate cyclase-cAMP system was involved as the mediator of cellular response, grew the overlap with a list, in which Ca^{2+} was involved in some coupling role, also grew (84). In addition, in a variety of systems in which cAMP was shown to mimic first-messenger action, one of its effects was shown to be that of invoking a change in cellular calcium metabolism (85).

Given these facts, we began nearly a decade ago to explore the possibility that both Ca^{2+} and cAMP served second-messenger functions and that in many systems their effects were interrelated. This exploration has expanded well beyond any bounds we initially conceived, has grown to include cGMP as a key element, and has led to the recognition that these three control elements serve as nearly universal interrelated second messengers and/or coupling factors in cell activation (10,81,84–86).

SECOND MESSENGERS VERSUS COUPLING FACTORS

Before describing some of their interrelationships, it is worth considering the two concepts: second messenger and coupling factor. At first glance, it would seem that Sutherland and co-workers merely took over the concept of coupling factor and renamed it second messenger. However, as used, there is an implied difference between these two concepts. Sutherland et al. (100) were explicit in defining a second messenger as a signal, generated at the cell surface in response to the interaction of a cell receptor with a first messenger, which then acted at one or more sites within the cell to generate a response. On the other hand, the concept of coupling factor, when applied to the role of calcium in contractile systems, emphasizes the fact that calcium is the ultimate mediator of the contractile response, i.e., it focuses on the ultimate response element. In different contractile systems, the source of calcium differs. In smooth muscle, it comes mainly from the extracellular calcium pool and that bound to the surface membrane. Hence, calcium serves both as coupling factor and second messenger in this tissue. In contrast, in skeletal muscle the calcium that serves to couple excitation to contraction is derived largely from the sarcoplasmic reticulum. In this instance, excitation leads to the generation of some second messenger (presumably ionic in nature) that in turn triggers calcium release from the sarcoplasmic reticulum. Hence, if we wish to discuss this system in the context of an information transfer hierarchy, calcium serves a third- rather than second-messenger function, although in both forms of muscle it functions to couple excitation to response.

CELLULAR CALCIUM METABOLISM

This distinction is of more than semantic interest because it points up the difference in complexity of cellular calcium metabolism compared to cellular cyclic nucleotide metabolism. The latter is a simple two-element system in which a surface source—the membrane-bound adenylate cyclase—and an intracellular sink—the soluble and/or membrane-bound phosphodiesterases—control events.

In contrast, cellular calcium metabolism is extremely complex as illustrated in an idealized cell in Fig. 2.

The mechanisms involved in the regulation of the free calcium ion concentration in the cell cytosol involve events at the plasma membrane (PM, Fig. 2) and the endoplasmic reticulum. In the plasma membrane, the processes operate as a pump-leak system with an inward leak of calcium down its concentration gradient (10^{-3} M extracellular calcium $\rightarrow 10^{-6}$ cytosolic calcium). However, there is considerable complexity in the organization of this pump-leak system (17,18,37,83,87). There are at least two separate channels by which Ca^{2+} leaks into the cells (① and ②, Fig. 2). The first ① is a relatively specific Ca^{2+} channel that is independent of the membrane potential and that may well be altered when certain hormones interact with their receptor but do not lead to a membrane depolarization. The second is a potential-dependent calcium permeability channel. In many cells, particularly those activated by acetylcholine, in which response is initiated by a rise in intracellular (cytosolic) calcium ion concentration, depolarization of the plasma membrane occurs as a consequence of first-messenger-receptor interaction. This membrane depolarization leads to an increase in the permeability of a potential-dependent calcium channel (②), and this leads to a rise in the intracellular calcium content. In addition, to these two pathways, some Ca^{2+} may also enter the cell, following membrane

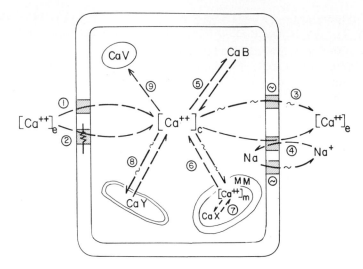

FIG. 2. Schematic representation of the factors regulating the calcium ion content of the cell cytosol $(Ca^{2+})_c$ in an idealized eukaryotic cell. $(Ca^{2+})_e$, extracellular calcium; **CaB**, calcium bound to cytosolic constituents; **CaV**, calcium deposited in endoplasmic reticulum; $(Ca^{2+})_m$, free calcium in mitochondria; **CaX**, calcium deposited in mitochondrial matrix space; ① to ⑨, the various processes regulating $(Ca^{2+})_c$ (see text for discussion).

depolarization, via the Na^+ channel. This is the "early calcium channel" in nerve, and the process depicted in ② in Fig. 1 is the "late calcium channel." It is quite likely that the relative importance of these two (or three) different processes by which the entry of calcium into the cell is regulated varies from one cell type to another, and different ones are of major importance in the activation of particular cells by extracellular messengers.

The mechanisms for controlling calcium efflux are just as complex as those involved in regulating influx. There are at least two different energy-dependent mechanisms for controlling efflux (③ and ④, Fig. 2). The first, ③ in Fig. 2, is a specific Ca^{2+}-activated ATPase or calcium pump. This has been most clearly identified and studied in the red blood cell, but there is evidence that it exists in other cells as well. The second, ④ in Fig. 2, is a Na^+-Ca^{2+} exchange (extracellular Na^+ exchanges for intracellular Ca^{2+}), which depends upon the energy-dependent maintenance of the Na^+ gradient by the classical Na^+/K^+-activated ATPase of Na^+ pump (6,7,17,87).

In addition to the complex events at the plasma membrane, control of $(Ca^{2+})_a$ also depends upon the binding of calcium (CaB) by cytoplasmic constituents (⑤ in Fig. 2); the energy-dependent uptake of calcium by the mitochondria and the passive leak of calcium out of the mitochondria (⑥ in Fig. 2); the deposition of a calcium-phosphate-ATP complex (CaX) within the mitochondrial matrix space (⑦ in Fig. 2); the energy-dependent uptake and passive release of calcium by the smooth endoplasmic reticulum and the deposition of calcium complexes (CaY) within this organelle (⑧ in Fig. 2); and the accumulation of calcium by secretory granules (CaV) in a variety of cell types (⑨ in Fig. 2).

All of the implications of this complexity of cellular calcium metabolism have not been thought out and discussed, but several are immediately obvious. The first is that the diffusion domain of calcium is limited. This has been shown most clearly by the microinjection studies of Rose and Loewenstein (91). They showed by using the luminescent protein, Aequorin, that Ca^{2+} injected into cells of the isolated salivary gland of *Chironomus* was confined to the local site of injection unless the capacity of mitochondria to accumulate calcium was blocked, in which case calcium diffused thoughout the cell. What this work means is that the calcium pump activities of the intracellular membranes are sufficiently active that they effectively and continuously control the calcium ion content of the cell cytosol. This, in turn, means that the number and distribution of mitochondria, and the extent and distribution of the endoplasmic reticulum will determine how rapidly any sudden influx of calcium from outside the cell will be buffered by these intracellular systems. On the other hand, for any sustained increase in the calcium ion content of the cell cytosol to occur, events at these intracellular membranes as well as those at the plasma membrane must be controlled.

The implications of these conclusions can be made obvious by considering three types of control systems in which calcium plays a coupling and/or second messenger role (see ref. 80 for review of these systems).

The first is a system such as histamine release from the mast cell (see below). This release is triggered by the sudden influx of Ca^{2+} into the cell across the plasma membrane. It is usually a very rapid but brief event. There are feedback controls at the plasma membrane that limit calcium entry. In addition, however, there must be a way to rapidly remove the calcium ions that have entered if response is to be terminated. This is presumably done by the intracellular organelles. In this system, then, there is, after cell activation, a high concentration of ionized calcium for a short time just beneath the cell surface, but likely very little change in calcium within the deeper regions of the cell cytosol.

The second is the secretion of insulin by the beta cell of the endocrine pancreas (see ref. 82). A physiological change in extracellular glucose concentration leads to a significant increase in calcium entry and an immediate release of insulin, a situation quite analogous to that of histamine release. However, if the high extracellular glucose concentration is maintained, then a rise in intracellular cAMP occurs, and this rise reduces the rate at which the newly entered calcium is sequestered by intracellular organelles, i.e., it acts to maintain the elevated cytosolic calcium ion concentration and thereby secretion by regulating transport events at one or more intracellular membranes.

The third is the fly salivary gland an exocrine gland with a polarity of cell surface function (see below). The receptors that interact with 5-hydroxy-tryptamine are located on the basal surface of the cell. One consequence of hormone-receptor interaction is an entry of calcium into the cell. However, these cells are rich in mitochondria. Hence, in the absence of any other signal the entering calcium would be rapidly accumulated by these organelles. This would mean that the luminal surface membrane of these cells would not see any change in $[Ca^{2+}]$. Yet, there is substantial evidence that such a change in $[Ca^{2+}]$ is seen by the luminal membrane, is sustained for long periods of time, and is a necessary event in hormone action. If this is so, the conclusion appears inescapable that changes in the activity of the calcium pump-leak system in the mitochondrial membranes must take place as well as those changes seen at the basal aspects of the plasma membrane. There is considerable evidence that cyclic nucleotides are important (10,18,82) regulators of calcium exchange across intracellular membranes.

These features of control of cell calcium metabolism are true not only for events in the fly salivary gland but must also hold for all exocrine glands and other polar cells where calcium is considered to serve a coupling role between excitation and response.

CALCIUM AND CYCLIC NUCLEOTIDES AS INTERRELATED SECOND MESSENGERS AND/OR COUPLING FACTORS

With this perspective, it is now possible to consider specific examples that illustrate the types of interrelationships one finds between calcium ions and cyclic nucleotides in control of cell function. The examples have grown rapidly in number and complexity, and it is not possible to catalog them all. However, several recent reviews give a fairly comprehensive account of this rapidly developing field (10,17,81–85). This discussion will be restricted to six examples that were chosen because they illustrate some of the particular variations of the universal theme that Ca^{2+} and cyclic nucleotide act as interrelated second messengers and/or coupling factors in cell activation. These six examples are: (a) the action of 5-hydroxytryptamine on fluid secretion in the fly salivary gland; (b) the actions of acetylcholine and epinephrine on contraction and relaxation of intestinal smooth muscle; (c) immune-mediated release of histamine from mast cells; (d) light activation of the amphibian rod; (e) the action of acetylcholine on the exocrine pancreas; and (f) acetylcholine-mediated epinephrine release from the chromaffin cells of the adrenal medulla.

The action of serotonin [5-hydroxytryptamine (5HT)] upon fluid secretion in the abdominal portion of the salivary gland of the blowfly (*Calliphora erythrocephala*) fits the general criteria outlined by Sutherland et al. (100) to establish that cAMP is the second messenger in the action of this hormone on this tissue: (a) the hormone stimulates the production of cAMP (79); (b) the effect of subnormal concentrations of hormone are enhanced by theophylline (9); and (c) exogenous cAMP mimics the effect of 5HT upon secretion (9). In addition, both 5HT and cAMP stimulate Ca^{2+} efflux (79). This efflux occurs even when no external calcium is present. Under these circumstances, there is a 5HT or cAMP-induced secretion. This secretory response decays with the same time course as the decline in rate of calcium efflux, so that it is reasonable to conclude that either 5HT or exogenous cyclic AMP mobilize calcium from an internal store causing a rise in the calcium ion content of the cell cytosol that leads to an activation of secretion and an increase in efflux of calcium from the gland. Stimulation of the gland for 1 hr in a medium containing no calcium but 5 mM EGTA leads to a 66% reduction in total tissue calcium, and by the end of this time fluid secretion rate is near the basal level. By the use of a combination of electron microscopy and electron probe X-ray analysis, Berridge et al. (11) have reported that the major intracellular calcium storage sites in this tissue are the mitochondria, and that following stimulation of the gland in calcium-free media the amount of intramitochondrial calcium is greatly reduced. These data are consistent with the concept that intramitochondrial calcium is mobilized during 5HT action, probably indirectly via the 5HT-induced rise in intracellular (cAMP).

When one compares the responses to 5HT and exogenous cAMP, clear differences become apparent: (a) 5HT stimulates Ca^{2+} uptake by the tissue, but exogenous cAMP does not (79); (b) 5HT causes the transepithelial potential to become more negative, but cAMP causes it to become more positive (12,79); (c) the effect of exogenous cAMP is greatly inhibited when the $[Cl^-]$ in the medium is lowered from 140 to 50 mM, but the effect of 5HT is only slightly reduced (12); and (d) the calcium ionophore, A23187, induces calcium-dependent secretion without causing a rise in cAMP, but causes changes in calcium influx and efflux and in transepithelial potential similar to those seen after 5HT (90).

The data on the mode of action of 5HT and of ionophore can be interpreted in terms of a system in which both Ca^{2+} and cAMP serve second messenger functions. When 5HT interacts with its receptor there are at least two immediate effects: an activation of adenylate cyclase and an increase in the permeability of the serosal membrane to calcium. The subsequent rise in intracellular cAMP concentration has at least two actions: an activation of a primary K^+ pump on the luminal membrane and an enhanced efflux of calcium from the mitochondria. The combination of this latter effect and the direct effect of 5HT upon Ca^{2+} entry into the cell leads to a rise in $[Ca^{2+}]$ of the cell cytosol. This has two effects: It increases the Cl^- permeability of the luminal membrane and inhibits further cAMP production and/or enhances its hydrolysis (a negative feedback loop in this intracellular control system).

It is reasonable to conclude that a rise in either $[cAMP]$ or $[Ca^{2+}]$ can lead to an increase in rate of fluid secretion in this tissue, but that under normal circumstances the natural hormone, 5HT, regulates fluid secretion by means of the interrelated effects of two intracellular second messengers: Ca^{2+} and cAMP. These two interact in two ways: (a) they regulate each other's concentrations conferring thereby feedback control, and (b) they act sequentially at the final effector organelle, the luminal membrane, to determine the magnitude of the eventual physiological response-fluid secretion.

THE SECOND-MESSENGER HYPOTHESIS IN SMOOTH MUSCLE

Applying the second-messenger model to a particular cellular system, one should expect some relationship between the magnitude of the physiological response and intracellular [cAMP]. However, the response of the intestinal smooth muscle to acetylcholine and epinephrine illustrate an apparent paradox (1–4). When stimulated by acetylcholine, the muscle contracts, and when stimulated by epinephrine, it relaxes. Nevertheless, both hormones cause an increase in [cAMP]. Further analysis reveals that the time course of change in [cAMP] differs with the two stimuli. The epinephrine-induced rise in [cAMP] is due to an activation of adenylate cyclase, whereas that induced by acetylcholine is due to an inhibition of phosphodiesterase.

To understand these data, it is necessary to consider the role of calcium. As in other muscles, calcium is the coupling factor between excitation and contraction. Acetylcholine causes calcium entry into the cell and calcium release from cell membranes, and epinephrine stimulates its membrane uptake and efflux from the cell (1–4). Based upon this information and the fact that Ca^{2+} is an inhibitor of phosphodiesterase in this tissue, it is possible to construct a closed-loop model of the hormonal control of contraction in this tissue. In this model, when acetylcholine acts it stimulates the release of membrane-bound calcium that induces, in turn, a contractile response, but also blocks cAMP hydrolysis by inhibiting phosphodiesterase activity. The rise in [cAMP] stimulates the reaccumulation of calcium by the membrane, thereby reducing the contractile response. The rise in [cAMP] is part of a negative feedback loop (closed loop) that modulates the initial acetylcholine response. In keeping with this conclusion, the rise in [cAMP] after acetylcholine addition is not immediate, but is seen only several minutes after addition of hormone and significantly later than the initiation of contraction. Conversely, when epinephrine acts there is an immediate rise in [cAMP] that precedes relaxation, and suggests that cAMP plays a direct role in the relaxation process. The epinephrine-induced rise in [cAMP] leads to a decrease in [Ca^{2+}] within the cell cytosol, and this, in turn, leads to an activation of phosphodiesterase with a resultant fall in [cAMP]; a second negative feedback loop within this closed system. However, the phase relationships between Ca^{2+} and cAMP in this tissue are opposite to those seen in the fly salivary gland. A rise in [cAMP] leads to a fall in [Ca^{2+}], and a rise in [Ca^{2+}] to a rise in [cAMP] rather than the situation in the fly salivary gland, where a rise in [cAMP] led to a rise in [Ca^{2+}], and a rise in [Ca^{2+}] to a fall in [cAMP] (see Table 1).

TABLE 1. *Patterns of calcium-cyclic nucleotide relationships in cell activation*

Pattern	Specific example	[cAMP] on $[Ca^{2+}]_c$	$[Ca]_c$ on [cAMP]	Effect of a rise in		
				[cGMP] on $[Ca^{2+}]_c$	$[Ca^{2+}]_c$ on [cGMP]	
I	5HT-fly salivary gland	Increase	Decrease	—	—	
II	Catecholamine, smooth muscle	Decrease	Increase	—	—	
III	Acetylcholine, exocrine pancreas	—	—	Increase	?Decrease	
IV	Light, amphibian rod	—	—	Decrease	Increase	

Under most circumstances some degree of muscle tone is maintained by a balance between cholinergic and adrenergic activity. Because of the dual feedback relationships between [cAMP] and Ca^{2+} uptake and between [Ca^{2+}] and cAMP hydrolysis, the contractile state of the smooth muscle depends upon the mutual interactions between these two second messengers, Ca^{2+} and cAMP, within this closed-loop system. Thus, it is possible to resolve the apparent paradox of a rise in [cAMP] during both hormonally-induced contraction and relaxation. The system does not function as a linear open-loop system, but as a closed-loop one. However, even this model is incomplete, because when acetylcholine acts a rise in cGMP concentration also occurs (60).

A major clue to the role of cGMP in regulating calcium metabolism is the recent observations of Andersson et al. (4). Working with isolated microsomal fractions from intestinal smooth muscle, they have shown a cAMP-mediated phosphorylation of these membranes and a cAMP-dependent calcium accumulation by these membranes. Addition of cGMP in a concentration one-tenth of that of cAMP completely antagonized the effect of cAMP on calcium accumulation. If this evidence is taken together with the evidence that the initial rise in cGMP in smooth muscle following cholinergic stimulation is calcium-dependent, it is then possible to construct a model in which the Ca^{2+}-induced rise in cGMP acts as a positive feedforward regulator of intracellular (cytosolic) [Ca^{2+}] by stimulating release of membrane-bound Ca^{2+}. Negative feedback control in the system is achieved by the Ca^{2+}-mediated inhibition of the cAMP-phosphodiesterase that leads to a rise in [cAMP] and a stimulation of calcium accumulation by the membranes.

HISTAMINE RELEASE FROM MAST CELLS

Interactions between cyclic nucleotides and calcium exist in the activation of histamine release by mast cell and circulating basophils (31,36,41,42,52). Natural secretion in these cells is triggered by the interaction of a soluble antigen with a specific immunoglobulin, IgE, bound to the cell surface. The IgE is analogous to a peptide hormone receptor. Interaction of antigen with membrane-bound IgE leads to a calcium-dependent release of histamine (63). A similar release can be induced by addition of the calcium ionophore, A23187 (31,36,42), or by the intracellular injection of Ca^{2+} (52). In addition, exogenous cAMP or endogenous cAMP, generated in response to isoproterenol or other β-catecholamine agonists, blocks the effects of antigen action (19). These data lead to a model of cell activation in which interaction of first messenger with receptor leads to an increased entry of calcium into the cell *without* an activation of adenylate cyclase, contrasting with the situation in the fly salivary gland in which the two are coupled. In the present case, the rise in intracellular [Ca^{2+}] has at least two effects: (a) activation of guanylate cyclase; and (b) activation of key steps in secretion. It is not yet known at what point cGMP acts to promote secretion. However, in analogy with the situation in smooth muscle, the rise in [cGMP] may act either as a positive feedforward regulator of Ca^{2+} release from internal stores, or as a negative regulator causing Ca^{2+} uptake into these stores. It is not presently known how cAMP blocks cell activation. It is possible that its effects are on either Ca^{2+} distribution within the cell or in blocking calcium entry. The recent observations of Foreman and Mongar (41) suggest the latter may be the case. They have shown that incubation of mast cells with appropriate natural activator in the absence of extracellular calcium leads to a progressive decline in histamine release upon subsequent addition of external Ca^{2+}. These data argue that the activator modifies response, in a negative feedback sense, independently of its effect upon Ca^{2+} entry. Foreman and Mongar have also shown that dibutyryl cAMP, added before antigen, blunts or abolishes subsequent antigen-induced histamine release in mast cells but does not block the histamine release caused by A23187. Thus, the action of cAMP may be to block Ca^{2+} entry or enhance Ca^{2+} efflux, and thereby modify secretion.

PHOTORECEPTOR ACTIVATION IN THE RETINA

The system in retinal rods that converts a photon (the first messenger) into cellular response, a decrease in Na^+ conductance of the rod outer segment is of great interest, because it represents a system in which the first messenger acts upon an intracellular membrane, the disk membrane, that is not attached to the outer plasma membrane. Photon absorption by the visual pigment in this membrane leads to plasma membrane hyperpolarization (46,49,56,102,107). This can only mean the action of light on rhodopsin must lead to the release of a second messenger from the disks that diffuses to the plasma membrane and thereby blocks the Na^+ channel. It also means that when light ceases to act there must be some way to decrease the concentration of this diffusible second messenger either by its catabolism or its reaccumulation by the disk membrane. Yoshikami and Hagins (107) first postulated that this internal transmitter was Ca^{2+}. This postulate was based upon the observation that membrane depolarization and the hyperpolarization response to light were greater in low versus high external calcium, and that high external calcium, presumably by increasing internal calcium, led to hyperpolarization of the membrane. However, the discovery of high levels of adenylate and guanylate cyclases and of a disk membrane-bound phosphodiesterase led to exploration of the possible role of cyclic nucleotides in photoreceptor activation (15,16,28–30).

Initial attention was directed toward photon activation and adenylate cyclase inhibition, and it was postulated that cAMP was a second messenger in this system (15,29). However, later evidence has shown there is a phosphodiesterase bound to the disk membrane, and this enzyme has a greater affinity for cGMP than cAMP (28,70). Upon exposure to light, this enzyme (in isolated disk membranes) is activated, and the action spectrum of this phosphodiesterase parallels the absorption spectrum of rhodopsin (52,69). Stimulation of the intact retina by light leads to a fall in both [cGMP] and [cAMP] with a greater relative fall of the former (15,16). Of particular interest is the observation that Ca^{2+} in the presence of low [Mg^{2+}] inhibits this same phosphodiesterase. This was the first established link between cyclic nucleotides and Ca^{2+} in this system.

More recently, several groups of investigators have suggested that light not only causes an activation of the cGMP phosphodiesterase but also the release of bound calcium from the disk, and calcium reaccumulation by the disk is controlled by a Ca^{2+}–Mg^{2+}-activated ATPase (20,47,65,73).

Direct evidence for a change in the concentration of free intracellular calcium during illumination of invertebrate photoreceptors has come from the recent work of Brown and Blinks (23). They microinjected aequorin into the cell as a means of measuring changes in intracellular [Ca^{2+}]. Upon illumination, there was a rise in [Ca^{2+}] within the cell, which did not reach a maximum until after the peak of the electrical response but returned to undetectable levels in the dark. Raising external [Ca^{2+}] increased the amplitude of the aequorin response, and removal of external calcium reduced the response to zero. Brown and Blinks concluded that both an external and internal pool of calcium were involved in the light-induced rise in [Ca^{2+}].

Working within the hypothesis that Ca^{2+} is the major regulator of the Na^+ conductance of the rod outer-segment membrane, that changes in [Ca^{2+}] within the cytosol occur after photon activation, and that cyclic nucleotides serve as regulators of intracellular [Ca^{2+}], Lipton et al. (64) carried out a series of experiments in the toad retina using the technique of electrical recording from single perfused rods (54). They examined the changes in membrane potential, after light stimulation, and light and dark adaptation in this cell type as: (a) a function of external [Ca^{2+}]; (b) after the addition of the calcium ionophore, A23187; (c) after addition of the dibutyryl derivatives of both cAMP and cGMP; and (d) after addition of isobutylmethylxanthine (IBMX), an inhibitor of phosphodiesterase.

Perfusion with 0.5 mM dibutyryl cGMP, 5 mM IBMX, or low Ca^{2+} (0.8 mM to 10^{-9} M) produced identical effects. Early effects were (a) membrane depolarization and (b) changes in amplitude waveform, time to peak response, and duration of response to brief flashes of light to a dark-adapted rod. Addition of A23187 led to membrane hyperpolarization. The effect of dibutyryl cGMP could be blocked either by high perfusate Ca^{2+} (3.2 mM), dibutyryl cAMP, or A23187. On the other hand, the effects of low Ca^{2+} could not be prevented by dibutyryl cAMP.

These results have been interpreted in terms of a model in which light interacting with rhodopsin leads to two simultaneous events: an activation of phosphodiesterase leading to a fall in [cGMP] and a release of disk membrane-bound calcium. The fall in [cGMP] leads to a decrease in Ca^{2+} reaccumulation of the disks, and so, in the operational sense, leads to a positive feedforward control loop, i.e., autocatalytic net release of Ca^{2+}. As [Ca^{2+}] rises, the Ca^{2+} has two effects: (a) by binding to the plasma membrane it blocks the Na^+ channels, leading to a decrease in Na^+ conductance and rise in membrane potential; and (b) it inhibits the activity of the phosphodiesterase, leading to a rise in [cGMP] that, in turn, stimulates the reuptake of calcium by the disk, i.e., a negative feedback loop in this control system. The evidence in support of this role of cGMP is indirect and is based upon the following observations. The addition of IBMX causes the same change in membrane potential and rod response to light as does lowering the external [Ca^{2+}]. It is postulated that the common denominator underlying these changes is a fall in [Ca^{2+}] in the cytosol of the rod. The addition of exogenous dibutyryl cGMP will mimic the effect of IBMX or low Ca^{2+}, but dibutyryl cAMP will not. Addition of cGMP plus IBMX had a synergistic effect. Thus, conditions in which a rise in intracellular [cGMP] would be expected or has been observed, mimic exactly the effects of lowering the [Ca^{2+}] in the medium. Conversely, either dibutyryl cAMP or high Ca^{2+} perfusion blocked the effect of dibutyryl cGMP, presumably because both led to a rise in the [Ca^{2+}] of the cell cytosol, but dibutyryl cAMP could not block the effects of low Ca^{2+}, presumably because under these conditions its addition could not cause a rise of [Ca^{2+}] in the calcium-depleted tissue.

ACETYLCHOLINE-MEDIATED SECRETION
IN THE EXOCRINE PANCREAS

A major problem in defining the calcium-cyclic nucleotide relationships in particular instances is that the source of calcium, which acts in stimulus-secretion coupling, varies from one system to the next, just as the source of calcium for stimulus-contraction coupling varies from one type of muscle to another (13,38).

When acetylcholine or pancreozymin act to stimulate pancreatic exocrine secretion, they increase the Na^+ but *not* the Ca^{2+} conductance of the plasma membrane (32,66,67,72), they increase the cGMP concentration within the cells but not the cAMP concentration (27,48,86) and they increase calcium efflux (from prelabeled glands) but do not increase calcium influx via a specific calcium channel (26,50,55). Calcium efflux is stimulated even if no external calcium is present (26), and preliminary data (*unpublished*) indicates [cGMP] increases even when no Ca^{2+} is present in the extracellular medium.

One of the most controversial aspects of the data concerned with acetylcholine action is whether or not an increase in Ca^{2+} uptake occurs in response to this stimulus. This particular problem points up the fact that there are two well-recognized pathways for Ca^{2+} entry into cells: (a) an early Na^+-dependent channel, and (b) a late potential-dependent via the calcium channel (17,87). As discussed above, in the case of the actions of 5HT on the fly salivary gland, cholinergic activation of the partotid, glucose activation of insulin secretion in the endocrine pancreas, and parathyroid hormone stimulation of renal gluconeogenesis, the hormone- or extracellular-messenger-induced increase in Ca^{2+} entry is thought to take place via specific calcium channels. However, in the case of acetylcholine action on the exocrine pancreas, there is no evidence for a rise in early Ca^{2+} conductance, but some Ca^{2+} may enter via the activated Na^+ channel, and it seems likely that Ca^{2+} is released from membrane sites (77). The present consensus is, however, that this small amount of calcium is not sufficient to activate secretion.

In the case of the exocrine pancreas, several points concerning calcium cyclic nucleotide relationships still need to be clarified. The first is the relationship between Ca^{2+} and cGMP. Although [cGMP] rises after acetylcholine action, it does not necessarily mean acetylcholine serves as a direct activator of guanylate cyclase. It is possible the acetylcholine-induced increase in Na^+ permeability leads to a sufficient rise in intracellular Na^+ to trigger Ca^{2+} release from the mitochondrial pool that, in turn, activates guanylate cyclase. Carafoli and co-workers (25) have shown that as little as a 5-mM increase in Na^+ can cause a significant efflux of Ca^{2+} from isolated liver mitochondria. Also, Meldolesi et al. (68) have suggested that the only major intracellular calcium pool in exocrine pancreas, other than the zymogen granules, is the mitochondrial pool and that this calcium can be released from these mitochondria by small amounts of cAMP, cGMP, or Na^+. Thus, either Na^+ or cGMP might be responsible for the mobilization of Ca^{2+} from this intracellular pool.

Although there is not yet complete unanimity about the role of calcium in acetylcholine action, nor the relationship of Ca^{2+} to cGMP in this tissue, the bulk of evidence favors the view that: (a) acetylcholine does not increase calcium uptake by specific calcium channels in this tissue; (b) acetylcholine does increase calcium efflux; (c) external calcium, although not required for initiation of secretion, is required for sustained protein secretion; (d) acetylcholine alters the Na^+ permeability of the plasma membrane, which may lead to an increase in the rate of Ca^{2+} entry into the cell; (e) acetylcholine stimulates guanylate cyclase even in the absence of extracellular calcium; and (f) either a rise in intracellular [Na^+] and/or a rise in [cGMP] mobilizes Ca^{2+} from an internal source (the mitochondria). These data can be incorporated into a model in which acetylcholine-receptor interaction leads to a simultaneous depolarization of the plasma membrane and an activation of guanylate cyclase. The membrane depolarization is associated with an increase in the Na^+ permeability of the membrane leading to an increase in entry of Na^+ into the cell. The rise in [cGMP] and/or [Na^+] within the cytosolic causes a release of calcium from an intracellular pool (probably the mitochondrial pool) leading to a rise in cytosolic [Ca^{2+}]. This increase in [Ca^{2+}] acts as the ultimate coupling factor between excitation and secretion. In addition, the rise in [Ca^{2+}] in the cytosol acts to repolarize the plasma membrane, i.e., it exerts a negative feedback control on the initial membrane events.

Operationally, this situation is similar in many respects to the proposed model of events in the action of 5HT on fluid secretion in the fly salivary gland. However, there are two features that are different in the exocrine pancreas. First, cGMP rather than cAMP is the second messenger generated by the first messenger-receptor interaction and is the prime regulator of intracellular Ca^{2+} metabolism. Second, a change in the Na^+ permeability of the membrane and thus the Na^+ gradient across the membrane is the major regulator of the calcium distribution across this membrane and may also play a second-messenger role.

PATTERNS OF CA^{2+}-CYCLIC NUCLEOTIDE RELATIONSHIPS

When one compares the model of events in the exocrine pancreas with that of light activation of the amphibian rod, that of serotonin activation of the fly salivary gland, and that of catecholamine action on the smooth muscle, an interesting set of patterns emerge (Table 1). These four different models represent the four different possible relationships between cytosolic calcium, on the one hand, and either cAMP or cGMP on the other. In the first relationship, exemplified by the action of serotonin on the fly salivary gland (Pattern I), cAMP stimulates a release of bound calcium and thus a rise in cytosolic [Ca^{2+}], and conversely Ca^{2+} causes a fall in [cAMP]. In the second relationship, exemplified by the action of catecholamines on intestinal smooth muscle (Pattern II), cAMP stimulates the uptake of free calcium and thus causes a fall in cytosolic [Ca^{2+}], and conversely Ca^{2+} causes a rise in [cAMP].

In the third relationship, exemplified by the action of acetylcholine on the exocrine pancreas (Pattern III), cGMP causes a release of bound calcium and thus a rise in cytosolic [Ca^{2+}]. In the fourth relationship, exemplified by light activation of the rod disc membranes in the amphibian eye (Pattern IV), cGMP stimulates the reaccumulation of calcium by the disc membrane and, hence, lowers cytosolic [Ca^{2+}], and Ca^{2+} inhibits PDE and thereby causes a rise in [cGMP]. Thus, all four of the possible phase relationships between Ca^{2+}, on the one hand, and the two cyclic nucleotides on the other, have been identified as patterns involved in activation of particular cell types by specific extracellular messengers.

ACETYLCHOLINE-MEDIATED RELEASE OF EPINEPHRINE
FROM THE ADRENAL MEDULLA

One of the earliest secretory systems in which Ca^{2+} was shown to play a second-messenger role was the acetylcholine-stimulated release of catecholamines from the adrenal medulla (34,35,93). Stimulation of cholinergic receptor leads to an increased uptake of calcium, and if stimulation is sustained, a 30% increase in total tissue calcium occurs. Removal of external calcium blocks cholinergic stimulation. Readdition of calcium leads to a prompt stimulation of catecholamine release (33,35,97). The addition of the calcium ionophore also stimulates catecholamine release (43). In all these circumstances, there is no direct evidence that acetylcholine causes an activation of tissue adenylate cyclase. It can be postulated that acetylcholine causes a calcium-dependent increase in [cGMP], but this remains to be established, as does the subsequent role of this nucleotide in the secretory process.

In spite of the fact that acetylcholine does not stimulate adenylate cyclase (98), it causes a rise in [cAMP] within the cell, and the addition of exogeous cAMP, dibutyryl cAMP, or isoproterenol causes a release of neurotransmitter (75). However, in the isolated perfused gland this cAMP-induced release is dependent upon the presence of acetylcholine. If the gland is not perfused with acetylcholine, then addition of either cAMP or dibutyryl cAMP causes no increase in catecholamine release.

If first perfused with acetylcholine alone and then cAMP is added, the nucleotide causes an enhanced release of catecholamine (96). If very large doses of exogenous cAMP are added to adrenal medullary slices *in vitro* in the absence of external Ca^{2+}, a release of catecholamine is still observed.

The effect of cAMP might be considered of pharmacological rather than physiological interest if it were not for the fact that the epinephrine-secreting cells of the adrenal medulla contain β-adrenergic receptors which are coupled to an adenylate cyclase and cause a rise in [cAMP] that potentiates the acetylcholine-stimulated epinephrine release (89). Thus, there is evidence in this tissue of a positive feedforward control loop in this system with the released product, epinephrine, enhancing its own release, i.e., serving as a sequential second messenger.

In considering the possibility that cAMP is a supplemental second messenger in the control of catecholamine release in the adrenal medulla, it is also necessary to consider that its main function may not be so much to enhance secretion as to stimulate hypertrophy and hyperplasia of the gland (24,45). It is a well-known physiological fact that the size and functional capacity of the adrenal medulla, adrenal cortex, or other endocrine tissues are regulated by the trophic influences of neurotransmitters or extracellular messengers. Denervation or removal of the appropriate first messenger leads to organ atrophy. Conversely, overstimulation leads to cell hypertrophy and limited hyperplasia.

In spite of the impressive evidence that initial cholinergic stimulation of the adrenal medulla does not involve any change in intracellular [cAMP], exposure of the rat to a cold environment, 4°C for 30 min, leads to a 10- to 20-fold increase in the cAMP content of the adrenal medulla. It seems likely that the rise in cAMP is due to β-adrenergic stimulation of these cells, i.e., the released epinephrine stimulates adenylate cyclase in the cell. However, it is also possible that an acetylcholine-induced second messenger may amplify the epinephrine-induced rise in [cAMP]. The rise in [cAMP] is followed by both an increase in specific catecholamine-synthesizing enzymes, tyrosine hydroxylase, and dopamine-β-hydroxylase, and by a dramatic increase in ornithine decarboxylase activity, one of the earliest events that occurs during tissue growth (45,94,95). Thus, the relationship between Ca^{2+} and cAMP in this tissue exemplifies an additional type of pattern wherein these two messengers are generated sequentially by different extracellular first messengers and act sequentially to integrate both short-term and long-term response.

The systems presented represent only a few of the many in which evidence shows a close relationship between calcium and one or both cyclic nucleotides in the activation of particular cell types by specific first messengers. They do, however, represent the spectrum of possible relationships between these three simple cellular control elements, and demonstrate that with just a few such elements a large number of possible particular variations upon a universal theme is possible.

EFFECTOR MOLECULES — RECEPTORS FOR SECOND MESSENGERS

In concluding this discussion of calcium-cyclic nucleotide relationships, it is worth taking note of one further aspect of their mutual interactions.

The discovery of the adenylate cyclase-cAMP system and the demonstration of its widespread occurrence led, as discussed above, to the second-messenger hypothesis. This, as noted, brought unity into the field of peptide and amine hormone action, but in a sense only moved the philosophical problem from outside to inside the cell. Just as previously, it seemed inconceivable that all peptide hormone brought about their diverse effects upon cell function by a common mechanism, it then seemed unlikely that the diverse effects of a rise in [cAMP] in different cells was mediated by a common cAMP receptor or control element. Initial studies into the mode of cAMP action were, thus, viewed as individual problems in cell biology. However, within a short time from the work of E. Krebs and co-workers (8) and that of Kuo and Greengard (44,59), unity was brought to this aspect of cAMP physiology by the discovery of the ubiquitous presence of cAMP-dependent protein kinases both in the soluble and particulate (membrane) parts of the cell. At present, the bulk of the evidence favors the view that the mode of action of cAMP depends upon its binding to specific receptor molecules within the cell, which in turn act as allosteric modifiers of a specific class of enzymes, protein kinases.

It is of more than passing interest, in the context of the present discussion, that some of the phosphoprotein products of these cAMP-dependent protein kinases are either calcium-activated enzymes or possibly regulators of calcium transport.

Although protein kinases are as yet the only class of enzymes known to interact with and be regulated by cAMP-binding proteins, it is quite possible that the cAMP-binding proteins will be found to interact with, and regulate the activity of, other types of protein molecules.

A similar philosophical problem had existed with the role of calcium in cell function. An increasing number of cell functions have been shown to be influenced by changes in $[Ca^{2+}]$ in one or more subcellular compartments. It seemed unlikely that a common calcium receptor protein was involved. However, within the past few years, this assessment has been changed by the demonstration of the widespread occurrence of a specific calcium-binding protein in nonmuscle cells.

The starting point of this work was the discovery that Ca^{2+} regulated phosphodiesterase activity in a variety of tissues (5,51,103,106). A homogenous, acidic, Ca^+-binding phosphoprotein from brain, heart, and other tissues that serves as an activator of partially purified brain or cardiac phosphodiesterase has been identified. This phosphoprotein, or calcium-dependent regulator (CDR) of PDE activity, thus functions as a regulatory subunit for PDE (a), but in the converse operational sense from the cAMP-dependent regulatory subunit of protein kinase (b),

$$\text{(a)} \quad Ca^{2+} + CDR \rightleftharpoons Ca\text{-}CDR + PDE_i \rightleftharpoons Ca\text{-}CDR\cdot PDE \quad \overset{\displaystyle cAMP|}{\underset{\displaystyle \downarrow}{}} \; 5'AMP$$

$$\text{(b)} \quad cAMP + RC \rightleftharpoons cAMP\cdot R + C \quad \underset{Pr\,P}{\overset{Pr}{|}} \left(\genfrac{}{}{0pt}{}{^{ATP}}{_{ADP}}\right.$$

Calcium, by binding to the regulator (CDR) of PDE, causes an association of CDR with PDE and, thus, an activation of the enzyme, i.e., a rise in V_{max} and a fall in K_m (103). In contrast, cAMP binding to the regulatory subunit of the protein kinase causes a dissociation of regulatory from the catalytic subunit, which now becomes enzymatically active.

The two findings with the broadest biological significance are that the calcium-dependent regulator protein is similar in structure to the calcium-binding subunit of troponin (TNC) and that it occurs in a variety of tissues (5,51,78,103–106). There is evidence that the regulator protein and the TNC evolved from a common ancestral protein through the mechanism of gene duplication (57,58,78,103,104). Both proteins have blocked NH_2 termini, similar UV spectra, similar calcium-binding properties, contain some identical peptide sequences, and have similar but not identical amino acid compositions. The CDR has a ninhydrin-positive component, not found in TNC (104,105).

In addition, the parvalbumins, a class of calcium-binding proteins in certain forms of muscle, also appear to be derived from the same ancestral protein (57,58,76). The parvalbumins are smaller proteins that probably represent incomplete copies of the primordial gene after its duplication. Nevertheless, they retain their calcium-binding ability.

From these and other data, Kretsinger (57,58) has proposed that in all processes in which calcium functions as a second messenger (or coupling factor) the molecular mechanism of calcium action is mediated by its binding to a calcium-binding protein containing one or several EF hands, i.e., regions homologous to the EF region of muscle parvalbumins, which are the sites of calcium binding to this protein. This hypothesis is in a sense the counterpoint of the hypothesis of Kuo and Greengard (59) that all cAMP effects are mediated via cAMP-binding proteins acting as regulatory subunits of protein kinase. It is already evident from the function of TNC, parvalbumins, and the CDR that during the evolution of these Ca^{2+}-binding proteins, they have taken on new functions. The TNC-calcium interaction is the initiator of the contractile response in mammalian muscle. The parvalbumins are thought to serve a calcium-buffering function in the contraction of fast muscles. The CDR is thought to regulate the activity of a crucial enzyme in cyclic nucleotide metabolism, phosphodiesterase. The fact that similar acidic calcium-binding proteins have been found in other tissues, and cells, e.g., sperm (22), raises the possibility that regulation of a variety of calcium-mediated processes involves a TNC-like protein as modulator. It will be of great interest to determine whether proteins of this type are involved in calcium-dependent secretion and other calcium-dependent cell responses.

ACKNOWLEDGMENTS

This work was supported by NIH Grants AM09650 and AM19813–01.

REFERENCES

1. Andersson, R. (1972): Role of cyclic AMP and Ca^{++} in metabolic and relaxing effects of catecholamines in intestinal smooth muscle. *Acta Physiol. Scand.*, 85:312–322.
2. Andersson, R. (1973): Cyclic AMP as a mediator of the relaxing action of Papaverine,

Nitroglycerine, diazoxide and hydralazine in intestinal and vascular smooth muscle. *Acta Pharmacol. Toxicol.,* 73:321–336.

3. Andersson, R. G. G. (1973): Relationship between cyclic AMP, phosphodiesterase activity, calcium and contraction in intestinal smooth muscle. *Acta Physiol. Scand.,* 87: 348–358.

4. Andersson, R., Nilsson, K., Wikberg, J., Johansson, S., and Lundholm, L. (1975): Cyclic nucleotides and the contractions of smooth muscle. *Adv. Cyclic Nucleotide Res.,* 5:491–518.

5. Appleman, M. M., and Terasaki, W. L. (1975): Regulation of cyclic nucleotide phosphodiesterase. *Adv. Cyclic Nucleotide Res.,* 5:153–162.

6. Baker, P. F., Hodgkin, A. L., and Ridgway, E. B. (1971): Depolarization and calcium entry in squid giant axons. *J. Physiol. (Lond.),* 218:708–755.

7. Baker, P. F., Meves, H., and Ridgway, E. B. (1973): Calcium entry in response to maintained depolarization of squid axons. *J. Physiol. (Lond.),* 231:527–548.

8. Beavo, J. A., Bechtel, P. J., and Krebs, E. G. (1975): Mechanisms of control for CAMP-dependent protein kinase from skeletal muscle. *Adv. Cyclic Nucleotide Res.,* 5:241–251.

9. Berridge, M. J. (1970): The role of 5-hydroxytryptamine and cyclic AMP in the control of fluid secretion by isolated salivary glands. *J. Exp. Biol.,* 53:171–186.

10. Berridge, M. J. (1976): The interaction of cyclic nucleotides and calcium in the control of cellular activity. *Adv. Cyclic Nucleotide Res.,* 6:1–96.

11. Berridge, M. J., Oschman, J. L., and Wall, B. J. (1975): Intracellular calcium reservois in *Calliphora* salivary glands. In: *Calcium Transport in Contraction and Secretion,* edited by E. Carafoli, F. Clementi, W. Drabikowsky, and A. Margreth, pp. 131–138. North Holland Publishing Co., Amsterdam.

12. Berridge, M. J., and Prince, W. T. (1972): The role of cyclic AMP and calcium in hormone action. *Adv. Insect Physiol.,* 9:1–49.

13. Bianchi, C. P. (1973): Drugs affecting excitation coupling mechanisms. In: *Fundamentals of Cell Pharmacology,* edited by S. Dikstein, pp. 454–468. Charles C Thomas, Springfield, Ill.

14. Birnbaumer, L. (1973): Hormone-sensitive adenyl cyclases: Useful models for studying hormone receptor functions in cell-free systems. *Biochim. Biophys. Acta,* 300:129–158.

15. Bitensky, M. W., Gorman, R. E., and Miller, W. H. (1971): Adenyl cyclase as a link between photon capture and charges in membrane permeability of frog photoreceptors. *Proc. Natl. Acad. Sci. U.S.A.,* 68:561–562.

16. Bitensky, M. W., Miki, N., Keirns, J. J., Keirns, M., Baraban, J. M., Freeman, J., Wheeler, M. A., Lacy, J., and Marcus, F. R. (1975): Activation of photoreceptors disk membrane phosphodiesterase by light and ATP. *Adv. Cyclic Nucleotide Res.,* 5:213–255.

17. Blaustein, M. P. (1974): The interrelationship between sodium and calcium fluxes across cell membranes. *Rev. Physiol. Biochem. Pharmacol.,* 70:33–82.

18. Borle, A. (1973): Calcium metabolism at the cellular level. *Fed. Proc.,* 32:1944–1950.

19. Bourne, H. R., Lichtenstein, L. M., Melmor, K. L., Henney, C. S., Wernstein, Y., and Shearer, G. M. (1974): Modulation of inflammation and immunity by cyclic AMP. *Science,* 184:19–28.

20. Bowinds, D., Gordon-Walker, A., Gaide-Huguenin, A. D., and Robinson, W. E. (1971): Characterization and analysis of frog photoreceptor membranes. *J. Gen. Physiol.,* 58: 225–237.

21. Bowyer, F., and Kitabchi, A. E. (1974): Dual role of calcium in steroidogenesis in the isolated adrenal cell of rat. *Biochem. Biophys. Res. Commun.,* 57:100–105.

22. Brooks, J. C. and Siegel, F. L. (1973): Calcium-binding phosphoprotein: The principal acidic protein of mammalian sperm. *Biochem. Biophys. Res. Commun.,* 55:710–716.

23. Brown, J. E., and Blinks, J. R. (1974): Changes in intracellular free calcium concentration during illumination of invertebrate photoreceptors detection with aequorin. *J. Gen. Physiol.,* 64:643–665.

24. Byers, C. V., and Russell, D. H. (1975): Ornithine decarboxylase activity: Control by cyclic nucleotides. *Science,* 187:650–652.

25. Carafoli, E., Malinstrom, K., Capano, M., Segel, E., and Crompton, M. (1975): Mitochondria and the regulation of cell calcium. In: *Calcium Transport in Contraction and Secretion,* edited by E. Carafoli, F. Clementi, W. Drabikowsky, and A. Margreth, pp. 53–64. North Holland Publishing Co., Amsterdam.

26. Case, R. M., and Clausen, T. (1973): The relationship between calcium exchange and enzyme secretion in the isolated rat pancreas. *J. Physiol. (Lond.)*, 235:75–102.

27. Case, R. M., Johnson, M., Scratcherd, T., and Sherratt, H. S. A. Cyclic adenosine 3′-5′-monophosphate concentration in the pancreas following stimulation by secretion, cholecystokinin-pancreozymin and acetylcholine. *J. Physiol. (Lond.)*, 223:669–684.

28. Chader, G., Fletcher, R., Johnson, M., and Bensinger, R. (1974): Rod outer segment phosphodiesterase: Factors affecting the hydrolysis of cyclic-AMP and cyclic-GMP. *Exp. Eye Res.*, 18:509–515.

29. Chader, G. J., Herz, L. R., and Fletcher, R. T. (1974): Light activation of phosphodiesterase activity in retinal rod outer segments. *Biochim. Biophys. Acta*, 347:491–493.

30. Chader, G. J., Johnson, M., Fletcher, R. T., and Bensinger, R. (1974): Cyclic nucleotide phosphodiesterase of the Fovine retina: Activity, subcellular distribution and kinetic parameters. *J. Neurochem.*, 22:93–99.

31. Cochrane, D. E., and Douglas, W. W. (1974): Calcium-induced extrusion of secretory granules (exocytosis) in mast cells exposed to 48/80 or the ionophores A23187 and X-537A. *Proc. Natl. Acad. Sci. U.S.A.*, 71:408–412.

32. Dean, P. M., and Matthews, E. K. (1972): Pancreatic acinar cells: Measurement of membrane potential and immature depolarization potentials. *J. Physiol. (Lond.)*, 224:1–13.

33. Douglas, W. W. (1974): Involvement of calcium in exocytosis and the exocytosis-vesiculation sequence. *Biochem. Soc. Symp.*, 39:1–28.

34. Douglas, W. W. (1968): Stimulus-secretion coupling: The concept and clues from chromaffin and other cells. *Br. J. Pharmacol.*, 34:451–474.

35. Douglas, W. W., and Rubin, R. P. (1961): The role of calcium in the secondary response of the adrenal medulla to acetylcholine. *J. Physiol. (Lond.)*, 159:40–57.

36. Douglas, W. W., and Veda, Y. (1973): Mast cell secretion (histamine) release induced by 48–80: Calcium-dependent exocytosis inhibited strongly by cytochalasin only when glycolysis is rate-limiting. *J. Physiol. (Lond.)*, 234:97–98P.

37. Dreifuss, J. J., Grau, J. D., and Nordman, J. J. (1975): Calcium movements related to neurophysical hormone secretion. In: *Calcium Transport in Contraction and Secretion*, edited by E. Carafoli, F. Clementi, W. Drabikowski, and A. Margreth, pp. 271–279. North Holland Publishing Co., Amsterdam.

38. Ebashi, S., and Endo, M. (1968): Calcium and muscle contraction. *Prog. Biophys. Molec. Biol.*, 5:123–183.

39. Exton, J. H. (1972): Gluconeogenesis. *Metabolism*, 21:945–990.

40. Fischer, E. H., Heilmeyer, L. M. G., and Haschke, R. H. (1971): Phosphorylase and the control of glycogen degradation. *Curr. Top. Cell. Res.*, 4:211–251.

41. Foreman, J. C., and Mongar, J. L. (1975): Calcium and the control of histamine secretion from mast cells. In: *Calcium Transport in Contraction and Secretion*, edited by E. Carafoli, F. Clementi, W. Drabikowsky, and A. Margreth, pp. 175–184. North Holland Publishing Co., Amsterdam.

42. Foreman, J. C., Mongar, J. L., and Gomperts, B. O. (1973): Calcium ionophores and movement of calcium ions following the physiological stimulus to a secretory process. *Nature*, 245:249–251.

43. Garcia, A. G., Kirpekar, S. M., and Prat, J. C. (1975): A calcium ionophore stimulating the secretion of catecholamines from cat adrenal. *J. Physiol. (Lond.)*, 244:253–262.

44. Greengard, P., and Kuo, J. F. (1970): On the mechanism of action of cyclic AMP. In: *Role of Cyclic AMP in Cell Function*, edited by P. Greengard, and E. Costa, pp. 287–306. Raven Press, New York.

45. Guidotti, A., Hanbauer, I., and Costa, E. (1975): Role of cyclic nucleotides in the induction of tyrosine hydroxylase. *Adv. Cyclic Nucleotide Res.*, 5:619–640.

46. Hagins, W. A. (1972): The visual process: Excitatory mechanisms in the primary receptor cells. *Annu. Rev. Biophys. Bioeng.*, 1:131–158.

47. Hagins, W., and Yoshikami, S. (1974): A role for Ca^{2+} in excitation of retinal rods and cones. *Exp. Eye Res.*, 18:299–305.

48. Haymovits, A., and Scheele, G. A. (1976): Cellular cyclic nucleotides and enzyme secretion in the pancreatic acinar cell. *Proc. Natl. Acad. Sci. U.S.A.*, 73:156–160.

49. Hechter, O. (1955): Concerning possible mechanisms of hormone action. *Vitam. Horm.*, 13:293–346.

50. Heisler, S., Fast, D., and Tenenhouse, A. (1972): Role of Ca^{2+} and cyclic AMP in

protein secretion from rat exocrine pancreas. *Biochim. Biophys. Acta,* 279:561–572.
51. Kakiuchi, S., Yamozaki, R., Teshiura, Y., Venishi, K., and Miyamoto, E. (1975): Ca^{2+}/Mg^{2+}-dependent cyclic nucleotide phosphodiesterase and its activator protein. *Adv. Cyclic Nucleotide Res.,* 5:163–178.
52. Kanno, T., Cochrane, D. E., and Douglas, W. W. (1973): Exocytosis (secretory granule extrusion) induced by injection of calcium in mast cells. *Can. J. Physiol. Pharmacol.,* 51:1001–1004.
53. Keirns, J. J., Miki, N., Bitensky, M. W., and Keirns, M. (1975): A link between rhodopsin and disk membrane cyclic nucleotide phosphodiesterase. Action spectrum and sensitivity to illumination. *Biochemistry,* 14:2760–2766.
54. Kleinschmidt, J. and Dowling, J. E. (1975): Intracellular recordings from geko photoreceptors during light and dark adaptation. *J. Gen. Physiol.,* 66:617–648.
55. Kondo, S., and Schulz, I. (1976): Calcium ion uptake in isolated pancreas cells induced by secretagogues. *Biochim. Biophys. Acta,* 419:76–92.
56. Korenbrot, J. E. (1973): Ionic flux and membrane characteristics of isolated rod outer segments. *Exp. Eye Res.,* 16:343–355.
57. Kretsinger, R. H. (1972): Gene triplication deduced from the tertiary structure of a muscle calcium binding protein. *Nature [New Biol.],* 240:85–88.
58. Kretsinger, R. H. (1975): Hypothesis: Calcium modulated proteins contain EF hands. In: *Calcium Transport in Contraction and Secretion,* edited by E. Carafoli, F. Clementi, W. Drabikowski, and A. Margreth, pp. 469–478. North Holland Publishing Co., Amsterdam.
59. Kuo, J. F., and Greengard, P. (1969): Cyclic nucleotide-dependent protein kinases. IV. Widespread occurrence of adenosine 3',5'-monophosphate-dependent protein kinase in various tissues and phyla of the animal kingdom. *Proc. Natl. Acad. Sci. U.S.A.* 64:1349–1355.
60. Lee, T. P., Kuo, J. F., and Greengard, P. (1972): Role of muscariuic cholinergic receptors in regulation of guanosine 3',5'-cyclic monophosphate content in mammalian brain, heart, muscle, and intestinal smooth muscle. *Proc. Natl. Acad. Sci. U.S.A.* 69:3287–3291.
61. Lefkowitz, R. I. (1975): Identification of adenylate cyclase coupled beta-adrenergic receptors with radiolabelled beta-adrenergic antagonists. *Biochem. Pharmacol.,* 24:1651–1658.
62. Lefkowitz, R. J., Roth, J., and Pastan, I. (1970): Effect of calcium on ACTH stimulation of the adrenal: Separation of hormone binding from adrenal cyclase activation. *Nature,* 228:864–866.
63. Lichtenstein, L. M., Ishizaka, K., Norman, P. S., Sobotka, A. K., and Hill, B. M. (1973): IgE Antibody measurements in ragweed hay fever. Relationship to clinical severity and the results of immunotherapy. *J. Clin. Invest.,* 52:472–482.
64. Lipton, S. A., Dowling, J. E., and Rasmussen, H. (1977): Adaptation of vertebrate photoreceptors. Similar effects of Ca^{2+}, cyclic nucleotides, and prostaglandins as intracellular messengers. *(Submitted.)*
65. Mason, W. T., Fager, R. S., and Abrahamson, E. W. (1974): Ion fluxes in disk membranes of retinal rod outer segments. *Nature,* 247:562–563.
66. Matthews, E. K., and Petersen, O. H. (1973): Pancreatic acinar cells: Ionic dependence of the membrane potential and acetylcholine-induced depolarization. *J. Physiol. (Lond.),* 231:283–295.
67. Matthews, E. K., Petersen, O. H., and Williams, J. A. (1973): Pancreatic acinar cells: Acetylcholine-induced membrane depolarization, calcium efflux and amylase release. *J. Physiol. (Lond.),* 234:689–701.
68. Meldolesi, J., Ramellini, G., and Clementi, F. (1975): Calcium compartmentalization in pancreatic acinar cells. In: *Calcium Transport in Contraction and Secretion,* edited by E. Carafoli, F. Clementi, W. Drabikowski, and A. Margreth, pp. 157–166. North Holland Publishing Co., Amsterdam.
69. Miki, N., Keirns, J. J., Marcus, F. R., and Bitensky, M. W. (1974): Light regulation of adenosine 3',5':cyclic monophosphate levels in invertebrate photoreceptors. *Exp. Eye Res.,* 18:291–297.
70. Miki, N., Keirns, J. J., Marcus, F. R., Freeman, J., and Bitensky, M. W. (1973): Regulation of cyclic nucleotide concentrations in photoreceptors: An ATP-dependent stimula-

tion of cyclic nucleotide phosphodiesterase by lights. *Proc. Natl. Acad. Sci. U.S.A.,* 70:3820–3824.

71. Moyle, W. R., Kong, Y. C., and Ramachandran, J. (1973): Steriodogenesis and cyclic adenosine 3′,5′-monophosphate accumulation in rat adrenal cells. Divergent effect of adrenocorticotropin and its o-nitrophenyl sulfenyl derivative. *J. Biol. Chem.,* 248:2409–2417.

72. Nishiyama, A., and Petersen, O. H. (1975): Pancreatic acinar cells: Ionic dependence of acetylcholine-induced membrane potential and resistance changes. *J. Physiol. (Lond.),* 244:431–465.

73. Ostwald, T. J., and Heller, J. (1972): Properties of magnesium- or calcium-dependent adenosine triphosphatase from frog photoreceptor outer segment disks and its inhibition by illumination. *Biochemistry,* 11:4679–4686.

74. Panko, W., and Kenney, F. (1971): Hormonal structure of hepatic ornithine decarboxylase. *Biochem. Biophys. Res. Commun.,* 43:346–350.

75. Peach, M. J. (1972): Stimulation of release of adrenal catecholamine by adenosine 3′:5′-cyclic monophosphate and theophylline in the absence of extracellular Ca^{2+}. *Proc. Natl. Acad. Sci. U.S.A.,* 69:834–836.

76. Pechere, J. F., Demaille, J., Capony, J. P., Dutruge, E., Baron, G., and Pina, C. (1975): Muscular parvalbumins. In: *Calcium Transport in Contraction and Secretion,* edited by E. Carafoli, F. Clementi, W. Drabikowski, and A. Margreth, pp. 459–468. North Holland Publishing Co., Amsterdam.

77. Petersen, O. H., and Veda, N. (1975): Ca^{2+} Control of pancreatic enzyme secretion. In: *Calcium Transport in Contraction and Secretion,* edited by E. Carafoli, F. Clementi, W. Drabikowski, and A. Margreth, pp. 147–156. North Holland Publishing Co., Amsterdam.

78. Potter, J., Leavis, P., Scidel, J., Lehrer, S., and Gergely, J. (1975): Interaction of divalent cations with troponin and myosin. In: *Calcium Transport in Contraction and Secretion,* edited by E. Carafoli, F. Clementi, W. Drabikowski, and A. Margreth, pp. 415–430. North Holland Publishing Co., Amsterdam.

79. Prince, W. T., Berridge, M. J., and Rasmussen, H. (1972): Role of calcium and adenosine 3′:5′-cyclic monophosphate in controlling fly salivary gland secretion. *Proc. Natl. Acad. Sci. U.S.A.,* 69:553–557.

80. Prince, W. T., Rasmussen, H., and Berridge, M. (1973): The role of calcium in fly salivary gland secretion analyzed with the ionophore A-23187. *Biochim. Biophys. Acta,* 329:98–107.

81. Rasmussen, H. (1970): Cell communication, calcium ion, and cyclic adenosine monophosphate. *Science,* 170:404–412.

82. Rasmussen, H., and Goodman, D. B. P. (1977): Calcium and cyclic nucleotides in cell activation. *Physiol. Rev. (in press).*

83. Rasmussen, H., Goodman, D. B. P., Friedmann, N., Allen, J. E., and Kurokawa, K. (1976): Ions and the control of metabolic processes. In: *Handbook of Physiology, Sect., 7, Vol. 7,* edited by G. D. Aurbach, pp. 225–264. Waverly Press, Baltimore.

84. Rasmussen, H., Goodman, D. B. P., and Tenenhouse, A. (1972): The role of cyclic AMP and calcium in cell activation. *CRC Crit. Rev. Biochem.,* 1:95–148.

85. Rasmussen, H., Jensen, P., Lake, W., Friedmann, N., and Goodman, D. B. P. (1975): Cyclic nucleotides and cellular calcium metabolism. *Adv. Cyclic Nucleotide Res.,* 5:375–394.

86. Rebhun, L. I. (1977): Cyclic nucleotides, calcium and cell division. *Int. Rev. Cytol.,* 49:1–54.

87. Rink, T. J., and Baker, P. F. (1975): The role of the plasma membrane in the regulation of intracellular calcium. In: *Calcium Transport in Contraction and Secretion,* edited by E. Carafoli, F. Clementi, W. Drabikowski, and A. Margreth, pp. 235–242. North Holland Publishing Co., Amsterdam.

88. Robberecht, P., Deschodt-Lanckman, M., DeNeff, P., Borgeat, P., and Christophe, J. (1974): *In vivo* effects of pancreozymin, secretion, vasoactive intestinal polypeptide and pilocarpone in the levels of cyclic AMP and cyclic GMP in the rat pancreas. *FEBS Lett.,* 43:139–143.

89. Robison, G. A., Butcher, R. W., and Sutherland, E. W. (1971): *Cyclic AMP.* Academic Press, New York.

90. Rodbell, M., Liu, M. C., Salomon, Y., Louclos, C., Harwood, J. P., Martin, B. R., Rendell, M., and Berman, M. (1975): Role of adenine and guanine nucleotides in the activity response of adenylate cyclase systems to hormones: Evidence for multisite transition states. *Adv. Cyclic Nucleotide Res.,* 5:3–29.

91. Rose, B., and Loewenstein, W. R. (1975): Calcium ion distribution in cytoplasm visualized by Aequorin: Diffusion in cytosol restricted by energized sequestering. *Science,* 190:1204–1206.

92. Roth, J. (1973): Peptide hormone receptors. *Metabolism,* 22:1059–1073.

93. Rubin, R. P. (1970): The role of calcium in the release of neurotransmitter substances and hormones. *Pharmacol. Rev.,* 22:389–428.

94. Russell, D. H., and Snyder, S. H. (1970): Amine synthesis in rapidly growing tissues: Ornithine decarboxylase activity in regenerating rat liver, chick embryo, and various tumors. *Proc. Natl. Acad. Sci. U.S.A.,* 60:1420–1427.

95. Russell, D. H., and Stambrook, P. J. (1975): Cell cyclic specific fluctuations in adenosine 3',5'-cyclic monophosphate and polyamines of Chinese hamster cells. *Proc. Natl. Acad. Sci. U.S.A.,* 72:1482–1486.

96. Serck-Hanssen, G. (1974): Effect of theophylline and propanolol on acetylcholine-induced release adrenal medullary catecholamines. *Biochem. Pharmacol.,* 23:2225–2234.

97. Serck-Hanssen, G., and Christiansen, T. (1973): Uptake of calcium in chromaffin granules of bovine adrenal medulla stimulated *in vitro. Biochim. Biophys. Acta,* 307:404–414.

98. Serck-Hanssen, G., Christoffersen, T., Morland, J., and Osnes, J. B. (1972): Adenyl cyclase activity in bovine adrenal medulla. *Eur. J. Pharmacol.,* 19:297–300.

99. Sutherland, E. W., and Rall, T. W. (1958): Formation of cyclic adenine ribonucleotide by tissue particles. *J. Biol. Chem.,* 232:1065–1076.

100. Sutherland, E. W., Robison, G. S., and Butcher, R. W. (1968): Some aspects of the biological role of adenosine 3',5'-monophosphate (cyclic AMP). *Circulation,* 37:279–306.

101. Tolbert, M. E. M., Butcher, F. R., and Fain, J. N. (1973): Lack of connection between catecholamine effects on cyclic adenosine 3':5'-monophosphate and gluconeogenesis in isolated rat liver cells. *J. Biol. Chem.,* 248:5686–5692.

102. Tomita, T., Miller, W. H., Hashiuioto, Y., and Saito, Y. (1973): Electrical response of retinal cells as a sign of transport. *Exp. Eye Res.,* 16:327–341.

103. Wang, J. H., Teo, T. S., Ho, H. C., and Stevens, F. C. (1975): Bovine heart protein activator of cyclic nucleotide phosphodiesterase. In: *Advances in Cyclic Nucleotide Research, Vol. 5.,* edited by P. Greengard and G. A. Robison, pp. 179–194. Raven Press, New York.

104. Watterson, D. M., Harrelson, N. G., Jr., Keller, P. M., Sharief, F., and Vanaman, T. C. (1977): Structural similarities between the Ca^{2+}-dependent regulatory proteins of 3':5'-cyclic nucleotide phosphodiesterase and actomyosin ATPase. *J. Biol. Chem. (in press).*

105. Watterson, D. M., Van Eldik, L. J., Smith, R. E., and Vanaman, T. C. (1977): Calcium-dependent regulatory protein of cyclic nucleotide metabolism in normal and transformed chicken embryo fibroblasts. *Proc. Natl. Acad. Sci. U.S.A. (in press).*

106. Wolff, O. J., and Brostrom, C. O. (1974): Calcium-binding phosphoprotein from pig brain: Identification as a calcium-dependent regulator of brain cyclic nucleotide phosphodiesterase. *Arch. Biochem. Biophys.,* 163:349–358.

107. Yoshikami, S., and Hagins, W. A. (1973): Control of the dark current in invertebrate rods and cones. In: *Biochemistry and Physiology of Visual Pigments,* edited by H. Langer. pp. 245–255. Springer-Verlag, New York.

Cell and Tissue Interactions, edited by
J. W. Lash and M. M. Burger. Raven
Press, New York, 1977.

Cell Interactions and DNA Replication in the Sea Urchin Embryo

B. De Petrocellis,* S. Filosa-Parisi,** A. Monroy,† and
E. Parisi*

*C.N.R. Laboratory of Molecular Embryology, Arco Felice, ** Institute of Histology and Embryology, the University of Napoli, and † Stazione Zoologica, Napoli, Italy*

The regulatory processes of DNA replication and of cell division during growth and differentiation are among the major problems of developmental biology. The sea urchin embryo appears to be a favorable object in which to explore some of these problems because it lends itself to combining morphological and biochemical studies.

In the first part of this report, we present the results of some current work on the pattern of cell divisions from the early cleavage stages to the late gastrula stage of the sea urchin, *Paracentrotus lividus*. The observations have been carried out on whole-mounts of gently compressed embryos and on histological sections stained with Feulgen reaction. In the second part, we discuss some observations on the behavior of the cells isolated at different stages of development, which may throw some light on the problem of the role of cell interactions in modulating DNA replication.

THE PATTERN OF CELL DIVISION

The first four cleavages in the sea urchin embryo follow each other at an interval of about 30 min and are synchronous in all blastomeres. Once the micromeres are segregated — at the fourth cleavage — synchrony is lost. Indeed, the micromeres not only divide at a much slower rate than the other blastomeres (20), but, as will be described presently, they appear to give rise to a gradient that influences the mitotic rate of the embryo in the vegetal-animal direction. From the 16-cell stage to the morula stage — with about 100 cells — the micromeres are easily identified in whole-mount preparations by their nuclei, which are smaller than those of the rest of the blastomeres; also, they have a denser appearance, and they are more uniformly stained than the nuclei of the rest of the embryo. This makes orientation of the embryo possible. An image we have frequently encountered in our preparations is that shown in Fig. 1 in which all the nuclei of the macro- and mesomeres, as well as the nuclei of the four center micromeres,

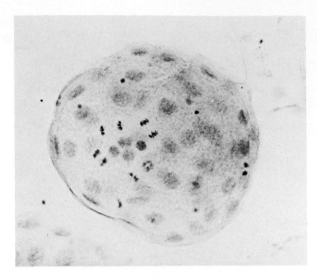

FIG. 1. Sixty-four-cell stage. The four outer micromeres in anaphase. ×320.

are in the resting stage. Four dividing nuclei stand out around the four micromeres. In embryos in which these four outer cells are in metaphase, the metaphase plates are clearly smaller than those observed in the macro- or mesomeres, and, hence, we are inclined to think that these are the four outer micromeres in the process of dividing. If this interpretation is correct, it would imply that the *outer row* of micromeres undergoes cell division, *while the inner row is in the resting stage.* Our interpretation, of course, needs to be substantiated by *in vivo* observations. However, we would like to refer to a photograph published by Zeuthen (20) (his Fig. 4, Plate VII), which shows that in the 32-cell stage embryo, the nuclei of the outer four micromeres are significantly larger than those of the micromeres of the inner row, as if they were in prophase. In anticipation of a hypothesis that will be discussed later, this observation may hint at the diffusion of a cell division stimulating factor (or factors) from the micromeres.

The next interesting point is that until the 32-cell stage, cell divisions — with the exception of the micromeres — are essentially synchronous in all blastomeres. At the sixth cleavage, however, the macromeres divide slightly ahead of the mesomeres. A slight asynchrony is observed in some embryos already at the fifth cleavage. The difference becomes much more pronounced at the seventh cleavage; by then, while the nuclei in the vegetal hemisphere of the embryo are in late anaphase, the nuclei at the animal pole are either in the resting stage or in prophase (Figs. 2a, 2b, and 3). Somewhat similar findings have been reported by Agrell (1). It thus appears that the segregation of the micromeres perturbs the synchrony of cell divisions and

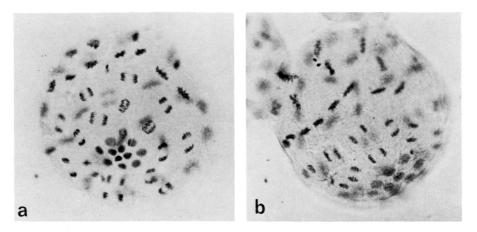

FIG. 2. a: Sixty-four-cell embryo; micromeres with nuclei in the resting stage. In **b** the micromeres are in the lower right-hand corner. The photographs show that while the nuclei closer to the vegetal pole are in anaphase, those just above the equator are in metaphase (especially clear in **b**), and those at the animal pole are in late prophase. ×320.

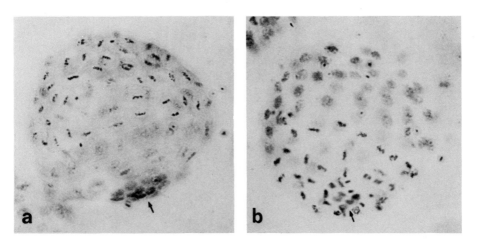

FIG. 3. Two embryos about 120-cell stage. In **a** the micromeres are in the lower right-hand corner **(arrow)**; dividing cells are confined to the animal half of the embryo. In **b** the micromeres are in the lower central part of the photograph **(arrow)**; dividing cells in the vegetal half of the embryo. ×195.

sets up a vegetal-animal gradient of cell division rate. We postulate that the gradient is due to a cell division modulating factor (or factors), which originates in the micromeres and diffuses toward the animal pole. We further suggest that the gradient becomes steeper between the fifth and the seventh cleavages. When the micromeres are first segregated they appear to be able to influence only the adjacent cells (i.e., the factor does not diffuse beyond the outer row of micromeres); later on, more and more factor is synthesized, thus resulting in a steeper concentration gradient and, hence, in a greater stage difference between the cells at the vegetal pole and those at the animal pole of the embryo. As yet we have no data regarding the rate of cell division in the different territories of the embryo, and, hence, we cannot make any guess as to the rate of diffusion of the hypothetical factor.

Thus, the micromeres, besides their well-known role as an organization center of the sea urchin embryo (see Hörstadius, ref. 10, for a review), appear to be responsible for the setting up of a gradient of asynchrony of cell divisions.

Some recent observations of van Dongen and Geilenkirchen (18) on the role of the polar lobe in the chronology of cell divisions in *Dentalium* are worth mentioning here. These authors have found that removal of the polar lobe at the trefoil stage (first cleavage) results in the disappearance of the normal division asynchrony between blastomeres in the different quadrants of the embryo. Furthermore, the development of these embryos is radialized due to the loss of the dorsoventral differences in the division pattern, which in normal development are responsible for the transition from radial to bilateral symmetry.

It is known (9) that the orientation of the cleavage spindles parallel to the equatorial plane during the first two cleavages, turns 90° at the third cleavage, thus resulting in the separation of four animal and four vegetal blastomeres. At the next cleavage, the spindle in the animal blastomeres is again parallel to the equatorial plane, and this gives rise to a crown of eight animal blastomeres (mesomeres). In the vegetal blastomeres, the spindle not only still lies perpendicular to the equatorial plane, but is markedly shifted toward the vegetal pole: this brings about the segregation of the four micromeres. It thus looks as though some material begins to concentrate at the vegetal pole at the four-cell stage, which influences the orientation of the spindle and makes it tilt 90° with respect to its original position during the first two cleavages. The strong influence that the hypothetical vegetal factor exerts is suggested by the shifting of the spindle toward the vegetal pole; at that time the factor must have reached a high concentration there.

The hypothesis has been recently put forward (Catalano, *in preparation*) that events of the early development of the sea urchin embryo can be interpreted on the basis of an asymmetrical radial gradient rather than of two gradients, one running from the animal to the vegetal pole and the other running in the opposite direction (15) (Fig. 4). According to this hypothesis, after fertilization, and in fact in the course of the second cleavage, the outer layer (zone 3, Fig. 4) should progressively flow toward the vegetal pole, eventually concentrating in the micromeres. For the time being, the hypothesis is based only on theoretical considerations, and, indeed, there is

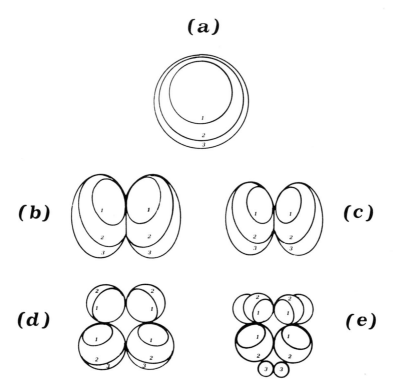

FIG. 4. Diagrammatic representation of the eccentric-fields hypothesis of Catalano. The diagram shows the flow of the subcortical material of layer 3 toward the vegetal pole and its eventual concentration in the micromeres. (Courtesy of Dr. G. Catalano.)

no experimental evidence that such a flow of subcortical material from the animal toward the vegetal pole does in fact occur; yet this is known to take place in the egg of *Beroe* (17).

The gradient is no longer detectable in the blastula; however, cell divisions now appear in clusters (Fig. 5); solitary dividing cells are seldom observed. At this stage, in the whole-mount preparations there are no visible markers to orientate the embryos; the embryo is indeed a hollow sphere, and there is no longer any size or staining difference between the derivatives of the micromeres and the other cells. Our impression thus far is that the clusters are randomly distributed in the embryo. The number and the size of the clusters are difficult to estimate; sometimes they are neatly localized and consist of 5 to 10 cells surrounded by resting cells (Fig. 5c); in other cases they are irregularly shaped (probably also as a result of the compression of the embryos), and they are confluent with neighboring clusters (Figs. 5a and b). There are two possibilities to consider: either each cluster is an independent unit resulting from a cell-division-inducing factor originating at one point in the embryo and remain-

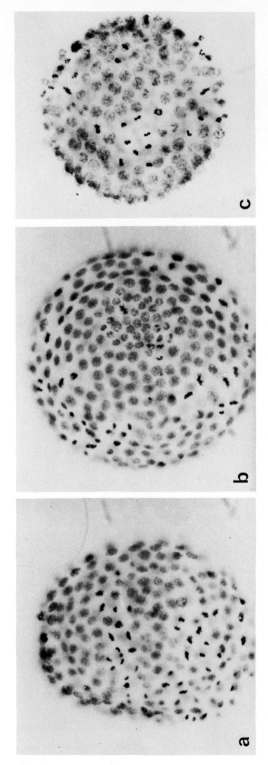

FIG. 5. a and b: Morulae. **c:** Young blastula. Different shapes and sizes of clusters of dividing cells are shown. ×320.

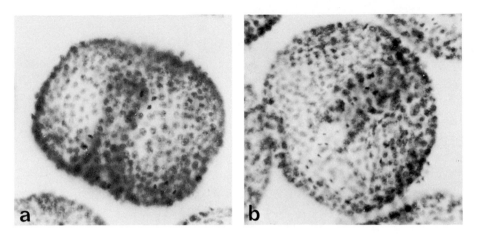

FIG. 6. Two gastrulae. **a:** Side view. **b:** The embryo is slightly tilted, thus showing a small cluster of dividing cells near the blastopore. ×320.

ing circumscribed to a small group of cells, or the factor spreads from its center of origin to a wider area, thus progressively inducing more and more cells to enter into mitosis. Experiments are in progress to answer these questions.

During gastrulation, the number of dividing cells at any one time is very small (Fig. 6). Dividing cells now appear mostly as solitary cells, but small clusters are visible in some embryos especially around the blastopore.

The rate of cell division changes dramatically in the course of development. Estimates of DNA content during development are summarized in Fig. 7. The earlier values obtained by Brachet (2) and the results of our own analyses are plotted in the diagram. Although a more detailed study is needed (especially for the time interval between the early and the mesenchyme blastula) and is in fact under way, the older and the newer data are essentially in agreement in suggesting that the increase of DNA content of the embryo follows a sigmoid curve in which three phases can be distinguished. During cleavage, the DNA content (and the number of cells) of the embryo increases very rapidly; at the time of hatching (10 to 12 hr after fertilization) the young blastula is made up of about 500 cells. At about this stage (the exact time remains to be established), the rate of the DNA increase slows down. An apparently higher rate of DNA synthesis is then resumed around the midgastrula stage. From the newly hatched blastula to the prisma stage (a period of about 24 hr) the DNA content of the embryo has increased roughly three times. These values agree with the old estimates of the increase of the cell number by Köhler (11).

It is interesting to compare these results with those of the changes in the percentage of dividing cells and of the absolute number of cell divisions

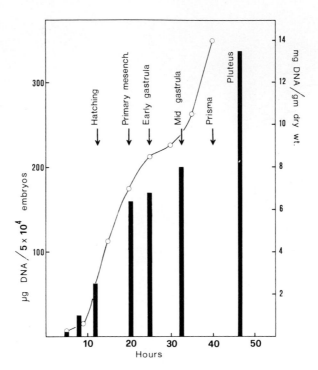

FIG. 7. The histogram shows our data on the DNA content per 5.10^4 embryos **(left ordinate)** at the indicated stages of development. The curve superimposed on the histogram is from Brachet (ref. 2) **(right ordinate).**

per embryo in the course of development. Although on this point also our observations are only preliminary, some of the data are worth being reported. In sections of morulae (about 7 hr of development) when the embryo is made up of about 150 cells, 80% of the cells are in mitosis. In the swimming blastula (10 to 12 hr after fertilization; the embryo is made up of about 400 cells), the percentage of cells in mitosis has dropped to 4% and remains at this level at least through the late gastrula (prisma) stage. Counts of cell divisions *per embryo* on whole-mount preparations, from the young blastula (prior to hatching) to the prisma stage, also show (Fig. 8) that in the young blastula there is a wide range of frequency distribution of mitosis per embryos, which ranges from 11–15 to 51–55 dividing cells per embryo, with a mode around 20 to 30. In the early gastrula, there occurs a shift toward the lower classes with a mode at 11 to 20, and this drops further down to 11 to 15 dividing cells per embryo in the mid- and late gastrula.

This mitotic index from the blastula stage on may appear too low to account for the increase of DNA content through the prisma stage. Yet, our calculations show that since cell growth is exponential, the curve of

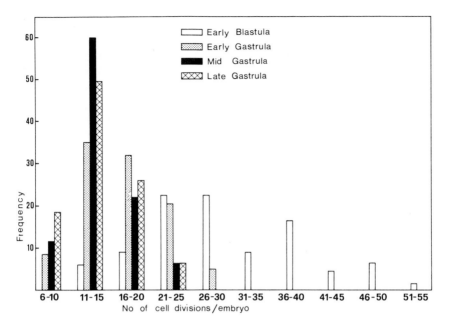

FIG. 8. Frequency distribution of the number of dividing cells in four stages of development of *Paracentrotus lividus.*

DNA increase during this time is compatible with the low percentage of dividing cells. This suggests that at some point around hatching, a control mechanism of cell division different from the one operating during cleavage is switched on.

It should be added that the rate of incorporation of thymidine (*per embryo*) into nuclear DNA in *Sphaerechinus* increases through the hatching blastula and then essentially levels off (Tosi, *personal communication*).

Studies on the activity changes of some of the enzymes of DNA biosynthesis also support the conclusion that the onset of the blastula stage is a critical turning point for the activity of the enzymes involved in DNA biosynthesis. For example, the activity of thymidylate synthetase undergoes an abrupt drop at the onset of the blastula stage (13). Also other enzymes of the DNA biosynthesis exhibit a more or less pronounced drop in the blastula (for a review, see Rossi et al., ref. 14).

DNA REPLICATION IN CELLS FROM DISSOCIATED EMBRYOS

We wish to propose here that cell interactions are at least one of the controlling factors of these changes.

It is known that cells obtained from dissociated early sea urchin embryos are able to reassociate and reconstitute embryos that can develop into larvae (8). On the other hand, if the dissociated cells are cultured at low cell density, they do not reassociate; this condition offers an opportunity to investigate the role of cell interactions in differentiation.

When late cleavage stages or early sea urchin blastulae (5 to 6 hr of development) are dissociated into individual cells and cultured under conditions preventing their reassociation, they continue to divide and to synthesize DNA for many hours and at an overall rate comparable to that of the stage from which they were prepared. (The observations were extended over a period of 50 hr; by that time the control embryos had reached the pluteus stage.) It looks as though the DNA replication and the cell division controlling mechanisms were "frozen" as they were at the time the cells were isolated. Disruption of the organization of the embryo, which results in the loss of cell interactions, thus appears to interfere with the sequence of events that in the embryo cause the cell-division-controlling mechanism operating during cleavage to change to that characteristic of the postblastula development.

Similarly, cells isolated from later blastula stages divide at a very low rate at the time the mitotic index has dropped. In this case, however, after a short treatment with trypsin, the cells divide at a high rate (19). This supports the view that cell-membrane-borne signals consisting of changes of the molecular organization of the cell membrane are a key factor in the control of the mitotic activity of the cells. We assume such signals originate from cell-to-cell interactions (see also the discussion by Edelman, ref. 7). So far no experiments have been carried out on blastomeres isolated before the fourth or fifth cleavage, and, hence, we do not know whether cell contact plays any role in the control of the synchrony of cell divisions before the segregation of the micromeres.

The role of cell interactions in the control of DNA synthesis is further shown by the recent work of De Petrocellis and Vittorelli (6) on the effects of dissociation of the sea urchin embryo on some enzymes of DNA biosynthesis.

The enzymes studied were deoxycytidylate-aminohydrolase (dCMP deaminase), thymidylate and thymidine kinase, an alkaline DNase, and DNA polymerase.

The activity of dCMP deaminase is high in the unfertilized egg and does not change during cleavage; at the blastula stage it begins to decline and in the pluteus larva its activity is only 25% of what it was in the unfertilized egg (16). These changes are abolished in the isolated cells: The enzyme activity remains "frozen" at the level it was in the embryo at the time of dissociation (Fig. 9). On the other hand, in the case of thymidylate kinase (Fig. 10) and thymidine kinase (Fig. 11) the response depended on the embryonic stage at which disaggregation had taken place. In cells isolated from cleavage stages the activities of these enzymes were the same as in the control embryos, whereas dissociation from hatching blastulae resulted

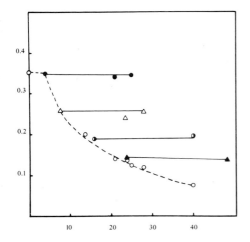

FIG. 9. Effect of isolation of cells on the activity level of dCMP deaminase during development of *Paracentrotus lividus.* **Abscissa:** Time after fertilization (hr). **Ordinate:** Micromoles of dCMP deaminated in 15 min per milligram of protein. Dissociation at ●——●, 32 to 64 blastomeres; △——△, early blastula; ◑——◑, mesenchyme blastula; ▲——▲, gastrula; ○——○, control. (From ref. 6.)

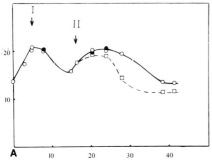

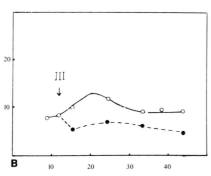

FIG. 10. Effect of cell isolation on the activity level of thymidylate kinase of the embryo of *Paracentrotus lividus.* **Abscissa:** Time after fertilization (hr). **Ordinate:** Activity expressed as nanomoles of dTDP + dTTP formed per milligram of protein per 20 min. **a:** Dissociation at 32 to 64 blastomeres **(I)** and at the mesenchyme blastula **(II) (arrows).** ○——○, control; ●——●, dissociated at 32 to 64 blastomeres; □----□, dissociated at the mesenchyme blastula. **b:** Dissociation at the early hatching blastula **(III).** ○——○, control; ●----●, dissociated. (From ref. 6.)

in the absence of the increase in the activities of these enzymes normally occurring after the blastula stage.

A similar stage-dependent effect is observed in the case of the alkaline DNase (Fig. 12): dissociation of the embryos into single cells during the preblastula stages had no effect on the enzyme level; dissociation at the early blastula did not permit the activity increase that normally occurs

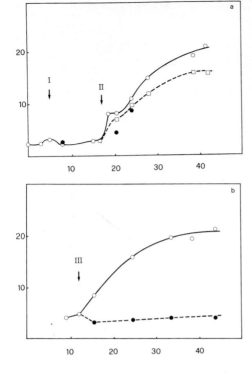

FIG. 11. Effect of dissociation on the activity level of thymidine kinase in the embryo of *Paracentrotus lividus*. **Abscissa:** Hours after fertilization. **Ordinate:** Activity expressed as nanomoles of dTMP formed per milligram protein in 20 min. **a:** Dissociation at 32 to 64 blastomeres **(I)** and at the mesenchyme blastula stage **(II) (arrows).** ○——○, control; ●----● and □----□, dissociated. **b:** Dissociation at the early hatching blastula **(III).** ○——○, control; ●----●, dissociated. (From ref. 6.)

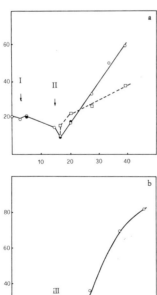

FIG. 12. Effect of dissociation on the activity level of DNase in the embryo of *Paracentrotus lividus*. **Abscissa:** Hours after fertilization. **Ordinate:** Activity expressed in U/mg protein. **a:** Dissociation at 32 to 64 cell stage **(I)** and at the mesenchyme blastula **(II) (arrows).** ○——○, control; ●——●, dissociated at 32 to 64 cells; □----□, dissociated at the mesenchyme blastula. **b:** Dissociation at the early hatching stage **(III).** ○——○, control; ●----●, dissociated. (From ref. 6.)

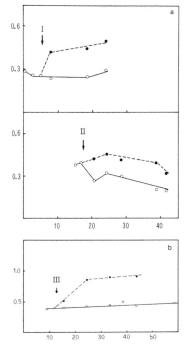

FIG. 13. Effect of dissociation on the activity level of DNA polymerase in the embryo of *Paracentrotus lividus*. **Abscissa:** Hours after fertilization. **Ordinate:** Activity expressed as nanomoles per milligram of protein per 20 min. O——O, control; ●——● dissociated. **a:** Dissociation at 32 to 64 cells **(upper curve)** and at the mesenchyme blastula **(lower curve).** **b:** Dissociation at the early hatching blastula **(arrow).** (From ref. 6.)

at the time of gastrulation. Previous work (3–5) had shown that the changes in the activity of these enzymes, which take place at the time of gastrulation, depend on transcriptional events occurring prior to the blastula stage. The observations on isolated cells add the notion that the phenotypic expression of the transcriptional events depends on cell-to-cell contacts.

A most striking effect is that on DNA polymerase. The overall activity of this enzyme does not change throughout development (12). Yet, in isolated cells, irrespective of the stage at which cell dissociation had taken place, the activity of the enzyme increases above control (Fig. 13).

At this point, it should be pointed out that experiments with isolated cells involve a number of unknowns. First, substances that might be contained in the blastula are lost when the embryo is dissociated, or undergo extreme dilution in the culture medium. The same is true for any substance that may be transferred directly from cell to cell in the intact embryo. It should also be borne in mind that only a portion of each cell surface is exposed in the embryo. These conditions may influence the metabolism and, hence, caution is required in evaluating such experiments.

In conclusion, the results we have presented suggest that cell interactions appear to play an important role in the modulation of cell division in the urchin embryo. First, we refer to the vegetal-animal gradient of cell divisions, which appears to be controlled by some factor(s) diffusing from the micromeres. The evidence for this is, however, circumstantial, and the hypothesis is now being submitted to experimental tests. The experiments on isolated cells point to a modulating role of the organization of the cell membrane which, in turn, depends on cell-to-cell interactions. The nature of the signals, how they originate, and how they are transferred from cell to cell and by which mechanism(s) they influence the rate of cell divisions is unknown. The same applies to the origin and role of the clusters of dividing cells in the blastula.

ACKNOWLEDGMENT

This work has been supported in part by the C.N.R. Project on the Biology of Reproduction.

REFERENCES

1. Agrell, I. (1956): A mitotic gradient as the cause of the early differentiation in the sea urchin embryo. *Zoology Papers in Honour of B. Hanström,* edited by K. G. Wingstrand, pp. 27–34.
2. Brachet, J. (1933): Recherches sur la synthèse de l'acide thymonucléique pendant le développement de l'oeuf d'oursin. *Arch. Biol.,* 44:519–576.
3. De Petrocellis, B., and Parisi, E. (1972): Changes in alkaline deoxyribonuclease activity in sea urchin during embryonic development. *Exp. Cell Res.,* 73:496–500.
4. De Petrocellis, B., and Parisi, E. (1973): Effect of actinomycin and puromycin on the deoxyribonuclease activity in *Paracentrotus lividus* embryos at various stages of development. *Exp. Cell Res.,* 82:351–356.
5. De Petrocellis, B., and Rossi, M. (1976): Enzymes of DNA biosynthesis in developing sea urchins. Changes in ribonucleotide reductase, thymidine and thymidylate kinase activities. *Dev. Biol.,* 48:250–257.
6. De Petrocellis, B., and Vittorelli, M. L. (1975): Role of cell interactions in development and differentiation of the sea urchin *Paracentrotus lividus*. Changes in the activity of some enzymes of DNA biosynthesis after cell dissociation. *Exp. Cell Res.,* 94:392–400.
7. Edelman, G. M. (1976): Surface modulation in cell recognition and cell growth. *Science,* 192:218–226.
8. Giudice, G. (1962): Restitution of whole larvae from disaggregated cells of sea urchin embryos. *Dev. Biol.,* 5:402–411.
9. Hörstadius, S. (1939): The mechanics of sea urchin developments studied by operative methods. *Biol. Rev.,* 14:132–179.
10. Hörstadius, S. (1973): *Experimental Embryology of Echinoderms.* Clarendon Press, Oxford.
11. Köhler, D. (1912): Uber die Abhängigkheit der Kernplasma-Relation von der Temperatur und vom Reifezustand der Eier. Experimentelle Untersuchungen an *Strongylocentrotus lividus. Arch. Zellforsch.,* 8:272–351.
12. Loeb, L. A., Fansler, B., Williams, R., and Mazia, D. (1969): Sea urchin nuclear DNA polymerase. I. Localization in nuclei during rapid DNA synthesis. *Exp. Cell Res.,* 57:298–304.
13. Parisi, E., and De Petrocellis, B. (1976): Sea urchin thymidylate synthetase. Changes of activity during embryonic development. *Exp. Cell Res,* 101:59–62.
14. Rossi, M., Augusti-Tocco, G., and Monroy, A. (1975): Differential gene activity and segregation of cell lines: An attempt at a molecular interpretation of the primary events of embryonic development. *Q. Rev. Biophys.,* 8:43–119.
15. Runnström, J. (1928): Plasmabau und Determination bei den Ei von *Paracentrotus lividus. W. Roux. Arch. Entw. Mech. Org.,* 113:556–581.

16. Scarano, E., and Maggio, R. (1959): The enzymatic deamination of 5'-deoxycytidylic acid and of 5-methyl-5'-deoxycytidylic acid in the developing sea urchin embryo. *Exp. Cell Res.,* 18:333–346.
17. Spek, J. (1926): Uber gesetzmässige Substanzverteilungen bei der Furchung des Ctenophoreneies und ihre Beiziehungen zu den Determinationsproblemen. *Arch. Entw. Mech. Org.,* 107:54–73.
18. van Dongen, C. A. M., and Geilenkirchen, W. L. M. (1975): The development of *Dentalium* with special reference to the significance of the polar lobe. IV. Division chronology and development of the cell pattern in *Dentalium dentale* after removal of the polar lobe at first cleavage. *Proc. Kon. Nederl. Ak. Wetensch. Amsterdam, Ser. C.,* 78:358–375.
19. Vittorelli, M. L., Cannizzaro, G., and Giudice, G. (1973): Trypsin treatment of cells dissociated from sea urchin embryos elicits DNA synthesis. *Cell Differ.,* 2:279–284.
20. Zeuthen, E. (1951): Segmentation, nuclear growth and cytoplasmic storage in eggs of echinoderms and amphibia. *Pubbl. Staz. Zool. (Napoli),* 23:47–69.

Cell and Tissue Interactions, edited by
J. W. Lash and M. M. Burger. Raven
Press, New York, 1977.

ACh Receptors Accumulate at Newly Formed Nerve-Muscle Synapses *In Vitro*

Eric Frank and Gerald D. Fischbach

*Department of Pharmacology, Harvard Medical School,
Boston, Massachusetts 02115*

There is now a considerable body of knowledge about synapses between nerve and muscle in adult animals. In vertebrates, motor nerve terminals are specialized for the synthesis, storage, and release of acetylcholine (ACh). The muscle membrane is also specialized in that ACh receptor molecules are concentrated immediately opposite sites of transmitter release. We are interested in how this subsynaptic clustering of receptors comes about.

Cholinergic synapses form between embryonic chick spinal cord and muscle cells *in vitro,* and receptor clusters or "hot spots" are located precisely at sites of transmitter release (7). Cultures in which fibroblasts have been eliminated with cytosine arabinoside are only one cell layer thick so the muscle fibers and nerve processes can be clearly visualized with interference-contrast optics. The same neurites and muscle cells can be identified and studied with microelectrodes over a period of several days, before, during, and after the formation of a synaptic contact. We have used this system to study when and how receptor clusters appear at newly formed synapses.

The methods we use for dissociating and plating mononucleated myogenic cells from 11-day embryonic pectoral muscle have been described (5). Spinal cord slices (200 to 300 μm transverse sections) cut from brachial segments of 14-day embryos are added to the cultures after the myoblasts have fused to form multinucleated myotubes. These relatively mature spinal cord explants send out far fewer processes than explants prepared from 4- or 7-day embryos (used in previous studies), and many of the neurites which exit from the ventral horns are relatively thick and unbranched. These features facilitate the identification of an individual nerve process as a motor axon. In addition, muscle fibers close to the explant do not become encrusted with neurites so individual sites of transmitter release (synapses) can be distinguished unambiguously.

The first synapses form within a few hours after the addition of a spinal explant to a muscle culture, and in a few days many muscle fibers contract repetitively as the result of spontaneous impulse activity originating within the explant. These first functional contacts are not transient. Individual synapses can be relocated over a period of several days. After synapses have formed, and ACh sensitivity is at least several-fold higher at the site of transmitter release than it is over surrounding regions. A graphic demonstration of the distribution of receptors in the vicinity of an identified synapse is shown in Fig. 1. The synapse was localized by recording synaptic

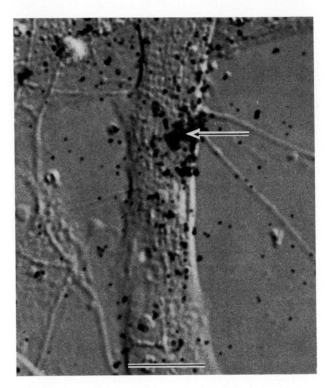

FIG. 1. Distribution of ACh receptors on an innervated muscle fiber *in vitro.* Muscle cells were plated 12 days earlier; a 14-day spinal cord explant was added 4 days after plating. The synapse **(arrow)** was located by recording spontaneous synaptic currents with an extracellular pipette (spatial resolution 2 to 5 μm). The culture was then incubated in 14 nM ^{125}I-α-BTX (113 cpm/fmole) for 1 hr at 37°C, washed, fixed, coated with NTB3 emulsion, and exposed for 9 days at 4°C. ACh receptor density (as revealed by silver grain density) is much higher in the immediate vicinity of the synapse than in extrasynaptic regions. Scale, 20 μm.

currents with an extracellular pipette. This technique affords a spatial resolution of 2 to 5 μm. The culture was then incubated with ^{125}I-α-bungaro-toxin, a small protein that binds specifically and essentially irreversibly to ACh receptors in skeletal muscle, and the distribution of receptors was determined by autoradiography. The density of silver grains in the immediate vicinity of the synapse is severalfold higher than it is over the surrounding, noninnervated membrane.

It is known that ACh receptors on noninnervated cultured chick muscle fibers tend to occur in clusters (6,8). Perhaps nerve fibers "seek out" these preexisting clusters as they grow in and innervate the muscle. Alternatively, ingrowing axons might induce the formation of new clusters as they "recognize" appropriate targets and form new synapses. To test these alternatives, we needed a nondestructive method of creating fine-grained maps of ACh receptor distribution on the surface of living muscle fibers before and after the formation of a synapse.

The method we used was ACh microiontophoresis. ACh was pulsed (0.5 to 1.0 msec, positive pulses) from a high resistance (300 to 500 MΩ) extracellular micropipette filled with 1 M ACh while recording the voltage response in the muscle with an intracellular microelectrode (3). The "sensitivity" of the muscle membrane, a measure of receptor density, at the point of ACh application is expressed as millivolts (mV) of depolarization per nanocoulomb (nC) of charge ejected from the pipette. To measure the sensitivity at a large number of points in a comparatively short period of time, we recorded the position of the ACh pipette at each test point on a frame of 16-mm movie film. At the same instant, the dose of ACh and the amplitude of the response was measured, and the sensitivity was computed with a small on-line computer. This semiautomated procedure facilitated the construction of fine-grain maps, so that 100 points separated by 10 μm or less could be tested in about 15 min. With care, the mapping process could be repeated several times with no apparent damage to the muscle fiber.

The results from one series of sensitivity measurements are shown in Fig. 2. Numbers in the diagram (traced from a photograph) indicate the sensitivity measured at each point. The scale relating integers to mV/nC is given in the legend to Fig. 2. Two areas of relatively high sensitivity are outlined with a dotted line for emphasis. Maps generated in this manner are reproducible. The same muscle cell mapped again a short time later shows the same distribution of sensitivity. For example, the map shown in Fig. 2b is of the same muscle fiber as Fig. 2a, but it was made 5 $\frac{1}{2}$ hr later.

The receptor distribution over uninnervated myotubes is apparently stable over longer periods of time. Repeated maps of selected muscle fibers —those that do not change shape or "grow" significantly—indicate that receptor clusters remain in the same vicinity and that few new hot spots appear over a period of 2 to 3 days. Our observations are limited to muscle fibers in 5- to 12-day-old cultures. Other studies indicate that hot spots increase in number on younger myotubes (8) and ultimately disappear in older cultures.

The situation is quite different on innervated fibers. In every case where we adequately mapped the same patch of muscle membrane before and after

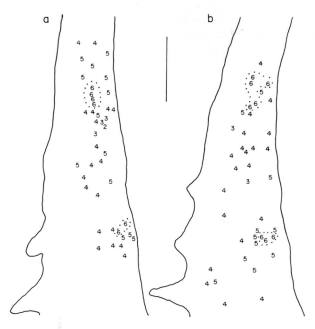

FIG. 2. Stability of ACh receptor distribution over a myotube. ACh was applied ion-tophoretically from one pipette while measuring the voltage response in the muscle cell with a second pipette. Each small number represents the sensitivity of the membrane to ACh at that spot. The numbers defined here apply to Fig. 3 as well: **1,** 10 − 22 mV/nC; **2,** 22 − 46 mV/nC; **3,** 46 − 100 mV/nC; **4,** 100 − 220 mV/nC; **5,** 220 − 460 mV/nC; **6,** 460 − 1,000 mV/nC; **7,** 1,000 − 2,200 mV/nC; **8,** 2,200 − 4,600 mV/nC; **9,** 4,600 − 10,000 mV/nC. The right-hand map was made 5¹/₂ hr after the one on the left, yet ACh sensitivity profiles are very similar. Muscle cells plated 7 days earlier. Scale, 50 μm.

the formation of a synapse, the sensitivity at the newly formed site of transmitter release increased severalfold. One example is shown in Fig. 3. During the interval between the two maps, a synapse was formed and the sensitivity in the immediate synaptic area (enclosed in a dotted circle) rose 8¹/₂-fold. The background, "extrasynaptic," sensitivity remained the same, or decreased slightly, during the same time period. This motor nerve terminal induced a new locus of high sensitivity and did not innervate a preexisting hot spot. As stated above, new hot spots rarely form on uninnervated myotubes in 5- to 12-day cultures, so the marked increase in sensitivity at new sites of transmitter release is unlikely to be coincidental.

These results are consistent with recent findings of Anderson and co-workers (1). They labeled ACh receptors on cultured *Xenopus* myocytes (mononucleated cells) with FITC-αBGT conjugates. Small hot spots were found on uninnervated cells. When neurons dissociated from the neural tube were added to previously labeled myocytes, elongated fluorescent patches (ACh receptor clusters) appeared along nerve-muscle contacts.

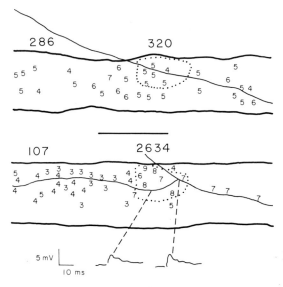

FIG. 3. Appearance of a new area of high ACh sensitivity at a newly formed neuromuscular synapse. Upper panel shows the distribution of ACh sensitivity before the synapse(s) had formed. The mean sensitivity within the dotted circle (320 mV/nC) was not significantly higher than background (286 mV/nC). Two days later **(lower panel),** synapses were present within this same dotted region. Synaptic potentials (recorded intracellularly from the muscle fiber) were evoked when the nerve terminals were focally depolarized with an extracellular pipette at the positions denoted by the dashed lines **(see traces below the map).** (The culture was bathed in 10^{-7} M TTX to block nerve impulses.) At this time, the average ACh sensitivity within the synaptic zone **(dotted circle)** had risen dramatically (2,634 mV/nC), while the background sensitivity was slightly lower than before (107 mV/nC). Myoblasts were plated 10 days earlier, and the spinal explant had been present for 6 days. Scale, 20 μm.

This suggests that clusters along neurites result from the migration of receptors within the plane of the membrane.

Regardless of whether subsynaptic receptors are "old" molecules that have migrated from extrasynaptic regions or whether they are newly synthesized (or inserted) molecules, it is important to ask whether they differ in any way from the remaining extrasynaptic receptors. In the adult rat, extrajunctional receptors (in denervated muscles) are degraded with a half-time of about 18 hr whereas junctional receptors are metabolized with a half-time of at least 5 days (2). The half-life of receptors in noninnervated chick muscle cultures is nearly identical to that of extrajunctional receptors in the adult rat (4). We have compared the stability of synaptic and extrasynaptic receptors on innervated myotubes by autoradiography. Several sites of transmitter release were located by extracellular recording and then the cultures were labeled for 1 hr with ^{125}I-αBGT, washed, and returned to the incubator. As judged from grain counts over myotubes fixed at various times after labeling, receptors "disappear" as rapidly at synapses (and also at uninnervated hot spots) as they do at extrasynaptic sites. This rapid disappearance of junctional receptors is surprising, since receptor clusters are fairly stable in size, position, and sensitivity. New receptors must appear at the same loci at the same rate. Taken together our results suggest that cluster formation and maintenance, at synapses and at noninnervated hot spots, do not simply involve stabilization of individual receptor molecules.

Experiments performed by Steve Burden in our laboratory indicate that embryonic junctional receptors also "turn over" rapidly *in vivo*. Receptors within clusters identified on muscle fibers between embryonic days 10 to 19 disappear at the same rate as extrajunctional receptors ($t_{1/2} = 32$ hr).

We plan to investigate subsequent stages in the maturation of this prototypic chemical synapse. Preliminary experiments indicate that motor axons can induce local patches of acetylcholinesterase (AChE) on cultured myotubes at sites of transmitter release within 1 to 2 days after the receptor clusters form.

In addition we plan to study the very first stages in nerve-muscle interaction. So far we have examined synapse formation over a many-hour time scale. Important stages may occur at earlier times. On one occasion we have evoked transmitter release by focal stimulation of an active growth cone. Eighteen hours later, the migrating tip had advanced several hundred microns along the muscle fiber, but a synaptic contact remained at the original site. The ACh sensitivity was extremely high beneath the growth cone when it was first assayed, so if hot spot induction at a new synapse is an invariant rule, it may occur rapidly. It will be important to determine if transmitter release begins before any postsynaptic changes can be detected. It is also possible that significant interactions between pre- and postsynaptic membranes occur before synaptic transmission is established.

Synapses are strikingly difficult to locate. They are not widely distributed over the muscle surface. One explanation for this finding is that only certain patches of muscle membrane are capable of forming synapses with ingrowing cholinergic axons. Our findings summarized in this discussion indicate that such hypothetical sites are not necessarily characterized by a high density of ACh receptor molecules. Future sites of synapse formation may be specialized in other ways, however. At this meeting, attention has been focused on surface molecules that mediate cell-cell adhesion. In some cases, specific adhesion has been found between cells dissociated from regions of embryonic retina and optic tectum, which are destined to communicate via chemical synapses in the adult. It is possible that such phenomena are relevant to the earliest stages of synapse formation between nerve and muscle.

REFERENCES

1. Anderson, M. J., Cohen, M. W., and Zarychta, E. (1976): Redistribution of acetylcholine receptors on cultured muscle cells. Abstracts of the Sixth Annual Meeting of Neuroscience Society, p. 707. Toronto, Canada.

2. Berg, D. K., and Hall, Z. W. (1975): Loss of α-bungarotoxin from junctional and extra-junctional receptors in rat diaphragm muscle *in vivo* and in organ culture. *J. Physiol. (Lond.)*, 252:771–789.
3. del Castillo, J., and Katz, B. (1955): On the localization of acetylcholine receptors. *J. Physiol. (Lond.)*, 128:157–181.
4. Devreotes, P. N., and Fambrough, D. M. (1975): Acetylcholine receptor turnover in membranes of developing muscle fibers. *J. Cell Biol.* 65:335–358.
5. Fischbach, D. F. (1972): Synapse formation between dissociated nerve and muscle cells in low density cell cultures. *Dev. Biol.,* 28:407–429.
6. Fischbach, G. D., and Cohen, S. A. (1973): The distribution of acetylcholine sensitivity over uninnervated and innervated muscle fibers grown in cell culture. *Dev. Biol.,* 31:147–162.
7. Fischbach, G. D., Berg, D. K., Cohen, S. A., and Frank, E. (1976): Enrichment of nerve-muscle synapses in spinal cord-muscle cultures and identification of relative peaks of ACh sensitivity at sites of transmitter release. *In: Cold Spring Harbor Symposium on Quantitative Biology. XL: The Synapse,* pp. 347–358.
8. Sytkowski, A. J., Vogel, Z., and Nirenberg, M. W. (1973): Development of acetylcholine receptor clusters on cultured muscle cells. *Proc. Natl. Acad. Sci. U.S.A.,* 70:270–274.

Cell and Tissue Interactions, edited by
J. W. Lash and M. M. Burger. Raven
Press, New York, 1977.

Abnormal Synaptic Connectivity Following UV-Induced Cell Death During *Daphnia* Development

E. R. Macagno

*Department of Biological Sciences, Columbia University,
New York, New York 10027*

Cell and tissue interactions play a central role in the formation of the nervous system. These range from the very general inductive interactions between tissues to the very specific recognition by a neuron of its synaptic targets. A "message" is exchanged between the cells or tissues, a message that carries information affecting their subsequent behavior. A molecular message may be secreted into the extracellular space, or it may be carried by a cell on its outer membrane surface. Or the message may be exchanged between cells after the formation of specialized contacts, such as gap junctions that allow electric currents and at least small molecules to pass freely from one cytoplasm to the other.

Many presentations at this meeting have been addressed to the isolation and characterization of the molecules involved in cell and tissue interactions. This is clearly one of the most interesting questions at present with respect to cell reaggregation and the interaction of cells with extracellular matrices. A related and equally interesting question is: What are cells telling each other through these interactions? In some cases the answer, at least superficially, seems to be straightforward: In reaggregation, cells tell each other that they belong together as a tissue. In other cases, however, even a detailed description of events following an interaction tells us little about what information is exchanged. This is especially true in neurogenesis.

For some years we have been studying the morphology and development of the *Daphnia* nervous system, principally of the compound eye and optic ganglion. By means of computer-aided analysis of serial section electron micrographs, we have obtained a precise knowledge of cellular architecture and synaptic connectivity and of their development. The number of cells is small and their characteristics are invariant from animal to animal to the extent that each can be uniquely identified. We have been particularly interested in the formation of synaptic contacts between photoreceptors and their postsynaptic neurons, the laminar cells. The morphological interaction between these two types of neurons during development (summarized in

293

the next section) has led us to suggest that the photoreceptors, which differentiate earlier, trigger the further differentiation of the laminar cells.

Various models can be constructed with respect to this hypothetical "triggering" action. Since the laminar neuroblasts subsequently develop into unique neurons in terms of morphology and synaptic connections, one can make the hypothesis that the trigger was merely a general signal to the neuroblasts to continue a predetermined developmental program that is different for each. Alternatively, one could conceive of the triggering signal as being a set of detailed instructions, different for each neuroblast, that is responsible for cell individuality. A third possibility is that the trigger is nonspecific and each neuroblast has the potential to be any laminar cell, and, hence, that uniqueness is a consequence of the very accurate temporal and spatial coordination of events following the triggering action: In other words, a laminar cell becomes L1 because it was at the right time at the right place to receive synapses from fibers X and Y.

In order to investigate the nature of this interaction and the validity of the models suggested above, we have used an ultraviolet (UV) beam to destroy some of the photoreceptors in the *Daphnia* embryo at stages prior to the occurrence of the interaction. After developing to early adulthood, the experimental animals were tested for simple photobehavioral responses and prepared for serial sectioning. The structure of the eye and the lamina were studied at both optical and electron microscopic levels.

Experiments related to this have been carried out extensively on vertebrate visual systems, although not at the level of single, identified cells. In addition, *Drosophila* mutations affecting the number of photoreceptors have been isolated and studied. We will discuss these results below after a description of our work on the *Daphnia* visual system.

SUMMARY OF ADULT *DAPHNIA* MORPHOLOGY

The *Daphnia* visual system consists of two organs, the compound eye and the optic ganglion, plus some cells and neuropil in the main or supra-esophageal ganglion (14). Within the compound eye there is one type of neuron, the photoreceptor or retinula, and at least two other kinds of cells, the glia and the secretory cells that form the lenses. The 176 photoreceptors are arranged in 22 groups of eight cells, 11 groups on each side of the midplane. There are 22 lenses, one for each group; the lens and the eight cells of a group form an ommatidium. Each receptor contains numerous opaque pigment granules; it contributes a set of microvilli to the fused rhabdom (these presumably contain the photopigment), and it sends a single unbranched fiber in a posterior direction to the optic ganglion (see Fig. 1). The eight fibers from each ommatidium, together with a glial process, form a bundle that extends as a unit from the compound eye to the lamina of the optic ganglion, a distance of about 50 μm. A bundle enters the optic ganglion

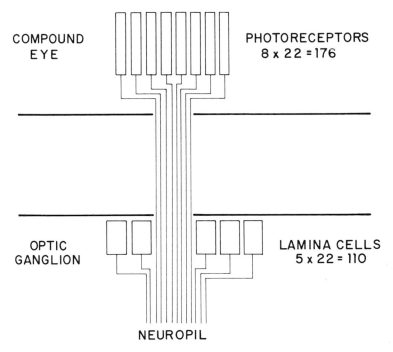

COMPOUND EYE

PHOTORECEPTORS
8 x 22 = 176

OPTIC GANGLION

LAMINA CELLS
5 x 22 = 110

NEUROPIL

FIG. 1. Schematic diagram of the photoreceptors and laminar cells. Each ommatidium is represented here by its eight photoreceptors. There are 22 ommatidia and, hence, a total of 176 photoreceptors. Each ommatidium also includes a lens that is not indicated in this figure. Each bundle of optic fibers associates with five laminar cells, forming an optic cartridge. The laminar cell somata lie in a cortical layer; the laminar fibers extend into the neuropil layer in parallel with the optic fibers. The fiber bundles travel directly from eye to optic ganglion without chiasmata or interchange of fibers between bundles.

through the cortical layer that contains the laminar somata and proceeds to the underlying neuropil, where each fiber branches and synapses onto processes of laminar cells. With each bundle is associated a group of five laminar cells whose somata form a ring around the entering bundle. The receptor fibers make most of their synapses with these five cells. We refer to the fiber bundle from one ommatidium plus the five associated laminar cells as a cartridge (see Fig. 1).

The optic ganglion consists of various types of neurons and a number of glia that branch extensively in the neuropil and separate the laminar cell somata from those of higher order neurons. Only the 176 photoreceptors and 110 laminar cells are of interest to us here. (No further mention of other neuronal types will be made.) The geometrical arrangement of both photoreceptors and laminar cells is simple: The somata in the eye are arranged as a hemispherical monolayer where each ommatidium occupies a specific position; similarly the somata of the lamina lie in a curved monolayer. The relative positions of the ommatidia and those of the corresponding cartridges are the same. Serial section reconstructions using a computer system (6,7,13) have permitted us to determine the geometry and synaptic connectivity of receptor and laminar cells. An important conclusion from such studies is that each cell has a characteristic morphology and connectivity pattern that, together with the simple geometry of the system, allows us to identify each cell uniquely and unambiguously under normal conditions.

SUMMARY OF *DAPHNIA* MORPHOGENESIS

The formation of synaptic contacts between receptor and laminar cells follows a precise sequence of events that occurs during the middle stages of embryogenesis (9,10). The photoreceptors differentiate earlier and send processes into the lamina, where they encounter laminar neuroblasts without processes. Within each future ommatidium one cell appears to mature earlier than the other seven: Its growing fiber precedes, or "leads," the other seven into the lamina. The lead fiber sequentially contacts five laminar neuroblasts, which then warp around this fiber and form transient gap junctions with it. As the following fibers of the bundle arrive, the laminar neuroblasts unwrap partially, form a ring around the bundle, and elaborate processes that grow in parallel with the receptor fibers into the forming neuropil. Synapses are then made between specific cells as branches of both receptor and laminar processes appear. The receptor processes terminate within the anterior neuropil area, whereas the primary laminar processes continue to grow into more posterior regions, where they synapse onto other cell types. Each cartridge is formed in this manner, in a temporal sequence in which those more laterally located mature earlier than those close to the midplane of the animal.

The occurrence of the events summarized here has led us to formulate the hypothesis that the lead fiber-laminar neuroblast interaction serves to trigger the further differentiation of the neuroblasts (8). The wrapping-around phenomenon can be construed as a mechanism for creating large areas of membrane apposition between fiber and neuroblast in order to facilitate either the exchange of a message through a mechanism involving intramembranous particles or the formation of transient gap junctions. If these junctions were to function here as the communication channels between the interacting cells, it would strongly imply that the message is carried by small molecules (12).

METHODS

The biochemical effects of ultraviolet radiation on living cells are varied and not completely understood. Of specific interest to us is the fact that high enough dosages are capable of inducing cell death. This is clearly the principle behind the use of mercury bulbs as germicidal lamps. Another effect of UV radiation that has been extensively studied is the induction of a growth delay in bacteria (5,16). Although such an effect has not been studied in maturing neurons, some of our preliminary observations (discussed below) indicate that such an effect does occur in the irradiated *Daphnia*.

Daphnia embryos were individually irradiated at the 29- to 30-hr stage using an American Optical H20 microscope fitted with a vertical illuminator for incident light fluorescence with a 50-W mercury vapor lamp. The light path was modified to allow the short wavelength spectrum to pass through to the specimen on the microscope stage by using quartz lenses and front surface mirrors, and by focusing the beam with a Beck reflecting objective. A set of apertures in the beam path allowed adjustment of the beam diameter at the focal plane to values from 1 to 8 μm. After irradiation, the embryos were individually placed in fresh medium in depression trays. Control embryos were put through the same procedure but with a filter in the beam that blocked the UV range of the spectrum. Once the development was completed and they began to swim, algae were added to the medium as food. After growing for a few days they were tested for simple behavioral responses to light and then fixed and prepared for sectioning. The specimens were subsequently cut serially into either 1-μm sections for optical microscopy or 1,000-Å sections for electron microscopy. The serial micrographs were then rephotographed into a movie of aligned sections, which was later analyzed using the CARTOS computer system (13). The number of photoreceptors and laminar cells was then determined, and the cellular morphology and synaptic connectivity were reconstructed for cartridges of interest.

At the time of irradiation, the pigmentation of the photoreceptor cells is beginning to appear. This makes the localization of the target cells quite straightforward. The receptor cells lie in a thin layer under the embryonic skin and over the more central developing optic ganglion. The position of the embryo during irradiation places the photoreceptor cell layer tangentially with respect to the beam. In this orientation, the optic ganglion is located laterally with respect to the receptors and, hence, well outside of the beam path. Since the amount of light scattering by the tissue (as seen under the microscope) is quite small, we are confident that the direct effects of the UV microbeam are entirely restricted to the photoreceptor layer.

RESULTS AND DISCUSSION

The first photoreceptors to mature in the compound eye elaborate visible pigmentation at about the 29-hr stage. There are two foci of maturation, one on each side of the midplane. As development proceeds, pigmentation increases both in extent and density, gradually and symmetrically forming the totally opaque hemisphere seen in the adult. In experimental animals that

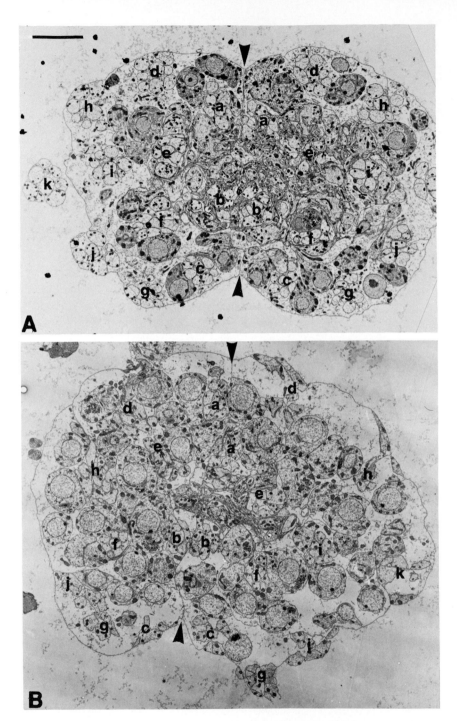

FIG. 2. Electron micrographs of the lamina of normal **(A)** and irradiated **(B)** animals. The midplane is indicated in each figure by arrowheads. The cartridges are individually identified by a letter, with the same letter used for corresponding ones on each side of the midplane. The experimental animal was irradiated on the left side of the developing eye, leading to the loss of two left cartridges, i and k. In addition, cartridges c, g, and j on the left side show less than the normal eight optic fibers. The bar on the upper left of **A** indicates 10 μm.

have been irradiated on one side, the pigmentation develops abnormally; the irradiated zone as seen under a dissecting microscope is an area devoid of or with little pigmentation. Depending upon the radiation dosage, one can then observe either a delayed production of pigment in this zone resulting in an eye that appears normal at a macroscopic level, or no further recovery of pigment with the eventual formation of an obviously asymmetrical eye. The type of asymmetry and the degree of malformation depends both upon dosage and location of the irradiated spot in the eye. Various interesting effects of this photosurgery have been observed. Here we will consider these: (a) hypoplasia of the lamina and (b) laminar cell sprouting and formation of abnormal synaptic connections.

Hypoplasia of the Lamina

Animals showing a reduced number of photoreceptors after irradiation also have a reduced number of laminar cells. Figure 2A is an electron micrograph of a normal lamina sectioned in a plane perpendicular to the midplane of the ganglion and close to the anterior surface. The 22-fiber bundles and some laminar cells can be seen. Figure 2B is a similar micrograph of an experimental animal showing a clearly asymmetrical lamina; the number of fiber bundles in the smaller side is only 9, whereas the normal side has 11. Bundles i and k on the left are missing. The same asymmetry is encountered in the number of laminar cells. These numbers for three experimental animals are given in Table 1. In these cases there are two bundles of fibers missing. The cartridges do not form without the fiber bundles, and their laminar cells have apparently degenerated. In addition, some bundles have less than the normal eight fibers; the corresponding cartridges include, in some instances, less than the normal five laminar cells, although not in all cases. More observations will be necessary before the pattern of these abnormal cartridges can be discerned, if such exists.

TABLE 1. *Values of various parameters in a normal and three irradiated animals*[a]

| | | Irradiated animals | | |
Parameters	Normal *Daphnia*	A	B	C
Photoreceptor fibers	176	153	163	151
Fiber bundles	22	20	20	20
Laminar cells	110	93	99	103[b]
Abnormal cartridges	0	3	3	4
Missing cartridges	0	2	2	2

[a] The three experimental animals were selected from a larger set because they all have two missing fiber bundles. Data including other cases will be published elsewhere.
[b] Some laminar cells have only a short fiber, are devoid of synaptic contacts, and appear to be at an early stage of chromatolytic degeneration.

A consequence of the reduction in numbers of photoreceptor fibers and laminar cells is that the volume occupied by the neuropil is smaller in the experimental animals. The elimination of a cartridge does not lead to an empty space in the neuropil, nor do other fibers proliferate to fill the unoccupied volume. The remaining cartridges coalesce into a smaller but normal-looking neuropil. When abnormal cartridges occur, on the other hand, some degree of sprouting of new laminar cell fibers does occur, as explained below. In such cases, the reduction in neuropil volume is less than would be expected from a purely linear relationship between neuropil volume and number of photoreceptor plus laminar cell fibers. The absolute value of the volume change of neuropil cannot be determined accurately because normal fluctuations of this parameter are not negligible; some specimens within a brood show considerably greater branching of fibers than do others (14). To this must be added variations due to fixation, distortions due to microtoming, and shrinkage of sections in the electron beam. On the other hand, as can be seen in Fig. 2B, an experimental specimen that is affected on only one side of the midplane has the normal opposite side as a built-in reference point; hence, a reliable relative measurement of the volume decrease is possible in those cases.

The degeneration of neurons deprived of their targets during development is a well-known phenomenon (2,4), although some cases have been studied where this does not happen (17). Target neurons deprived of their main synaptic inputs have also been studied extensively, largely in the visual system of vertebrates (2,4). The observation of relevance to this discussion is that unilateral eye enucleation early in development leads to hypoplasia of the optic tectum. Although this is generally in agreement with the *Daphnia* results reported here, the observations are at a fairly gross level, and it is unclear which elements of the tectum are responsible for the hypoplasia. Also of great interest to us here is the observation by Power (15) that genetic mutants in *Drosophila,* which have fewer ommatidia than the wild type, also have proportionately smaller optic glomeruli. In the case of eyeless mutants, Power found the external glomerulus to be completely absent. Power also argues that this reduced glomerular volume (equivalent to our reduced neuropil) is not the direct result of genetic factors but rather a secondary effect of the ingrowth of reduced numbers of ommatidial fibers. His data, however, are not inconsistent with a genetic, rather than neurogenic, explanation. Our results on the *Daphnia* clearly support his conclusions, although our observations show specifically that the hypoplasia results not only from the reduced volume occupied by ommatidial fibers but also from the disappearance of laminar cells. We have not at this time determined whether there is also a reduction in higher order cells in the *Daphnia* optic ganglion, which would correspond to Power's observation of a reduced volume of the inner glomeruli. In addition to those studied by Power, mutations that eliminate specific types of photoreceptors from each ommatidium have been reported (3), but without a description of the effects upon the lamina.

Sprouting and Abnormal Synaptic Connectivity

In normal *Daphnia,* the fibers of one ommatidium make synapses only onto the five cells in their own cartridge and not onto other adjacent laminar cells. This is quite unlike the case in flies, where specific photoreceptors within an ommatidium send their fibers to different cartridges in the lamina ganglionaris (1). In addition, most laminar cell processes in the *Daphnia* remain within their cartridge area in the lamina; the small processes that sometimes can be found extending into other cartridges are quite short and devoid of synapses. We have found that in animals showing abnormal ommatidia after UV irradiation, laminar cells sprout new lateral branches that form anomalous synaptic connections in adjacent cartridges. Two interesting cases will be discussed here.

The first case is illustrated in Fig. 3, which shows three micrographs from the serial sections of an experimental animal, UVA-9. Figure 3A shows the fiber bundles as they appear immediately after exiting from the eye. The labeled bundle includes 14 fibers rather than the normal 8. This particular bundle is shown again as it enters the lamina in Fig. 3B. Five laminar cells that associate only with this supernumerary bundle are labeled as L1 to L5. Figure 3C shows a section further into the lamina neuropil; we see in this figure a large branch from a different laminar cell (LX) than the original five. The principal branches of L1 to L5 at this level are also identified. Synapses onto this extraneous process from three photoreceptor terminals are indicated by arrowheads. In addition to the one seen in this figure, we have found postsynaptic processes at other levels which belong to other laminar cells from adjacent cartridges.

Although we do not know the actual mechanism by which this supernumerary bundle was formed, we can make the reasonable guess that the UV treatment caused the disappearance of the lead fiber and one other cell of one ommatidium, and that the remaining six fibers followed the lead fiber of an adjacent group, giving rise to a bundle of $6 + 8 = 14$. Since only one lead fiber is present, the bundle picks up the normal number of laminar cells, 5. However, as the 14 fibers mature, there isn't enough postsynaptic area available to complement the growing presynaptic membrane surface. The choices are: degeneration or retraction of presynaptic membrane; growth of the photoreceptor terminal branches to other cartridges; or sprouting of postsynaptic processes. The evidence from our analysis points to the third choice and specifies that laminar cells external to a cartridge participate in this sprouting. Whether there is significant sprouting by laminar cells within the cartridge is difficult to estimate, since there is a great deal of variability in numbers of branches of any particular laminar cell (14). The extension of new branches and the formation of anomalous synaptic connections appears to us to be nonspecific (in terms of which cells make connections) on the basis of preliminary data. More observations will be required before a firm statement can be made.

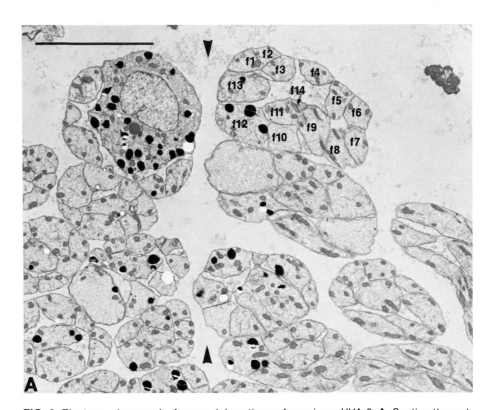

FIG. 3. Electron micrographs from serial sections of specimen UVA-9. **A:** Section through the fiber bundles in the region between the compound eye and the optic ganglion. The midplane of the animal is indicated by the arrowheads. The bundles are each held together by a complex glial process. Most bundles contain the normal eight fibers, except for the two in the upper center of the picture. The abnormal bundle on the left of the midline includes not only six fibers, but also has a photoreceptor soma, something we never observe in normal animals but see quite frequently in those that are irradiated with the UV microbeam. The abnormal bundle on the right of the midplane contains 14 fibers that have been numbered arbitrarily as f1 to f14. The bar on the upper left indicates a length of 10 μm. **B:** Section through the cortical layer of the optic ganglion. The 14 fibers in the abnormal right-side bundle have been indicated by the letter f but not numbered as in **A** due to ambiguities resulting from the loss of a few intervening sections of the series. The five associated laminar cells have been labeled as L1 to L5. The midplane of the animal is indicated by an arrowhead; the bar corresponds to 1 μm. **C:** Section in the laminar neuropil. The five original laminar cells have branched extensively, and only their principal fibers are labeled, L1 to L5. A laminar cell (labeled **LX**) from an adjacent cartridge is seen to send a large process into this region. This branch is postsynaptic to three different optic fibers; the synaptic sites are indicated by small arrowheads. Such branches are not found in normal animals. The midplace of the animal is indicated by the large arrowhead; the bar corresponds to 1 μm.

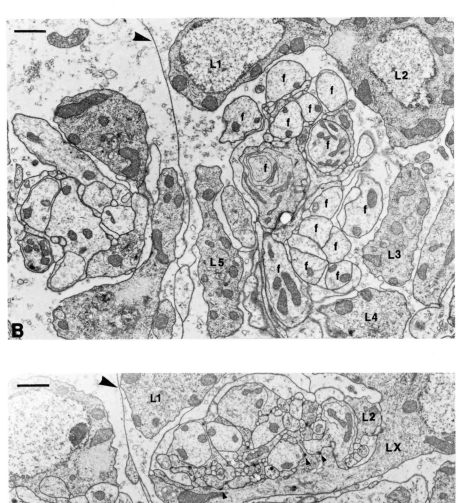

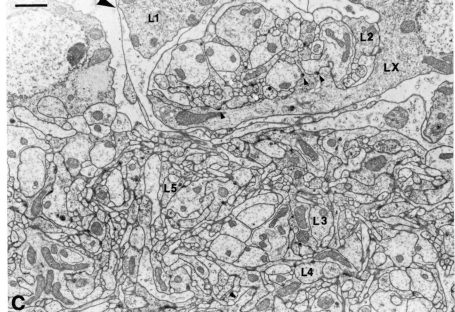

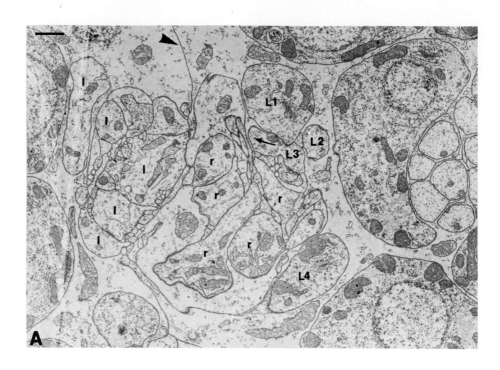

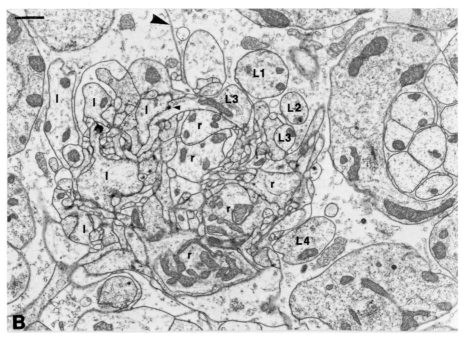

The second case (UVA-10) is illustrated in Fig. 4. The micrograph in Fig. 4A shows two subnumerary bundles, each with five fibers, on opposite sides of the midline. The bundle on the right has four laminar cells associated with it; the one on the left has none. The laminar cell labeled L3 appears to be about to split into two branches. It does so a few sections later, as seen in Fig. 4B. The left branch crosses the midline and forms synapses with fibers in the bundle on the left. One such synapse is illustrated at higher magnification in Fig. 4C. We have reconstructed this laminar cell with the computer and located and identified most of its synaptic contacts. The computer reconstruction is shown in Fig. 5. The synaptic connectivity is given in Table 2. Cells L2 and L4 also cross the midline to make anomalous synaptic connections.

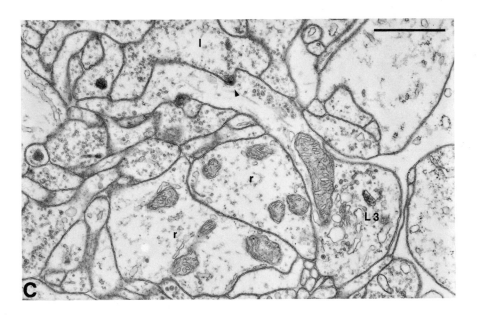

FIG. 4. Electron micrographs from serial sections of specimen UVA-10, illustrating two abnormal cartridges on opposite sides of the midplane. **A:** Section at the beginning of the neuropil. The five fibers in the left cartridge are labeled with the letter *l*, the five on the right with the letter *r*. The left cartridge does not show any laminar cells specifically associated with it; the right cartridge has four, labeled as L1 to L4. At this level L3 is about to split into two branches, one of which will send a secondary branch across the midplane, as seen in **B.** Other cells are visible, and a normal cartridge appears partially at right. **B:** Section in the neuropil, 1μm further than the section shown in **A.** The fibers are labeled as before. A subbranch of the left branch of L3 is seen to cross the midplane into the left cartridge where it receives a synaptic contact **(small arrowhead)** from one of the *l* fibers. Under normal conditions, such crossing of the midplane to synapse is not observed. **C:** Higher magnification electron micrograph of the same section as shown in **B,** showing the branch of L3 that crosses the midplane. Synaptic vesicles and a presynaptic density can be seen clearly. Left (*l*) and right (*r*) optic fibers have been labeled. The bar in each figure represents 1 μm.

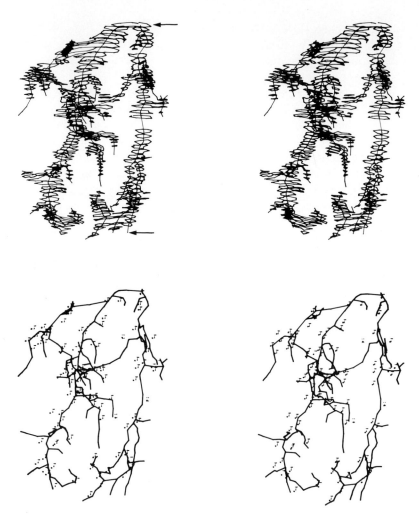

FIG. 5. Computer reconstruction of cell L3 from specimen UVA-10. The reconstruction begins with the section shown in Fig. 4A and continues through 200 serial sections, a distance of about 20 μm. The computer-generated image is shown rotated about 80° with respect to the sectioning plane, in an approximate sagittal plane. The upper stereo pair shows the cellular membrane contours plus a stick figure drawn through the center of each branch. The lower stereo pair shows the same stick figure plus the location of synaptic sites. Each site is denoted by a point and a character. In our convention, a (+) character makes the site postsynaptic, a (−) character presynaptic, and an (o) ambiguous. Since this cell at this neuropil level is essentially postsynaptic, most of the characters are (+). The upper arrow points at the section in Fig. 4A; the lower arrow points at the main branch of this cell, which continues further in the neuropil to the medulla. In this figure the top corresponds to anterior, the bottom to posterior. The branches on the left receive synaptic input from left optic fibers and those on right from right optic fibers, a condition not seen in normal tissue.

TABLE 2. *Synaptic sites between photoreceptor fibers (presynaptic) and laminar cells (postsynaptic) in two abnormal cartridges of animal UVA-10*[a]

Photoreceptor fiber	Laminar cells					Unidentified postsynaptic cells
	L1	L2	L3	L4	LX	
l1		3				23
l2		8	37	8		21
l3			8	11	2	9
l4				5	11	17
l5				5	3	23
r1		29	59	1		22
r2		23				14
r3				8	2	10
r4				23	6	9
r5				3		7

[a] The volume included is the anterior 25 μm of the laminar neuropil. Laminar cell L1 receives synapses further into the neuropil. LX are identified laminar cells from other cartridges. Unidentified postsynaptic cells include medullar cells. In a normal animal the synapses from l1–l5 to L2, L3, and L4 are never found.

The morphology and synaptic connections of these laminar cells are quite unlike those of such cells in a normal animal. And again, as in the case of animal UVA-9, it is the laminar cells that sprout new branches that seek out the presynaptic fibers. We do not know how the deletion of some cells from these ommatidia gave rise to the particular abnormal cartridges shown here. A reasonable conjecture is that the radiation slowed the maturation of the surviving photoreceptors to the extent that they arrived too late to pick up any or enough laminar cells, if one assumes that laminar neuroblasts have a limited period of availability. On the other hand, it could be suggested that an effect of the irradiation is to cause in some instances the loss of the capacity to trigger laminar cell maturation.

It is apparent from these results that developing laminar cells are capable of reacting to changed conditions in abnormal ways: by sprouting new branches, by making new synaptic contacts, and by degenerating. The new synapses are made only with photoreceptor terminals. The extra branches are found only locally where there are extra presynaptic fibers. We see no evidence of extensive sprouting into other regions, although such sprouting followed by retraction may have occurred prior to the time of fixation.

CONCLUSIONS

What do these observations suggest to us about the nature of the interaction between the growing photoreceptor fibers and the laminar neuroblasts?

The Disappearance of Laminar Cells

Since our observations are made on the adult nervous system, we do not know at what point the laminar cells disappear. One alternative is that the neuroblasts cannot mature and elaborate their neurites without the interaction with the receptor fibers; such neuroblasts would remain in an arrested stage of development and eventually degenerate. A different possibility is that the maturation of laminar neuroblasts can proceed without the wrapping around or the transient gap junctions occurring, and that the laminar cells degenerate much later because of insufficient synaptic input. In the first case, the interaction is a necessary signal; in the second it is not. We expect to resolve this question in the future by looking at experimental animals soon after irradiation and determining whether untouched neuroblasts mature, at least partially, without associated receptor fibers. At this time, however, the disappearance of laminar cells only tells us that the fiber bundle, or at least part of it, must be present in order for the laminar cells of a cartridge to survive. This trophic function apparently cannot be performed by bundles from adjacent cartridges.

Sprouting and Formation of Abnormal Synapses

The observation that laminar cells can proliferate new branches that seek out and receive synaptic contacts from other than their normal input cells suggests that the "triggering" signal does not carry a detailed set of instructions for the specific subsequent behavior of a laminar neuroblast. Nor is it a signal that triggers a rigid and specific developmental program in these cells. The additional branches seek out receptor fibers that have available presynaptic sites; the receptor fibers, however, do not travel into adjacent cartridges. If we assume on the basis of this observation that receptor fibers generally are capable of attracting laminar cell branches, then the formation of specific synaptic connections can be the result of the very accurate timing of sequential events occurring in different spatial locations. In this scenario, the signal transferred during the early fiber-neuroblast interaction is the trigger that starts the internal clock of the neuroblast at the proper time for the events to follow. The signal can be very general, even a small molecule, or perhaps an electrical signal passing across a gap junction.

In summary, we have obtained some evidence that we interpret as supporting the view that the interaction between these two cell types in early development is a general means for coordinating the events that follow and not a mechanism for labeling a cell as a unique individual with a specific and restricted repertoire for its developing morphology and synaptic connectivity. The evidence is suggestive, not conclusive. Other experiments now in progress will hopefully change this qualification in the near future.

ACKNOWLEDGMENTS

I wish to thank Ms. Bonnie Peng for her invaluable technical assistance. Thanks are also due to Cyrus Levinthal, Robert Schehr, and Neil Bodick for their contributions to the design and construction of the UV microbeam apparatus. This work was supported by NIH Grants NS11738 and NS09821. The computer work was performed at the Columbia Biological Sciences Computer Facility, which is supported by NIH Grant RR00442.

REFERENCES

1. Braitenberg, V. (1967): Patterns of projection in the visual system of the fly. I. Retina-lamina projections. *Exp. Brain Res.,* 3:271–298.
2. Gaze, R. M. (1970): *The Formation of Nerve Connections.* Academic Press, New York.
3. Harris, W. A., Stark, W. S., and Walker, J. A. (1976): Genetic dissection of the photoreceptor system in the compound eye of *Drosophila melanogaster. J. Physiol. (Lond.),* 256:415–439.
4. Jacobson, M. (1970): *Developmental Neurobiology.* Holt, Rinehart and Winston, New York.
5. Jagger, J. (1976): Effects of near-ultraviolet radiation on microorganisms. *Photochem. Photobiol.,* 23:451–454.
6. Levinthal, C., Macagno, E. R., and Tountas, C. (1974): Computer-aided reconstruction from serial sections. *Fed. Proc.,* 33:2336–2340.
7. Levinthal, C., and Ware, R. (1972): Three-dimensional reconstructions from serial sections. *Nature,* 236:207–210.
8. Levinthal, F., Macagno, E. R., and Levinthal, C. (1976): Anatomy and development of identified cells in isogenic organisms. *Cold Spring Harbor Symp. Quant. Biol.,* 40:321–331.
9. Lo Presti, V. (1975): The development of the eye-optic lamina neuroconnections in *Daphnia magna:* A serial section electron microscope study. Ph.D. dissertation, Columbia University, New York.
10. Lo Presti, V., Macagno, E. R., and Levinthal, C. (1973): Structure and development of neuronal connections in isogenic organisms: Cellular interactions in the development of the optic lamina of *Daphnia. Proc. Natl. Acad. Sci. U.S.A.,* 70:433–437.
11. Lo Presti, V., Macagno, E. R., and Levinthal, C. (1974): Structure and development of neuronal connections in isogenic organisms: Transient gap junctions between growing optic axons and lamina neuroblasts. *Proc. Natl. Acad. Sci. U.S.A.,* 71:1098–1102.
12. Lowenstein, W. R. (1973): Membrane junctions in growth and differentiation. *Fed. Proc.,* 32:60–64.
13. Macagno, E. R., Levinthal, C., Tountas, C., Bornholdt, R., and Abba, R. (1976): Recording and analysis of 3-D information from serial section micrographs: The CARTOS system. In: *Computer Technology in Neuroscience,* edited by P. B. Brown, pp. 97–112. John Wiley & Sons, New York.
14. Macagno, E. R., Lo Presti, V., and Levinthal, C. (1973): Structure and development of neuronal connections in isogenic organisms: Variations and similarities in the optic system of *Daphnia magna. Proc. Natl. Acad. Sci. U.S.A.,* 70:56–61.
15. Power, M. E. (1943): The effect of reduction in numbers of ommatidia upon the brain of *Drosophila melanogaster. J. Exp. Zool.,* 94:33–71.
16. Ramabhadran, T. V. (1975): Effects of near-ultraviolet and violet radiations (313–405 nm) on DNA, RNA and protein synthesis in *E. coli* B/r: Implications for growth delay. *Photochem. Photobiol.,* 22:117–123.
17. Sanes, J. R., Hildebrand, J. G., and Prescott, D. J. (1976): Differentiation of insect sensory neurons in the absence of their normal synaptic targets. *Dev. Biol.* 52:121–127.

Subject Index

Ablation, in neural crest cell migration research, 11
Acetylcholine
  adrenal medulla epinephrine release and, 260–261
  exocrine pancreas secretion and, 257–259
  microiontophoresis, 287
  receptors, at new nerve muscle synapses *in vitro,* 285–290
Actin, in fibroblasts, 61
Adenylate cyclase, second messenger concept and, 244–246
Adhesion, between fibroblasts, 59–62. *See also* Cell adhesion
Adrenal medulla, acetylcholine-mediated epinephrine release from, 260–261
Adrenergic cells of sympathetic nervous system, origin of, 14
Agglutination, *see* Cellular agglutination
Alkaline DNase, in sea urchin embryos, 279–280
Antibodies
  to interstitial collagens and procollagens, 77–78
  to teratoma cells, 48–49
Antigen(s)
  collagen as, 76–78
  on teratoma cells, 52–53
Autonomic ganglia formation, 14
  coculture of truncal neural crest and aneural hind gut and, 19
  differentiation of cephalic neural crest and, 17–21

Basement membrane
  capillary, secondary invasion by meta-stizing tumor cells, 228
  collagen extraction, 74
  collagen-like molecules in, 72
Baseplate
  purification of, 161–164
  role in sponge cell aggregation, 160–164
Bone marrow, thoracic duct lymphocytes and, 48–49

Calcium. *See also* Calcium-cyclic nucleotide relationships

acetylcholine-mediated exocrine pancreas secretion and, 257–259
  cellular metabolism of, 247–250
  in contractile systems, 246–247
  as coupling factor in muscle, 243–244
  cyclic nucleotides and, 250–252
  -dependent regulator, 262–263
  fly salivary and exocrine glands and, 249–250
  histamine release and, 249
  insulin secretion and, 249
  photoreceptor activation in retina and, 255–257
  as second messengers, 243–263
Calcium-cyclic nucleotide relationships
  acetylcholine-mediated epinephrine release from adrenal medulla and, 260–261
  effector molecules as receptors for second messengers and, 261–263
  patterns of, 259
CAM, *see* Chick chorioallantoic membrane
"Capping" of fibroblasts, 64
Capillary endothelium and basement membrane,
  secondary invasion by metastasizing cells of, 228
Cartilage. *See also* Cartilage proteoglycans
  biochemical changes and osteoarthrosis, 95–96
  collagen, 87
  developing, cell-surface extracellular matrix interactions and, 121–122
  embryogenesis, 7, 101
  physical chemistry and function of, 87
Cartilage proteoglycans, 87–96
  cell-matrix interactions and, 87
  extraction of, 88–89
  influence of cellular environment on, 93–94
  structure and aggregation of, 88–93
Catecholaminergic cell differentiation, 21–22
Catecholamines, detection of, 22–24
Cell adhesion. *See also* Cell binding; cellular agglutination
  multicomponent model for, 187–194
Cell-aggregating factors, 177–178. *See also* Cell aggregation